Vaccine
Evaluation

Pocket Guides To

Biomedical Sciences

The *Pocket Guides to Biomedical Sciences* series is designed to provide a concise, state-of-the-art, and authoritative coverage on topics that are of interest to undergraduate and graduate students of biomedical majors, health professionals with limited time to conduct their own searches, and the general public who are seeking for reliable, trustworthy information in biomedical fields.

Series Editor
Lijuan Yuan
Department of Biomedical Sciences and Pathobiology, Virginia Tech
Blacksburg, Virginia, USA

Editorial Board
Lisa Gorski, U. S. Department of Agriculture

Baochuan Lin, U. S. Naval Research Laboratory

Russell Paterson, University of Minho, Portugal

Yiqun Zhou, Roche Pharmaceuticals

Science Publisher
Charles R. Crumly
CRC Press/Taylor & Francis Group

Tumors and Cancers: Endocrine Glands – Blood – Marrow – Lymph,
by Dongyou Liu

A Guide to Cancer: Origins and Revelations, *by Melford John*

Pocket Guide to Bacterial Infections, *edited by K. Balamurugan*

A Beginner's Guide to Using Open Access Data,
by Saif Aldeen Saleh Airyalat and Shaher Momani

Pocket Guide to Mycological Diagnosis,
edited by Rossana de Aguiar Cordeiro

Genome Editing Tools: A Brief Overview, *by Reagan Mudziwapasi and Ringisai Chekera*

Vaccine Efficacy Evaluation: The Gnotobiotic Pig Model,
by Lijuan Yuan

For more information about this series, please visit:

https://www.crcpress.com/Pocket-Guides-to-Biomedical-Sciences/book-series/CRCPOCGUITOB

Vaccine Efficacy Evaluation

The Gnotobiotic Pig Model

Lijuan Yuan

CRC Press
Taylor & Francis Group
Boca Raton London New York

CRC Press is an imprint of the
Taylor & Francis Group, an **informa** business

Cover art by Peilian Yang, China

First edition published 2022
by CRC Press
6000 Broken Sound Parkway NW, Suite 300, Boca Raton, FL 33487-2742

and by CRC Press
4 Park Square, Milton Park, Abingdon, Oxon, OX14 4RN

CRC Press is an imprint of Taylor & Francis Group, LLC

Library of Congress Cataloging-in-Publication Data

Names: Yuan, Lijuan, author.
Title: Vaccine efficacy evaluation : the gnotobiotic pig model / Lijuan Yuan.
Description: First edition. | Boca Raton : CRC Press, 2022. | Series: Pocket guides to biomedical sciences | Includes bibliographical references and index.
Identifiers: LCCN 2021050135 | ISBN 9781032234748 (hardback) | ISBN 9780367486341 (paperback) | ISBN 9781003277828 (ebook)
Subjects: LCSH: Vaccines. | Vaccines--Research.
Classification: LCC QR189 .Y83 2022 | DDC 615.3/72--dc23/eng/20211113
LC record available at https://lccn.loc.gov/2021050135

ISBN: 9781032234748 (hbk)
ISBN: 9780367486341 (pbk)
ISBN: 9781003277828 (ebk)

DOI: 10.1201/b22816

Typeset in Frutiger Light
by Deanta Global Publishing Services, Chennai, India

Contents

Preface

The successful rearing of pigs under germfree (gnotobiotic) conditions was first reported in 1960 from the Department of Immunology in the Institute of Microbiology of the Czech Academy of Sciences, Prague, Czechoslovakia. Dr. Jaroslav Sterzl and fellow immunologists used germfree piglets to study the ontogeny of B cell and antibody immune responses. Gnotobiotic pigs are also an excellent model system for the study of viral pathogenesis. Due to the lack of confounding commensal microbes that are present in all specific pathogen-free (SPF) animals, gnotobiotic pigs can be used to identify the pathological changes directly as the result of a particular virus infection. Virus replication, dissemination, shedding, and the resulting tissue damage and clinical signs can be more accurately recapitulated compared to the SPF animal models. Because microbiota and pathogens (i.e., wild-type rotavirus) are absent, immune responses to a single pathogen or vaccine can be assessed under a clean immunological background in gnotobiotic pigs.

The neonatal gnotobiotic pig model of Wa HRV (G1P[8]) infection and diarrhea was first established in 1996 by my Ph.D. advisor Dr. Linda J. Saif at The Ohio State University. This model has been used in her laboratory and then, since 2008, in my laboratory at Virginia Tech for the evaluation of numerous human rotavirus (HRV) candidate vaccines including live attenuated (Wa), reassortant (RotaTaq, Lanzhou trivalent reassortant vaccine), inactivated (Wa, CDC-9), recombinant fusion protein (P2-VP8*), nanoparticle (P24-VP8*, S60-VP8*), DNA plasmid (VP6), virus-like particle (VLP) (2/6 and 2/6/7), and mRNA-based P2-VP8* vaccines, with different adjuvants (LT-R192G, ISCOM, QS21, aluminum phosphate, aluminum hydrogel), immunostimulating supplements (probiotics and rice bran), and immunization routes (oral, intramuscular, intranasal, intradermal by gene gun), and with or without maternal antibodies.

Human norovirus (HuNoV) vaccine development is still in its early stage, the frontrunner is a VLP vaccine developed by Takada. The gnotobiotic pig model of GII.4 HuNoV infection and diarrhea has been developed and used to evaluate efficacies of HuNoV P particle and VLP vaccines with different immunization routes and adjuvants. To better mimic humans, human gut microbiota (HGM)-transplanted Gn pig models for HRV and HuNoV infection have also been developed. These models are useful in studying the influence of HGM on the immunogenicity and protective efficacy of HRV and HuNoV vaccines.

Writing this book allows me to look back at over 28 years of research in the field and at the body of work that I participated in, starting from being a visiting scientist (1993–1996), a Ph.D. student (1996–2000), a tenure-track assistant professor (2007), and a tenured full professor (2019–). I say it half-jokingly at scientific meetings and to my students that "In my entire career, I only did one thing—perfecting the gnotobiotic pig model for the study of enteric viruses."

In this book, I mainly focus on studies of the development of gnotobiotic pig models of human rotavirus and human norovirus infection and all the vaccine studies using gnotobiotic pigs in which I participated in with a major role. Through our studies, the gnotobiotic pig model has been firmly established as the most reliable animal model for the preclinical evaluation of HRV and HuNoV vaccines.

It is my wish that this book will inspire more researchers, vaccine developers, and graduate students to appreciate the extraordinary value of gnotobiotic pig models in biomedical research and vaccine development.

Acknowledgments

First, I would like to thank everyone who has been involved in the establishment and the maintenance of the Gn pig program at the Virginia-Maryland College of Veterinary Medicine. I am deeply grateful to the people who drove the success of the creation of the Gn pig facility at our college and the ongoing success we are having in running the program and providing the Gn pigs needed for all my research projects. The deans of our college, Gerhardt Schurig, Cyril Clarke, Interim Dean Gregory Daniel, Senior Associate Dean for Research Roger Avery, the department heads Ludeman Eng and Ansar Ahmed, and my faculty mentor X. J. Meng provided leadership during the time we worked to establish, and later, to expand the facility. Drs. Schurig, Avery, and Eng have since retired. Now Dr. Daniel Givens is our new dean, Dr. Ahmed is the Associate Dean for Research, and Dr. Margie Lee is the department head and they have continued the strong support of my research using Gn pigs. After two expansions, our Gn pig facility became one of the largest in the country. It has nine Gn pig holding rooms and 24 four-place isolators and two sets of surgery and transfer units.

I owe Mr. Peter M. Jobst a tremendous amount of gratitude. He was the new manager of TRACSS (Teaching and Research Animal Care Support Service) at Virginia-Maryland College of Veterinary Medicine and led the heroic effort in creating the Gn pig facility from scratch. During the process, he demonstrated such great resourcefulness and creativity using his vast and detailed knowledge in swine gained from his many years of work experience at Revivicor (PPL Therapeutics Inc.), where they created the first cloned pig in the world (Phelps et al., 2003). He is currently the Director of College Facilities and still lends his helpful hands to our Gn pig program whenever we need him. Steve Dehart, Sharon Carbaugh, and Andrea Aman Pulliam helped me with ordering the parts and pieces from at least seven different companies to assemble the isolators. Steve Dehart, Peter Jobst, and Andrea Pulliam led the building renovations to turn two unused animal facilities into the Gn pig facility as well as the renovation of a necropsy room for the need of hysterectomy procedure. Andrea Pulliam, Shannon Viers, Mariah Weiss, Karen Hall, Kimberly Allen Skroupa, and Rachel McNeil are all TRACSS members who were or are in charge of managing the Gn pig program. Karen Hall has been the TRACSS manager until December 2021. Now she takes on the new role of Associate Director, Animal Care and Operations. They are the most competent, skillful, and devoted animal care professionals and I am so fortunate to have worked with them. To derive Gn piglets, the sow hysterectomy surgeries are performed by large animal surgeons Dr. Kevin Pelzer, Dr. Sherrie Clark-Deener, and Dr. Jamie Stewart. They are the reason that a Gn pig facility has to be established within a college of veterinary medicine, not in a medical school. Veterinary services to Gn pigs provided by Dr. Marlice Vonck, Betsy Midkiff, Dr. Amy Rizzo, and Dr. Calvin Lau are greatly appreciated. I am grateful to Dr. Juliette Hanson, the agricultural veterinarian on Wooster Campus, The Ohio State University for providing guidance and training for the surgical procedure to derive germ-free piglets. I also want to thank the staff, especially Rickard McCormick, at Wooster Gn pig facility for providing consultation and training to our staff when we were setting up our Gn pig facility at Virginia Tech.

Secondly, I want to thank my Ph.D. advisor Dr. Linda J. Saif, fellow students and colleagues in Saif Lab, all my collaborators, my former and current graduate students, post-docs, visiting scholars, and undergraduate research volunteers in Yuan Lab, who contributed to the research projects discussed in this book. The individual who led each of the research topics and the funding agency that sponsored each of the research projects have been acknowledged at the beginning of each chapter or section. However, all the projects using Gn pig models are teamwork. One person can't undertake a Gn pig experiment. In Yuan Lab, every member is a team member in all our research projects. While I only gave the name of the lead author of the papers discussed in the chapters, the contributions of all co-authors are equally acknowledged here. Our work discussed in this book covers the time span from 1993 to the present. I am grateful to all the colleagues I have interacted with for research more than a quarter-century.

Lastly, I want to thank all the funding agencies that have supported my research over the many years, and especially NIH and Bill and Melinda Gates Foundation, and the strategic leaders for enteric vaccines in the enteric and diarrheal diseases team at the foundation, Dr. Duncan Steel and Dr. Carl Kirkwood, for their appreciation of the importance of using Gn pig models for preclinical evaluation of rotavirus and norovirus vaccines. And finally, I want to thank my dear husband Rick Orick and my mother Yufang Gao for all of their love, support, and attention to my work over the years. My mother lives with us and she takes care of me. She is 87 years old and still able to plant a large vegetable garden and cooks for me with the fresh vegetables she collects from the garden every day. Since the beginning of the COVID-19 pandemic, I have been working mostly from home. She would say this to me: "Go to work on your book; when I have lunch ready I will call you to eat"; this makes me the luckiest and happiest daughter in the world and makes the completion of this book in this trying time an enjoyable task.

Lijuan Yuan
Professor of Virology and Immunology
Department of Biomedical Sciences and Pathobiology
Virginia Polytechnic Institute and State University
Blacksburg, VA

About the Author

Lijuan Yuan is a Professor of Virology and Immunology in the Virginia-Maryland Vet Med's Department of Biomedical Sciences & Pathobiology. Dr. Yuan studies the interactions between enteric viruses and the host immune system. Her lab's research interests are focused on pathogenesis and innate and adaptive immune responses induced by enteric viruses, especially noroviruses and rotaviruses, and on the development of safer and more effective vaccines as well as passive immune prophylaxis and therapeutics against viral gastroenteritis. These studies utilize wild type, gene knock-out and human gut microbiota transplanted gnotobiotic pig models of human rotavirus and norovirus infection and diseases and cell culture model of rotavirus infection. Currently, the Yuan lab is evaluating the immunogenicity and protective efficacy of several candidate novel rotavirus and norovirus vaccines and engineered probiotic yeast *Saccharomyces boulardii* secreting multi-specific single-domain antibodies as a novel prophylaxis against both noroviruses and *Clostridioides difficile* toxins. See: https://vetmed.vt.edu/people/faculty/yuan-lijuan.html

1

Introduction

1.1 Importance of pig models in human vaccine development

Testing the immunogenicity, protective efficacy, and safety in animal models is one of the most important steps in vaccine development after the construction and formulation of the protective antigens and before human clinical trials. Regulatory authorities require vaccine candidates to undergo preclinical assessment in animal models before being permitted to enter the clinical phase in human subjects, and much of the success in bringing a new vaccine to market depends on choosing the most appropriate animal model for preclinical testing (Gerdts et al., 2015). Appropriate animal models are critical for establishing safety and efficacy, identifying the mechanisms of immune protection, and determining the optimal route and formulation of new vaccines. The selection of animal models for human vaccine evaluation should be based on the following considerations: 1) modes of infection and disease similar to those in humans, 2) background knowledge of biological properties including genetics and physiological and immune system similarity, 3) transferability of information and generalizability of the results, 4) size, availability, ease of care, and breed, 5) adaptability to experimental manipulation, 6) ecological consequences, and 7) ethical considerations (Davidson et al., 1987; Gerdts et al., 2015).

One of the most appropriate and reliable models for the evaluation of vaccines against human enteric pathogens is the pig (*Sus scrofa*), which has high similarities in gastrointestinal anatomy, physiology, nutritional/dietary requirements, and mucosal immunity (Gerdts et al., 2015). The first pig genome sequence was completed in August 2011 and published in 2012 (Groenen et al., 2012). To date, there are several hundred complete individual pig genome sequences available in public databases (Groenen, 2016). The pig genome has high sequence homology to humans, 60%, compared with a 40% sequence homology of rodents to humans (Thomas et al., 2003; Humphray et al., 2007). The pig chromosomal structure has a higher similarity to humans than those of the mouse, rat, dog, cat, horse, or cattle (Meyers et al., 2005; Murphy et al., 2005). Pigs are monogastric and closely resemble humans physiologically and immunologically (Butler et al., 2009a; Sinkora and Butler, 2009). The ontogeny of the porcine immune system, especially the development of immunoglobulins, antibody repertoire, and B cells has been extensively reviewed by Butler and colleagues (Butler et al., 2009b, 2009c; Sinkora and Butler, 2009). The development of the lymphoid organs from birth to adult pigs and T and B cell immune responses have been documented and show great similarity to humans (Bianchi et al., 1992). All the immune cell populations identified in humans are present in pigs. Most proteins of the immune system of pigs share structural and functional similarities with their human counterparts, except some host defense polypeptides. Therefore, there is a large number of cross-reactive anti-human monoclonal antibodies that can be utilized in research using pig models. Compared to other laboratory animals, pigs have large body sizes to ensure ample sampling of tissues and blood for analysis. For vaccine evaluation, the ability to isolate large numbers of lymphocytes following disruption of solid organs offers outstanding advantages for the study of the intestinal and systemic immune responses induced by the vaccination (Wen et al., 2012a, 2012c; Yuan et al., 2001a). Moreover, pigs are born in large litter sizes (usually >10 piglets) with enough animals to provide for age-matched controls in study groups. Pigs are widely produced as food animals in most of the world and therefore there are unlimited supplies, and their use as tools of science is generally more acceptable to the public; compared to non-human primates, there are much fewer ethical concerns. Because of all these factors, the pig is an increasingly important large animal model for human biomedical research (Butler, 2009; Butler et al., 2009a; Meurens et al., 2012).

For human rotavirus (HRV) and human norovirus (HuNoV) vaccines, mice models are useful for testing immunogenicity; however, mice cannot be infected by HRV or HuNoV. Murine rotaviruses and noroviruses are very different from HRV and HuNoV in host-age restriction and

DOI: 10.1201/b22816-1

pathogenesis. The most dominant host factor in rotavirus pathogenesis is age. Murine rotaviruses and noroviruses can induce diarrhea after infection only in neonatal mice, not in adult mice (Boshuizen et al., 2003; Roth et al., 2020; Ward et al., 2007). Adult mice can be infected by murine rotavirus and norovirus and shed virus for a long period, but these infections are asymptomatic in wild-type mice. Therefore, mouse models are not useful for the evaluation of vaccine-induced adaptive immunity associated with protection against rotavirus or norovirus disease. Many rotavirus immunization schemes that have succeeded in adult mice have failed when evaluated in neonatal gnotobiotic (Gn) pigs (Yuan and Saif, 2002). For example, the rotavirus DNA vaccine (Chen et al., 1997) and 2/6 virus-like-particle (2/6-VLP) vaccine (Bertolotti-Ciarlet et al., 2003) conferred almost completed protection against virus shedding upon challenging vaccinated mice with murine rotavirus; however, these same candidate vaccines did not confer any clinical protection in Gn pigs challenged with HRV (Yuan et al., 2005, 2000). For the preclinical testing of the protective efficacy, an animal model that can exhibit the same or similar clinical sign of disease like that in humans are critical for assessing protection against disease upon challenge infection. Gn pig models fulfill this need for testing HRV and HuNoV vaccines. No immunization scheme that fails in Gn pig models would be considered for human clinical trials.

1.2 Definition of gnotobiotic pigs

The detailed technical instructions on how to establish and maintain a Gn pig program, including standard operating procedures for the derivation of Gn pigs and their rearing on sterile commercially available milk, have been provided elsewhere (Yuan et al., 2017). The word gnotobiotic is from Greek roots *gnostos* means "known" and *bios* means "life." A gnotobiotic pig is a pig in which only certain known viruses or bacteria are present. Technically, the term also includes germ-free pigs, as the status of their microbial communities is also *known*. Germ-free pigs are derived by the surgical procedure hysterectomy one day before the expected delivery date (112–113 days of pregnancy) and moved into a transfer isolator, after which they are transported into rearing isolators where they are kept for the remainder of the experiment. Germ-free pig isolators provide a sterile environment free of bacteria, mycoplasma, molds, fungi, viruses, etc. Once an agent, such as a virus, is introduced into the germ-free isolator, it is no longer truly germ-free and is referred to as being gnotobiotic. In the literature, however, the words germ-free pig, gnotobiotic pig, and isolator pig are often used interchangeably.

1.3 Brief history of the first studies of HRV and HuNoV infection in Gn pigs

The Gn pig model of G1P[8] HRV infection and diarrhea was established in 1996 using the virus isolate the HRV Wa strain from Wyatt and Kapikian's group (Saif et al., 1996; Ward et al., 1996a; Wyatt et al., 1980). Dr. Albert Kapikian's group conducted the first study showing that Gn pigs were productively infected with HRV and developed diarrhea and virus shedding (Torres-Medina et al., 1976a, 1967b). This study reported that 19 of 21 piglets developed diarrhea after being inoculated at one to four days of age with the rotavirus isolate acquired from human infants with acute gastroenteritis in 1974, which is the original Wa strain HRV that belongs to G1P1A[8] serotypes with the genome constellation of G1-P[8]-I1-R1-C1-M1-A1-N1-T1-E1-H1 (Wentzel et al., 2013). The total duration of rotavirus shedding was two to six days after the onset of diarrhea. The onset of diarrhea was two to seven days after virus inoculation. The same infectivity was reproduced in Gn pigs for up to five serial passages (Torres-Medina et al., 1976a). Subsequent studies showed that Gn pigs can be infected with G1P[8] (Wa strain), G3P[8] (M strain), and G4P[6] (Arg strain) (Yuan et al., unpublished data) HRV and develop diarrhea, but not G2P[4] (DS-1 strain) HRV (Hoshino et al., 1995; Torres and Ji-Huang, 1986). The virulence of the Wa HRV in Gn pigs was maintained by passaging the virus solely through the intestinal contents of infected neonatal Gn pigs and the virus inoculum stock is generated from the pooled intestinal contents. Hence, Wa HRV inoculum stock can be easily replenished using Gn pigs. Over the next two and half decades since 1996, the Gn pig model of Wa HRV infection and diarrhea has been the key element for many studies published in over 50 peer-reviewed journal articles, testifying to the usefulness of this unique animal model in rotavirus research (Azevedo et al., 2013, 2010, 2004, 2005, 2006, 2012; Chang et al., 2001; Chattha et al., 2013; Gonzalez et al., 2010, 2004; Hodgins et al., 1999; Iosef et al., 2002a, 2002b; Kandasamy et al., 2014; Liu et al., 2013, 2014; Nguyen et al., 2003, 2006a, 2006b, 2007; Parreno et al., 1999; Ramesh et al., 2019; Saif et al., 1997; To et al., 1998; Vega et al., 2012, 2013; Wang et al., 2010; Ward et al., 1996a, 1996b; Wen et al., 2009, 2012a, 2012b, 2012c, 2011, 2015, 2014a, 2014b; Wu et al., 2013; Yang et al., 2014; Yang and Yuan, 2014; Yuan et al., 2005, 2000, 2001a, 2001b, 1998, 1996, 2008; Yuan and Saif, 2002; Zhang et al., 2014, 2008a, 2008b, 2008d). The Gn pig model of Wa HRV infection and diarrhea has been used in evaluating numerous rotavirus candidate vaccines including live attenuated (Wa), reassortant (RotaTaq, Trivalent Lanzhou Reassortant vaccine), inactivated (Wa, CDC-9), recombinant protein (P2-VP8*, P24-VP8*), DNA plasmid (VP6), and virus-like particle (VLP) (2/6 and 2/6/7) vaccines with different adjuvants (LT-R192G, ISCOM, aluminum phosphate, aluminum hydroxide), immunostimulating supplements (probiotics and rice bran), and immunization routes (oral, intramuscular, and intranasal), and with or without maternal antibodies. Through these studies, the Gn pig model has established its role as the most reliable animal model for the preclinical evaluation of HRV vaccines.

The Gn pig model of HuNoV infection was first attempted for the studies of the pathogenesis and humoral immune responses in 2006 (Cheetham et al., 2006) and 2007 (Souza et al., 2007a) with a genogroup II, genotype 4 (GII.4)/HS66/2001 variant. This challenge model, without determining the median infectious dose (ID50), was then used to evaluate a HuNoV VLP vaccine with two different adjuvants (Souza et al., 2007b). However, due to the insufficient virus challenge dose used (only 57% of control Gn pigs shed virus), the protective efficacy of the vaccines could not be calculated. In 2013, with the determination of the ID50 of a GII.4/2006b variant in newborn Gn pigs (four to five days of age) and older Gn pigs at 33–34 days of age, the Gn pig model of HuNoV infection and diarrhea was formally established (Bui et al., 2013). The model was utilized first for the evaluation of a HuNoV P particle vaccine (Kocher et al., 2014). Since then, the Gn pig model of GII.4/2006b infection and diarrhea has been used in several studies of HuNoV infection and immunity (Lei et al., 2016a, 2016b, 2016c, 2019). Recently, with the determination of ID_{50} and median diarrhea dose (DD_{50}) and the optimal challenge dose of a GII.4/2003 variant in older Gn pigs (Ramesh et al., 2020), the Gn pig challenge model is further standardized for the accurate evaluation of protective efficacies of therapeutics, vaccines, and other prophylactics against HuNoV. Unlike Wa HRV, passaging of HuNoV in Gn pigs has not been successful. The virus titers in the intestinal contents of HuNoV infected Gn pigs in general only reach the level of ~10^4 RNA genome copies per gram feces, not high enough to be used as the source of virus inoculum which is about 2×10^5 to 6.43×10^5 RNA genome copies/pig (Bui et al., 2013; Ramesh et al., 2020). Therefore, the source of virus inoculum has to rely on collecting the stool samples of HuNoV challenged human volunteers and the stool pool needs to go through a stringent screening process to exclude the presence of any other viruses infecting pigs. Different approaches have been attempted to increase the virus titer in the intestinal contents of neonatal Gn pigs after HuNoV infection, including simvastatin treatment (Bui et al., 2013) and using severe combined immunodeficient (RAG2/IL2RG knockout) Gn pigs (Lei et al., 2016b), but without substantial improvement. It was postulated that different immunodeficient Gn pigs, such as STAT1–/– might allow HuNoV to replicate to higher titers in the pig intestine. However, unlike STAT1-deficient mice that are viable and display no developmental defects, STAT1–/– was lethal to pig fetuses. Multiple attempts using CRISPR/Cas9 technology failed to establish or maintain pregnancy in surrogate sows aimed to derive STAT1–/– piglets. A recent study found that GII.4 HuNoV did not show increased replication and virus spread in STAT1–/– human intestinal enteroids (Lin et al., 2020). In another study, we found that human gut microbiota (HGM) significantly enhanced HuNoV replication in Gn pigs (Lei et al., 2019). Compared to germ-free Gn pigs, HuNoV inoculation of HGM Gn pigs resulted in increased HuNoV shedding on post-inoculation day (PID) 3, 4, 6, 8, and 9. The peak titer in a few HGM pigs reached 10^5 genome RNA copy numbers per gram feces. This is a promising direction but further efforts are still needed to generate a high titer HuNoV inoculum pool in Gn pigs.

2
Establishment of Gn Pig Model of HRV Infection and Diarrhea
Infectivity and Pathogenesis of HRV in Gn Pigs

2.1 Origin of the virulent Wa HRV inoculum

To establish the Gn pig model of HRV infection and diarrhea, an infant stool filtrate containing Wa HRV (from R. G. Wyatt, Laboratory of Infectious Diseases, National Institute of Allergy and Infectious Diseases, Bethesda, Maryland) was used to orally inoculate newborn Gn pigs whose intestinal contents were collected and serially passaged (orally) in additional Gn pigs (Ward et al., 1996a). The passage history is summarized in Figure 2.1. Pooled intestinal contents obtained from the 16th Gn pig passage were used as the virulent Wa HRV stock inoculum in Ward et al.'s study (Ward et al., 1996a). The passage of the Wa HRV strain in Gn pigs provided an amplified source of virulent Wa HRV. In subsequent studies, the virulence of the virus inoculum is maintained by passaging only in neonatal Gn pigs. Meanwhile, the Wa HRV strain was successfully adapted to cell culture (Wyatt et al., 1980) and was attenuated during the passages in Gn pigs and then in cell culture. After 11 Gn pig passages, the Wa HRV was adapted to culture in fetal Rhesus monkey kidney (MA104) cells and plaque purified (Wyatt et al., 1980). The plaque-purified, cell culture–adapted Wa HRV was subsequently cloned six times in MA104 cells by limiting dilution. An MA104 cell culture lysate containing the 27th passage of the cloned cell culture–adapted Wa HRV was used as the attenuated Wa HRV (AttHRV) stock inoculum in the Ward et al. study (Ward et al., 1996a) (Figure 2.1) and later used as a prototype HRV vaccine in many subsequent studies in Gn pigs. The genomes of the Wa HRV variants were sequenced for the elucidation of the evolutionary history of the Wa HRV strain (Wentzel et al., 2013). The GenBank access numbers for the sequences of all the 11 gene segments of the VirHRV genome are FJ423113–FJ423123. For the AttHRV, the GenBank access numbers are FJ423135–FJ423145. Important to note, the AttHRV passage that was sequenced in the study (Wentzel et al., 2013) contains rearrangements in gene segments 7 and 8, due to passaging under high MOI. The rearrangements were detected at 33th to 34th passage in MA104 cells. To prevent the occurrence of gene rearrangement, we started from the 31st passage in Yuan lab, only passage the AttHRV at low MOI, and monitor the virus by genomic sequencing.

An MA104 cell culture adapted Wa HRV variant with lower than 27 passages (the exact number is unknown), namely parental Wa HRV (ParWaHRV) from Dr. Yasutaka Hoshino's lab at NIH was also sequenced (GenBank access numbers FJ423124–FJ423134). When comparing the infectivity and pathogenesis of the three Wa variants, VirHRV, AttHRV, and ParWaHRV in Gn pigs (Table 2.1), we can see that, unlike AttHRV, ParWaHRV is not fully attenuated and caused diarrhea in 67% of the pigs and viremia in 17% of the pigs.

2.2 Determination of infectious dose of virulent Wa HRV inoculum in Gn pigs

A comprehensive study was carried out to determine the pathogenesis and infectious doses of the VirHRV and AttHRV variant pair in Gn pigs in Linda Saif's lab. Lucy Ward led the study, who was a post-doctoral research associate and a veterinary pathologist in Saif Lab. Blair Rosen and I

DOI: 10.1201/b22816-2

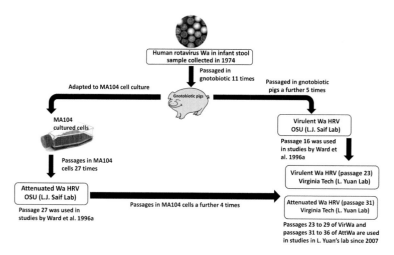

Figure 2.1 Schematic diagram summarizing the passage history of the Wa human rotavirus (HRV) strains. The Wa HRV was originated from the 1974 infant stool sample and was obtained following passaging in Gn pigs (virulent Wa HRV strain) and MA104 cells (attenuated Wa HRV strain).

participated in the study. Blair Rosen just graduated with his PhD degree in Saif Lab and stayed on as a post-doc. I was a visiting scientist from November 1993 to August 1996 in Saif Lab. The research project was funded by grants from the World Health Organization (GPV/V27/181/24) and the National Institute of Allergy and Infectious Diseases (NIAID), National Institutes of Health (NIH) (R01A133561 and R01A137111, PI: Linda Saif).

Near-term pigs were derived and maintained under germ-free conditions. A total of 45 Gn pigs were used for determining the infectivity of the VirHRV in Gn pigs. The VirHRV inoculum was serially diluted in MEM (~10^6 to 10^{-1} focus-forming units [FFU]/ml). Five ml diluted inoculums were fed to four- to five-day-old Gn pigs 10 min after feeding them 5 ml of 100 mM-sodium bicarbonate (to neutralize the gastric acidity). Pigs were examined daily for diarrhea and their faces scored as follows: 0, normal (solid); 1, pasty; 2, semi-liquid; and 3, liquid. Diarrhea was present if the fecal consistency score was ≥2. Rectal swabs were collected daily and blood samples were drawn weekly. Infectivity was defined as the dose of live VirHRV required to infect 50% of inoculated pigs as determined by fecal virus shedding (median infectious dose, ID$_{50}$). Different dilutions of the Vir HRV were administered orally to eight groups (n = 2–4) of Gn pigs (<1, ~1, 25, 550, 3×10^4, 8.5×10^4, 3×10^5, and 1×10^6 FFU) in the attempt to determine the ID50 and median diarrhea dose (DD$_{50}$) using the Reed and Muench method (Reed and Muench, 1938). However, all of the dilutions resulted in 100% of the pigs developing virus shedding and no less than 50% of the pigs developing diarrhea at any given dose group. Therefore, both ID$_{50}$ and DD$_{50}$ were designated to be at ≤1 FFU (Ward et al., 1996a). The severity and duration of virus shedding and diarrhea were similar in all the eight dose groups, thus appearing to be dose-independent. Of pigs inoculated with VirHRV at the dose of ~10^5 FFU, 80–100% developed diarrhea for an average duration of four days and 100% shed virus for an average of 5.5 days. This dose of ~10^5 FFU VirHRV was used as the challenge dose for Gn pigs at 4–7 and 33–34 days of age in all the subsequent studies using this Gn pig model and it has consistently induced 80–100% of virus shedding and diarrhea in all the naïve Gn pigs when the model system is properly calibrated.

2.3 VirHRV fecal shedding pattern in Gn pigs

Daily rectal swabs from Gn pigs are collected for HRV antigen detection by ELISA and infectious virus particle detection by cell-culture immunofluorescent (CCIF) assay (Twitchell et al., 2016). The ELISA is done using goat anti-bovine rotavirus polyclonal antibody as detector antibody,

Table 2.1 Comparison of Virus Shedding, Viremia, and Diarrhea in Gn Pigs Inoculated with Wa HRV Variants

Wa HRV variant inoculum*	n	Fecal virus shedding		Viremia	Diarrhea			Reference
		% shedding	Mean duration (days)		% diarrhea	Mean duration (days)	Mean cumulative score	
VirHRV	43	100%	6.0 (0.3)**	100%	100%	4.1 (0.4)	10.4 (0.5)	(Ward et al., 1996a)
AttHRV	29	17%	1.0 (0)	0%	13%	2.0 (0)	6.1 (0.3)	(Zhang et al., 2008b)
ParWaHRV	6	67%	1.0 (0)	17%	67%	2.0 (0.8)	7.6 (1.4)	Unpublished data
Mock control	18	0%	0 (0)	0%	13%	0.5 (0.2)	4.0 (1.0)	(Ward et al., 1996a)

Note:
* The inoculation dose for AttHRV and ParWaHRV were 2–5 × 10⁷ FFU. VirHRV was 1 × 10⁵ FFU.
** Standard error of the mean.

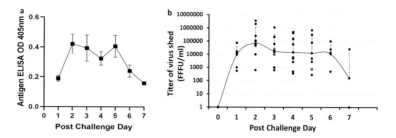

Figure 2.2 The pattern of VirHRV shedding in fecal samples. ELISA mean OD values for rotavirus antigen (a) and infectious virus particles measured by cell culture immunofluorescence (b) in fecal samples of Gn pigs. Naïve Gn pigs in placebo control groups were challenged with 1×10^5 FFU virulent Wa HRV at 33–34 days of age. Fecal rotavirus antigen was measured by ELISA and results are expressed as OD units. The OD values were adjusted based on the average OD values of the negative controls from different ELISA plates. Data are presented as mean OD values or FFU/mL. Error bars are standard errors of the mean (n = 3–15). ELISA: enzyme-linked immunosorbent assay; FFU: fluorescent focus-forming units; Gn: gnotobiotic; OD: optical density.

that reacts mainly with the most abundant rotavirus capsid protein VP6, which works for all the group A rotaviruses. The CCIF assay is done on 96-well microtiter plates using coded samples to determine the titer of infectious HRV. Fluorescing cells were visualized by fluorescent microscopy and the number of fluorescent focus-forming units counted in each well.

The highest titer of virus (2×10^7 FFU/ml) was detected in fecal samples of Gn pigs at 24–48 hours post-inoculation (hpi) with the VirHRV, which forms the first peak of virus shedding. Two peaks of fecal rotavirus shedding are often observed in naïve Gn pigs infected with VirHRV (Azevedo et al., 2005; Wen et al., 2016), as well as in suckling mice infected with murine rotavirus (Boshuizen et al., 2003; Maffey et al., 2016). The second peak in Gn pigs happens at post-inoculation day (PID) 5–6 (Figure 2.2). The two peaks of rotavirus shedding in mice were observed at PID 1–2 and 4 (Boshuizen et al., 2003; Maffey et al., 2016). It was suggested that because epithelial renewal occurs within four days, the second peak of viral replication is most likely due to infection of newly matured villus epithelial cells since rotavirus only infects mature villus epithelial cells at the apices of small intestinal villi (Rollo et al., 1999).

2.4 Pathogenesis and distribution of HRV antigens in tissues of VirHRV infected Gn pigs

In the pathological study (Ward et al., 1996a), four-day-old Gn pigs were orally inoculated with ~10^5 FFU (~10^5 ID$_{50}$) VirHRV; age-matched pigs were inoculated with diluent as controls. Pigs were euthanized at 13, 24, 48, 72, and 96 hpi and 7 days pi. After virus inoculation, pigs developed diarrhea within 13 hpi, which correlated with detectable rotavirus antigen in villous epithelial cells. Villous atrophy was observed at 24–48 hpi coincident with the peak of virus replication. Viral antigens were readily found within the gastrointestinal tissues of the virulent Wa HRV-inoculated pigs. Relatively large amounts of antigen were present throughout the small intestine of the virulent Wa HRV-inoculated pigs at 13 hpi with smaller amounts of viral antigens additionally detected in gastric, colonic, rectal, and mesenteric lymph node tissues at 24 hpi. From 48–96 hpi, viral antigen was again limited to the duodenum, jejunum, and ileum. The greatest number of antigen-positive cells were in the epithelial cells of jejunal and ileal villi at 24 hpi (Ward et al., 1996a). In another study that tested the presence of rotavirus in serum of Gn pigs after inoculation with the virulent Wa HRV, rotavirus antigen was detected by ELISA from PID 1–7 (Azevedo et al., 2005). There was one peak at PID 3. Although this viremia was only detectable by ELISA, not CCIF, the infectivity of sera from the viremic VirHRV-inoculated pigs was confirmed by inoculating Gn pigs orally with pooled HRV-positive serum. Serum-inoculated pigs developed diarrhea and fecal virus shedding and seroconverted with serum and intestinal HRV antibodies (Azevedo et al., 2005).

2.5 Attenuated Wa HRV as a prototype HRV vaccine in Gn pigs

The MA104 cell culture adapted Wa HRV at more than 27 passages (Figure 2.1 and Table 2.1) is attenuated for infectivity and pathogenesis in Gn pigs; it has been used as a prototype G1P[8] HRV live oral vaccine, mimicking the Rotarix™, in many of our studies. After oral inoculation of Gn pigs with 5×10^7 to 3×10^8 FFU of the AttHRV, 6% to 25% of pigs shed virus in feces for one to two days with the mean peak titers ranging from 1×10^3 FFU/ml to 3.4×10^3 FFU/ml and had transient loose stools (fecal score = 2 for one to two days) (Azevedo et al., 2005; Ward et al., 1996a; Zhang et al., 2008b). After oral or intranasal inoculation, AttHRV nasal shedding is observed more frequently (50–95%) than fecal shedding. It lasts for one to two days and the peak titers were in the range of 7×10^3 to 3×10^4 FFU/ml. AttHRV does not cause viremia/antigenemia as does after inoculation with the VirHRV, so direct contact of the nasal or oral mucosal surfaces with the virus delivered orally or intranasally may be required for infection of epithelial cells at the upper respiratory tract. There is no sign of respiratory disease in Gn pigs after AttHRV infection.

There is no significant villous atrophy in any of the intestinal sections from the AttHRV inoculated pigs (Ward et al., 1996a). Minor microscopic changes were observed within the villous epithelial cells at 13 and 24 hpi. Scattered duodenal and jejunal villous absorptive cells lost normal cytoplasmic vacuolation but the affected cells remained attached to their respective basement membrane and normal cellular morphology returned by 48 hpi (Ward et al., 1996a). No further morphological changes or lesions were observed in tissues from the AttHRV inoculated pigs except for the presence of mild lymphoid hyperplasia in the Peyer's patches and MLN at 72–96 hpi and at PID 7, suggesting the induction of immune responses by the AttHRV. After two doses of AttHRV inoculation at doses $\geq 10^7$ FFU, 100% pigs seroconverted. The geometric mean titers (GMT) of virus-neutralizing (VN) antibodies in serum at PID 21 are higher after oral inoculation with higher AttHRV inoculation doses (Table 2.2). The ID_{50} of the Wa AttHRV was determined to be 1.3×10^6 FFU based on seroconversion using the Reed and Muench method, not virus shedding since the fecal virus shedding rates were lower than 50% even after the highest inoculation dose used in the study (Ward et al., 1996a).

The knowledge of replication, dissemination, and virus shedding patterns of VirHRV and AttHRV are important for the evaluation of the design and delivery effectiveness of HRV vaccines. Oral live rotavirus vaccines are effective in inducing mucosal IgA antibodies at the intestinal induction site (mainly ileum); these antibodies can block rotavirus entry of epithelial cells in the small intestine to confer a high degree of protection rate against infection. On the other hand, intramuscular non-replicating rotavirus vaccines (NRRV) only induce systemic IgG, not mucosal IgA. By neutralizing the rotavirus in blood during viremia, the serum IgG antibodies induced by NRRV can shorten the duration of virus shedding and diarrhea, thus conferring a certain degree of protection, even though they do not prevent initial infection of the small intestine. For evaluation of vaccine effectiveness, in addition to fecal virus shedding, detection of viremia may be another parameter to assess protection against infection induced by the candidate vaccines (Azevedo et al., 2005).

Table 2.2   Rotavirus Neutralizing Antibody Responses in Gn Pigs Inoculated with AttHRV at PID 21

AttHRV2x dose/pig (FFU)	8×10^5	4.5×10^6	2×10^7	7×10^7	3×10^8
Percent of VN positive	33%	80%	100%	100%	100%
Serum VN GMT (range)	4.2 (<4–19)	32 (4–86)	62 (14–86)	42 (18–280)	69 (12–256)

Note: Virus-neutralizing (VN) antibody titers were determined by plaque reduction assay. Serum VN antibody titers were calculated as the reciprocal of the serum dilution that reduced the number of plaques by $\geq 80\%$; VN titers of <4 were considered to be negative and assigned a value of 2 for determining the GMT (n = 3–7). Data are summarized from Ward et al. (1996a).

3

Establishment of Gn Pig Model of HuNoV Infection and Diarrhea

3.1 Median infectious dose of HuNoV GII.4/2006b in two different age groups of Gn pigs

The median infectious dose (ID_{50}) of two inocula containing HuNoV GII.4/2006b and GII.4/2003 variant, respectively, was determined in Gn pigs in my laboratory (Bui et al., 2013; Ramesh et al., 2020). For the Gn pig HuNoV challenge model to be used for an accurate evaluation of vaccine efficacy, ID_{50} at the time of challenge (33–34 days of age) is an important parameter to determine. In the first study, the ID_{50} of a GII.4/2006b variant (GenBank accession number KC990829) was determined in both neonatal (4–5 days of age) and older pigs (33–34 days of age). This study was led by my PhD student Tammy Bui Castellucci and was funded by an R01 sub-award I received from NIAID, NIH titled "Novel vaccine against norovirus" (R01AI089634, 2010–2015, PI Xi Jiang).

The pool of human stool samples (ID 092895) was collected in Dr. Xi Jiang's laboratory in Cincinnati Children's Hospital Medical Center from a child with HuNoV gastroenteritis in 2008. The stool sample was diluted to 10% and treated by high-speed centrifugation to remove bacteria. The absence of contaminating viruses was confirmed by testing on a Virochip Microarray (University of California, San Francisco Viral Diagnostics and Discovery Center). Endotoxin in the inoculum may cause diarrhea in Gn pigs thus needs to be measured. The inoculum contains an endotoxin level of 1 EU/ml, which is below the recommended level for live attenuated vaccines (<200 EU/ml) (Brito and Singh, 2011), therefore it is safe for neonatal Gn pigs.

Before virus inoculation, Gn pigs were confirmed to be A+ or H+ in blood type by PCR and immunofluorescence assay using antibodies against human A, H1, and H2 antigens that are cross-reactive with porcine A and H antigens. It has been shown that histo-blood group antigen (HBGA) phenotype influences susceptibility to norovirus infection and that A+ or H+ pigs are more likely to be infected than A– or H– pigs (Cheetham et al., 2007; Huang et al., 2003; Huang et al., 2005; Hutson et al., 2002; Marionneau et al., 2002; Rockx et al., 2005; Tan and Jiang, 2005a; Tian et al., 2007). In all the subsequent studies using the Gn pig model of HuNoV infection, only A+ or H+ pigs were used. In Bui et al study, pigs (n = 3–6) were orally inoculated at either 4–5 or 33–34 days of age with 2.74×10^3, 2.74×10^4, 2.74×10^5, or 2.74×10^6 viral RNA copies of GII.4 variant 092895. Four ml of 200 mM sodium bicarbonate were given 10 minutes prior to inoculation to neutralize stomach acids. Pigs were euthanized at PID 3, 4, or 7 to collect intestinal contents and tissues. HuNoV virus shedding titers in daily fecal swab samples are measured with a TaqMan® real-time RT-PCR (Bui et al., 2013).

To assess diarrhea induced by HuNoV, the fecal consistency scoring system was amended depending on the age of the pigs due to the change in consistency of the feces (become looser) with maturity. In addition, the sign of diarrhea in Gn pigs inoculated with GII.4/2006b at 4–5 days of age was less severe compared to that after the virulent Wa HRV infection. Pigs inoculated at 4–5 days of age with HuNoV were scored based on the following system: 0) solid, 1) semi-solid, 2) pasty, 3) semi-liquid, and 4) liquid. Pigs inoculated at 33–34 days of age with HuNoV were scored based on the more stringent system that is the same for the Gn pig model of Wa HRV diarrhea: 0) solid, 1) pasty, 2) semi-liquid, and 3) liquid. For both scoring systems, a score of 2 or greater was considered diarrhea. Therefore, after HuNoV infection, a pasty stool is considered diarrhea in 4–5-day-old pigs, but not in 33–34 days old pigs.

The ID_{50} of the GII.4/2006b in neonatal (4–5 days of age) pigs was determined to be $\leq 2.74 \times 10^3$ viral RNA copies and in older pigs (33–34 days of age), the ID_{50} was 6.43×10^4 using

DOI: 10.1201/b22816-3

the Reed and Muench method (Reed and Muench, 1938). The higher dose required to infect the older pigs is probably due to the more developed innate immune system of these pigs. Moderate to severe cytopathological changes were observed in the small intestine, including irregular microvilli, necrosis, and apoptosis, and viral antigen was detected in the tip of villi in the duodenum (Bui et al., 2013). Bui et al.'s study also investigated the impact of a choles-terol-reducing drug, simvastatin on increasing the susceptibility of HuNoV infection in older pigs. The ID_{50} in simvastatin treated Gn pigs was reduced to $<2.74 \times 10^3$ viral RNA copies.

In subsequent five studies in Gn pig using this GII.4/2006b inoculum, 10 ID_{50} (2.74×10^4 viral RNA copies for 4–5-day-old pigs and 6.43×10^5 viral RNA copies for 33–34 day old pigs) has consistently resulted in virus shedding in 83–100% and diarrhea in 78% to 100% of naïve con-trol Gn pigs (Kocher et al., 2014; Lei et al., 2016a; Lei et al., 2016b; Lei et al., 2016c; Lei et al., 2019). The successful establishment of the Gn pig model of GII.4/2006b infection and diarrhea greatly facilitated our studies to understand the factors in norovirus infection and immunity (Lei et al., 2016b; Lei et al., 2016c; Lei et al., 2019) and provided the first well-standardized animal model for evaluation of vaccine (Kocher et al., 2014) and prophylaxis (Lei et al., 2016a) against HuNoV diarrhea.

3.2 Median infectious dose and median diarrhea dose of HuNoV GII.4/2003 in Gn pigs and dose-response models

The primary objective of this study was to determine the ID_{50} and DD_{50} of another pandemic strain of HuNoV GII.4/2003 inoculum pool in Gn pigs (Ramesh et al., 2020). This study was led by my PhD student Ashwin Ramesh and Dr. Viviana Parreno who is my life-long friend and col-laborator. Viviana was studying toward her master's degree in biometrics and statistics. Data analyses by different methods for this project benefited greatly from her statistical knowledge; the work also provided the contents of her master thesis. The project was funded by a HuNoV vaccine evaluation contract I received from Anhui Zhifei Longcom Biopharmaceutical, China (July 2017–December 2018).

ID_{50} values are often estimated using classical methods, such as the Reed and Muench method that we used for determining the ID_{50} of the GII.4/2006b (Bui et al., 2013). In recent decades, several mechanistic models have been developed to describe plausible phenomena that are inherent to experimentally derived dose-response data (Schmidt, 2015; Teunis et al., 2008). Critically, such models provide a basis for inference about the probability of infection at any dose level (i.e., not just ID_{50}) (Messner et al., 2014). Contemporary dose-response models may offer greater flexibility and accuracy (i.e., managing inconsistent dilution factors between doses or numbers of subjects per dose group or small numbers of subjects) in estimating ID_{50}. The detailed significance of this study includes 1) experimental evaluation of GII.4/2003 HuNoV dose-response in 33–34 day old Gn pigs; 2) comparison of different dose-response analyses used for estimating the ID_{50} and DD_{50} of HuNoV in Gn pigs to determine the best-fit model; 3) comparison of the ID_{50} obtained in this study with the ID_{50} used in previous human volunteer challenge studies; 4) identification of the most appropriate challenge dose in 33–34 day old Gn pigs to standardize the model for HuNoV vaccine evaluation. Comparing the infectiousness of the GII.4 variant challenge pools in Gn pigs and humans will lead to a better understanding of the zoonotic potential (Villabruna et al., 2019) of HuNoVs between differ-ent species and further validate the Gn pig model of HuNoV infection as a proper predictive tool of the future efficacy of vaccines and other antiviral strategies to control HuNoV diarrhea in humans.

The GII.4 HuNoV challenge inoculum pool (ID 103041) used in this study, identified hence-forth as Cin-2, was the unfiltered 10% suspension of stool samples collected from a volunteer who was challenged with HuNoV GII.4/2003 (Hu/GII.4/Cin-1/2003/US), a 2002 Farmington Hills–like variant (GenBank number JQ965810), as part of a vaccine study conducted by Xi Jiang's laboratory at Cincinnati Children's Hospital Medical Center. The volunteer developed a three-day illness characterized by diarrhea, vomiting, nausea, abdominal cramps, and fever (Bernstein et al., 2015; Frenck et al., 2012). The HuNoV concentration of the challenge pool was determined by RT-qPCR to be approximately 2×10^6 viral genomic RNA copies/mL of stool (Ramesh et al., 2020).

A+/H+ Gn pigs at 33–34 days of age were divided into seven groups and inoculated orally with different doses of HuNoV GII.4/2003 Cin-2. Diarrhea and fecal virus shedding were monitored daily until PID 7. Fecal consistency scores and presence of HuNoV genomes in rectal swab samples from PID 1–7 were determined based on previous studies carried out by our group (Lei et al., 2016a; Lei et al., 2016b; Lei et al., 2016c). The percentage of affected animals, mean days to onset, duration, peak titer, and AUC of virus shed in feces of pigs in each of the seven dose groups are summarized in Table 3.1. An increase in inoculation dose was positively associated with a shorter incubation period, which coincided with the observed increased duration as well as increased overall virus shedding in feces measured by AUC. All pigs belonging to dose groups 1 to 4 shed detectable titers of virus in their feces, while 67% of pigs in dose groups 5 and 6 and 25% of the pigs in dose group 7 shed viruses. It is important to note that pigs in dose group 3 shed a significantly higher ($>1 \times 10^6$ RNA copies) amount of viruses in their feces ($p < 0.0001$) and shed viruses for a significantly longer duration (>3 days) than pigs in the other groups ($p < 0.0001$).

Using various mathematical approaches (Reed–Muench, Dragstedt–Behrens, Spearman–Karber, logistic regression, and approximate beta-Poisson dose-response models), we estimated the ID_{50} and DD_{50} to be between 2,400–3,400 RNA copies, and 21,000–38,000 RNA copies, respectively (Table 3.2). Contemporary dose-response models offer greater flexibility and accuracy in estimating ID_{50}. In contrast to classical methods of endpoint estimation, dose-response modeling allows seamless analyses of data that may include inconsistent dilution factors between doses or numbers of subjects per dose group, or small numbers of subjects. While the different data analysis approaches used in this study feature varying assumptions, limitations, and degrees of statistical complexity, the somewhat tight clustering of results indicates that the simple classical methods perform adequately for this particular dataset.

Using logistic regression, the ID_{50} of the original Norwalk virus (GI.1, 8fIIa isolate) challenge pool was predicted to be around 2.8×10^3 RNA copies in secretor-positive individuals (Atmar et al., 2014). Similarly, in the current study, we estimated the ID_{50} of Cin-2 to be around 2.51×10^3 RNA copies (Table 3.2). These data show a similar infection potential between the prototypic GI.1 Norwalk virus and the Cin-2.

We compared the percentage, mean duration, and peak day of virus shedding between humans and Gn pigs after inoculation with GII.4/2003 variants of HuNoV (Table 3.3). In the human adult challenge study (Frenck et al., 2012), 16 out of 23 secretors (70%) showed signs of symptomatic illness along with the presence of HuNoV RNA in stool samples detected by RT-qPCR, when orally inoculated with 5×10^4 RNA copies of the challenge strain. In Gn pigs, a dose of 2×10^4 RNA copies infected 67% of pigs, whereas 8×10^4 and 2×10^5 RNA copies infected 100% of pigs (Table 3.1). It is observed that pigs inoculated with 2×10^5 RNA copies shed virus for a similar duration of days compared to those of the human volunteers in the study conducted by Frenck and colleagues (Frenck et al., 2012).

With the availability of more dose-response studies and the relatability between human studies and Gn pig studies, it could eventually be possible to compare the pathogenesis of different HuNoV genogroups to help further identify similarities and/or differences between different HuNoVs.

Based on the virus shedding and diarrhea status of infected Gn pigs in this study and on the virus shedding data in humans, we approximated the optimal dose for Cin-2 to be 2×10^5 RNA copies. Our data show that pigs infected at this dose had a mean onset day of 1.3 with viruses shed in feces in large quantities (measured by AUC) for almost the whole duration of the infection period (6.3 days out of 7). Moreover, pigs in this dose group also experienced the highest diarrhea burden among all dose groups, with diarrhea starting at 2.8 days after inoculation and occurring for a duration of four days. Pigs in this group also had the highest mean cumulative diarrhea score of 9.31 (Table 3.1).

The ID_{50} of GII.4/2006b variant was identified to be 6.43×10^4 RNA copies in 33–34 days old Gn pigs using the Reed and Muench method (Bui et al., 2013). In the current study, using the Reed and Muench method, we estimated the ID_{50} of Cin-2, a variant of the Farmington Hills virus, to be 2.51×10^3 RNA copies in similarly aged pigs (Table 3.2). This analysis suggests that at least a 25-fold higher virus titer is required to establish infection among 50% of the pigs inoculated with the GII.4/2006b virus as compared to the Cin-2, indicating that Cin-2 is a more infectious strain in Gn pigs (Ramesh et al., 2020).

Table 3.1   HuNoV Fecal Shedding and Diarrhea after Inoculation of Gn pigs with Different Doses of Cin-2

Dose group	# of viral genome copies	n	Virus shedding				Diarrhea			
			(%)[a]	Mean duration days (SEM)[c-e]	AUC[d,f]	Mean onset day (SEM)[c,d]	(%)[b]	Mean duration days (SEM)[c-e]	AUC[d,f]	Mean onset day (SEM)[c,d]
1	2×10^6	4	4 (100%)	2.5 (0.6)^{BC}	9,506^B	2 (0.4)^B	4 (100%)	5.0 (0.3)^A	9.06^A	1.5 (0.5)^D
2	4×10^5	4	4 (100%)	1.3 (0.3)^{CD}	5,232^B	4 (1.2)^{AB}	3 (75%)	1.3 (0.9)^{ABC}	7.04^A	5.5 (1.2)^{ABC}
3	2×10^5	4	4 (100%)	6.3 (0.5)^A	126,774^A	1.3 (0.3)^B	4 (100%)	4.0 (0.9)^{AB}	9.31^A	2.8 (0.5)^{CD}
4	8×10^4	6	6 (100%)	2.8 (0.5)^B	13,495^B	1.5 (0.3)^B	6 (100%)	3.8 (1)^{AB}	7.46^A	3.2 (0.9)^{CD}
5	2×10^4	3	2 (67%)	1 (0.6)^B	93^B	3.3 (2.3)^{AB}	0 (0%)	0.0 (0)^{ABC}	1.50^B	6.3 (1.7)^{AB}
6	3.20×10^3	3	2 (67%)	1.3 (0.9)^{BCD}	2,667^B	4 (2.1)^{AB}	1 (33%)	1.0 (0)^{BC}	3.17^{AB}	5 (2.1)^{BC}
7	8×10^2	4	1 (25%)	0.5 (0.5)^D	2,972^B	6.8 (1.3)^A	0 (0%)	0.0 (0)^C	2.75^{AB}	8 (0)^A

Note:

a) Gn pigs were orally inoculated with HuNoV GII.4/2003 variant (Cin-2) at 33–34 days of age. Rectal swabs were collected daily after inoculation from PID 1–7 to determine virus shedding by RT-qPCR.

b) Fecal consistency was assessed from PID 1–7 as 0: solid; 1: pasty; 2: semi-liquid; 3: liquid. Pigs with daily fecal consistency scores of 2 or greater were considered to be diarrheic.

c) SEM, standard error of the mean.

d) Numbers in the same column followed by different capital letters (A, B, C, D) differ significantly (Tukey-Kramer HSD, $p < 0.05$); while shared letters indicate no significant difference.

e) As pigs were sacrificed on PID 7, some data are right-censored (>7). A value of 7 was substituted in mean calculations.

f) AUC, the area under the curve.

Table 3.2 ID_{50} and DD_{50} Calculations of Cin-2

Method	ID_{50}	DD_{50}
Reed–Muench		
Hand calculation "skrmdb" R script	2.51×10^3	3.80×10^4
Dragstedt–Behrens *"skrmdb" R script*	2.45×10^3	3.80×10^4
Spearman–Karber *Hand calculation "skrmdb" R script online calculator*	3.31×10^3	3.09×10^4
Logistic regression *online calculator*	2.51×10^3	2.18×10^4
Approximate beta-Poisson		
R script [a]	2.57×10^3	2.13×10^4

Note:
[a] Exponential and beta-Poisson were determined based on the R script used by Weir et al. (2017).

Table 3.3   Comparison of Virus Shedding in Unvaccinated Gn Pigs and Humans after Inoculation with GII.4/2003 HuNoV Cin-2 Inoculum

Host	Age	Challenge dose	n	Virus shedding (%)	Mean duration days (range) [c]	Peak virus shedding day (PID)
Human [a]	19–48 years	5.00E + 04	23	70	5.2 (2–30)	3
Human [b]	18–50 years	4.40E + 03	34	76.5	—	—
Human [b]	18–50 years	4.40E + 03	48	62.5	—	—
Gn pig	33–34 days	2.00E + 04	3	67	1.0 (1–2)	2
Gn pig	33–34 days	8.00E + 04	6	100	2.8 (2–4)	3
Gn pig	33–34 days	2.00E + 05	4	100	6.3 (5–7)	4

Note:
[a] Data reported by Frenck et al. (2012).
[b] Data reported by Bernstein et al. (2015).
[c] Virus shedding in human stools was monitored for up to 30 days; and in Gn pigs for up to seven days post-inoculation.

In conclusion, Gn pigs have a similar susceptibility to HuNoV infections as in humans. These pigs are also capable of exhibiting the full course of HuNoV infection and developing the disease as it is presented in humans. Determining the optimal HuNoV dose would help establish consistency in terms of the challenge dose for the evaluation of novel vaccine candidates between preclinical and clinical trials. To this end, we have established a reliable model that can be used to test candidate prophylactics and therapeutics before clinical trials in humans, and the model is ready to be used for the evaluation of immunogenicity and the protective efficacy of candidate HuNoV vaccines.

4

Evaluation of Live Oral and Inactivated Intramuscular HRV Vaccines in Gn Pigs

4.1 Protective efficacy conferred by natural infection versus live attenuated HRV vaccines and correlates of protective immunity

With the establishment of the Gn pig model of Wa HRV infection and diarrhea in 1996 (Saif et al., 1996; Ward et al., 1996a), the pig model was used in the first studies for the identification of correlates of protection with cell-mediated and antibody immune responses induced by prior infection with the virulent Wa HRV (VirHRV) and with those induced by the live attenuated Wa HRV (AttHRV), to mimic host response to natural infection versus attenuated oral vaccines, respectively, after homotypic challenge with the VirHRV (Ward et al., 1996b; Yuan et al., 1996).

At three to five days of age, Gn pigs were inoculated orally with 10^5 FFU of Wa VirHRV or 2×10^7 FFU of live Wa AttHRV. All pigs are confirmed seronegative for anti-rotavirus antibodies and germ-free prior to HRV exposure in all our studies using Gn pigs. Pigs were fed 5 ml of 100 mM sodium bicarbonate to neutralize stomach acid approximately 10–20 min before inoculation. Pigs receiving AttHRV were reinoculated with the same dose 10 days later (PID 10) and a subset of the pigs received the third dose AttHRV at PID 21. Pigs given equal volumes of diluent served as mock-inoculated controls. Three or four weeks after the first inoculation (PID 21 or PID 28), virus-inoculated and mock-inoculated pigs were challenged orally with 10^6 FFU of Wa VirHRV. The pigs were examined daily from post-challenge day (PCD) 0-7 for the severity of diarrhea. Virus shedding from PCD 1–7 was assessed by ELISA and CCIF from the daily rectal swab samples. Blood samples were collected weekly to measure virus-specific IgM, IgA, and IgG antibody and virus-neutralizing (VN) antibody titers (To et al., 1998). A subset of each group of pigs (n = 4 to 6 at each time point) was euthanized at PID 0, 8, 21, 25, and 28 and PCD 7. The small intestines (duodenum and ileum), mesenteric lymph nodes (MLN), spleens, and blood samples were collected at euthanasia for isolation of mononuclear cells (MNC) for measuring antibody-secreting cell (ASC) and memory B cell responses by ELISPOT assay (Yuan et al., 2001a; Yuan et al., 1996) and T cell responses by lymphocyte proliferation assay (LPA) (Ward et al., 1996b). The protection rate was calculated as (1 – [% Wa HRV-inoculated pigs in each group with diarrhea/% diluent-inoculated control pigs with diarrhea]) × 100.

Pigs recovered from VirHRV infection were completely protected from virus shedding upon challenge (Table 4.1). Two or three doses of AttHRV inoculated pigs were partially protected against virus shedding. VirHRV inoculated pigs developed significantly higher numbers of HRV-specific IgA and IgG ASC and memory B cells in intestinal lamina propria (average numbers from duodenum and ileum) than the AttHRV inoculated pigs at PCD 0. The protection rate from the VirHRV, AttHRV2x, and AttHRV3x inoculation groups against virus shedding and diarrhea were positively correlated with the numbers of IgA ASC and memory B cells in the intestinal lamina propria. The VirHRV also elicited higher levels of HRV-specific serum and intestinal IgA antibodies compared to AttHRV (To et al., 1998). There is a positive correlation existed between protection and serum IgA and intestinal IgA antibody titers. These findings suggest that serum IgA antibodies to HRV could act as an indicator of IgA antibodies in the intestine after rotavirus infection.

Lymphocyte proliferation assay (LPA) was a method to assess cell-mediated immunity, mostly T helper cell function before flow cytometry become widely used to measure subtype-specific

DOI: 10.1201/b22816-4

Table 4.1 Intestinal and Systemic IgA and IgG ASC and Memory B cells at PCD 0 and Protection Induced by Oral Virulent and Attenuated Wa HRV Inoculation in Gn Pigs[a]

Virus inoculation group	Protection rate against challenge		Mean no. ASC/5 × 10⁵ MNC				Mean no. memory B cells/5 × 10⁵ MNC			
	Diarrhea (%)	Shedding (%)	Intestinal lamina propria		Spleen		Intestinal lamina propria		Spleen	
			IgA	IgG	IgA	IgG	IgA	IgG	IgA	IgG
Virulent Wa HRV	87	100	171	114	4	13	364	3637	58	2,288
Attenuated Wa HRV 3x	62	67	8	3	2	10	8	55	135	3,700
Attenuated Wa HRV 2x	34	19	6	41	1	2	ND[b]	ND	ND	ND
Diluent-inoculated control	0	0	0	0	0	0	0	0	0	0

Note:
a. Data are summarized from Yuan et al. (1996) and (2001a).
b. Not determined.

T cell responses induced by infection and vaccination. In VirHRV and AttHRV2x inoculated pigs and the controls, the magnitude of LPA responses in intestinal and systemic lymphoid tissues (blood, spleen, MLN, and small intestinal lamina propria) at PCD 0 correlated positively with the degree of protection from VirHRV challenge (Ward et al., 1996b). The magnitude of a tissue's LPA response is also positively correlated with the numbers of virus-specific ASCs in that tissue (Yuan et al., 1996). The Ward et al. study also showed that peripheral blood MNC LPA activity transiently (at PID 5–9 and PCD 3–4) reflected that of intestinal MNC, supporting the use of blood as a temporary window to view intestinal T cell activity and providing a possible tool by which mucosal immunity may be studied indirectly using blood MNC.

In a subsequent study, IFN-γ-producing CD4+ and CD8+ T cell responses in the intestinal (ileum) and systemic (spleen) lymphoid tissues and peripheral blood of Gn pigs induced by VirHRV infection and AttHRV2x, AttHRV3x, and AttHRV2x/virus-like particle (VLP)1x vaccine (using a combined regimen with oral AttHRV priming and intranasal VLP boosting) were evaluated (Yuan et al., 2008). Production of IFN-γ after 17 hours of antigenic stimulation is a functional characteristic of virus-specific effector-memory T cells (Mayer et al., 2005). We used semi-purified Wa AttHRV as a stimulating antigen in flow cytometry and intracellular staining assay to detect intracellular accumulation of IFN-γ by CD4+ and CD8+ T cells after re-activation. Studies of the interaction of rotavirus with human myeloid dendritic cells (MDC) demonstrated that intact peripheral blood MNC populations contain antigen-presenting cells are equally efficient compared to rotavirus-infected MDC, in stimulating IFN-γ-producing rotavirus-specific effector-memory T cells (Narvaez et al., 2005). Furthermore, the frequencies and patterns of cytokines produced by effector-memory CD4+ T cells after stimulation of peripheral blood MNC with the purified rhesus rotavirus (RRV) or RRV-infected MDC were similar. Thus, intracellular IFN-γ staining assays using semi-purified intact Wa AttHRV as stimulating antigen with MNC populations can provide comparative information regarding the tissue distribution and magnitude of rotavirus-specific CD4+ and CD8+ effector-memory T cell responses elicited by rotavirus infection and vaccines.

After VirHRV infection or HRV vaccine immunizations, the frequencies of HRV-specific intestinal IFN-γ producing CD4+ and CD8+ effector-memory T cells in Gn pigs at the time of challenge (PID 28/PCD 0) correlated significantly with protection rates against VirHRV-induced diarrhea. However, there is no correlation between splenic IFN-γ producing T cell responses and protection. Thus, this study demonstrated a direct correlation between the existence of effector-memory T cells and protection at the site of virus entry (small intestine) (Yuan et al., 2008) (Table 4.2).

Table 4.2   Correlation between Mean Frequencies of IFN-γ Producing T Cells Induced by HRV Infection or Vaccines at PCD 0 and Protection Rates against VirHRV Diarrhea

Group	Protection rate against HRV diarrhea (%)[a]	Mean frequencies of effector-memory T cells in the ileum	
		IFN-γ+CD3+CD4+ T cells	IFN-γ+CD3+CD8+ T cells
VirHRV	87[A b]	1.40[A c]	1.49[A]
AttHRV2/6VLP	71[A]	0.82[AB]	1.07[A]
AttHRV3x	62[AB]	0.28[B]	0.07[B]
AttHRV2x	34[B]	0.18[B]	0.07[AB]
Mock	0[B]	0.02[C]	0.01[B]
		$r = 1.0$[d]	$r = 0.975$[d]
		$p < 0.0001$	$p = 0.0048$

Note:
[a] Summarized from historical published data (Nguyen et al., 2006a; Yuan et al., 2001b; Yuan et al., 1996).
[b] Numbers in the same column with different superscript letters differ significantly (Fisher's exact test, $p < 0.05$).
[c] Numbers in the same column with different superscript letters differ significantly (Kruskal-Wallis test, $p < 0.05$).
[d] Correlation coefficient between protection rates and mean frequencies of IFN-γ+ T cells (Spearman correlation).

At PID 28/PCD 0, blood MNC had the lowest frequencies of IFN-γ producing T cells compared to ileum and spleen in VirHRV infected or AttHRV vaccinated pigs, and the frequencies did not differ among treatment groups; therefore, they did not reflect the magnitude of HRV-specific T cell responses in ileum or spleen at PID 28. Together with Ward et al.'s study discussed earlier (1996b), it clearly indicated the importance of timing in measuring T cells responses using MNC from blood samples.

Our findings on the magnitudes of intestinal B cell and T cell responses after VirHRV infection and AttHRV vaccination reinforce the notion that the increased intensity of an intestinal immune response is directly related to the increased magnitude of viral replication in the intestine, presumably reflecting, in part, substantially increased virus quantities. These studies of magnitude, tissue distribution, and dynamics of humoral and cellular immune responses induced by VirHRV infection and AttHRV vaccines have established basic parameters related to immune protection in the Gn pig model of VirHRV-induced disease, verifying the usefulness of this model to examine new strategies for the design and improvement of HRV vaccines.

4.2 Evaluation of immunogenicity and protective efficacy of inactivated HRV vaccines

Inactivated whole virus vaccine is the most mature vaccine development technology. It contains physically or chemically killed, cultured viruses mixed with an adjuvant. Compared to live attenuated rotavirus vaccine, inactivated vaccines are safer, without the risk of reassorting with wild type rotaviruses or reverting to the virulent virus, and will not post the potential risk of prolonged virus shedding in immunocompromised infants as it does by live attenuated rotavirus vaccines (Kaplon et al., 2015; Patel et al., 2010). Compared to subunit rotavirus vaccines, the inactivated vaccine contains all rotavirus structural proteins, including the inner capsid protein (VP6) and the two outer capsid proteins (VP4 and VP7), with a large variety of antigenic epitopes that may lead to a broader, cross-reactive immune response (Jiang et al., 2013).

Two inactivated whole rotavirus vaccine candidates have been evaluated in the Gn pig model of Wa HRV infection and diarrhea using the intramuscular (IM) injection route (Wang et al., 2010; Yuan et al., 1998) or the skin immunization route (Wang et al., 2016b). In the first study, inactivated AttHRV was used as a prototype vaccine and was inactivated using binary ethylenimine (BEI) (Yuan et al., 1998). To prepare the vaccine, the supernatant of AttHRV-infected MA104 cell lysates (the virus titer before inactivation was approximately 10^7 FFU/ml) was treated with 10% BEI for 18 h at 37° C with continuous agitation. Sodium thiosulfate solution (1 M) was added to the virus-BEI mixture to a final concentration of 10% to inactivate residual BEI. Inactivation was verified by loss of infectivity in MA104 cells.

Gn pigs were inoculated twice or three times with 5 ml of the BEI inactivated Wa strain AttHRV mixed with an equal volume of incomplete Freund's adjuvant (IFA) and challenged with ~10^6 FFU of VirHRV at PID 20 or PID 24. Neither two nor three doses of the vaccine conferred significant protection against the VirHRV challenge (0 to 6% protection rate) (Table 4.3).

Few virus-specific IgA ASC or memory B cells were detected in any tissues of the inactivated HRV vaccinated pigs prior to challenge (Table 4.4). But these pigs had high numbers of virus-specific IgG ASC and IgG memory B cells in spleen and blood and high serum virus neutralizing (VN) antibody titers at challenge (Tables 4.3 and 4.4) and had significant anamnestic virus-specific IgG ASC and memory B cell responses post-challenge. Hence, high numbers of IgG ASC or memory IgG ASC in the systemic lymphoid tissues and the serum VN antibody titers at the time of challenge did not correlate with protection in this study (To et al., 1998; Yuan et al., 1998).

There are several drawbacks of this study, which may help to explain the lack of protective efficacy of the inactivated HRV vaccine in Gn pigs. The time between the second and third dose vaccine was seven days and the time of challenge from the first dose vaccine administration was only 21 days. These times are very compact. Longer time intervals may give the immune system more time to develop higher secondary immune responses. The process of inactivating HRV with BEI may have damaged the epitopes on the outer capsid proteins VP4 and VP7, both are important virus-neutralizing antigens, leading to the immune responses mostly against the inner capsid protein VP6, which is not a virus-neutralizing antigen.

Table 4.3 Diarrhea and Virus Shedding in Inactivated HRV IM Inoculated and Control Pigs after Oral Challenge with VirHRV

Treatments (inactivated HRV[a])	Diarrhea[b]				Virus shedding[b]				Serum VN GMT at PCD 0[a]
	% with diarrhea	Protection rate against diarrhea[c]	Mean duration (days)	Mean cumulative fecal score	% of shedding virus	Mean days to onset	Mean duration (days)	Mean peak titer (FFU/g of feces)	
HRV2x	83%	6	2.8 (0.4)	8.4 (0.8)	100%	1.7 (0.2)	3.2 (0.3)[b]	1.3×10^5	4,000[a]
HRV3x	100%	0	2.5 (0.7)	7.3 (1.1)	100%	2.0 (0)	3.5 (0.7)[a,b]	3.0×10^4	10,951[a]
Diluent control	88%	0	3.2 (0.5)	9.6 (0.9)	100%	2.0 (0.2)	4.4 (0.3)[a]	3.5×10^5	<4[b]

Note:

a. Summarized from published data (Yuan et al., 1998). Pigs were immunized twice or three times IM with inactivated Wa AttHRV or placebo at five days (post-inoculation day [PID] 0), 15 days (PID 10), and 22 days (PID 17) of age. IFA was used as an adjuvant. On PID 21 or 24, all pigs were orally challenged with ~10^6 FFU of virulent Wa HRV and monitored for diarrhea and virus shedding for seven days post-challenge.

b. Fecal consistency scores were used to assess diarrhea; scores are defined as 0) solid, 1) pasty, 2) semi-liquid, and 3) liquid. Scores of 2 or higher are considered diarrheic. Rotavirus shedding titers were determined by cell culture immunofluorescent infectivity assays.

c. Protection rate = [1 − (percentage of Wa HRV-inoculated pigs in each group with diarrhea/percentage of diluent-inoculated control pigs with diarrhea)] × 100.

d. Different letters indicate significant differences between groups ($p < 0.05$), while shared letters or no letters indicate no significant difference.

Table 4.4 Peak Intestinal and Systemic IgA and IgG ASC and Memory B Cell Responses in Gn Pigs after Inactivated HRV IM Inoculation and VirHRV Oral Challenge[a]

Treatments (inactivated AttHRV[a])	PCD 0 Mean no./5 × 10⁵ MNC								PCD 4 Mean no./5 × 10⁵ MNC							
	ASC				Memory B cells				ASC				Memory B cells			
	Intestinal lamina propria		Spleen		Intestinal lamina propria		Spleen		Intestinal lamina propria		Spleen		Intestinal lamina propria		Spleen	
	IgA	IgG	IgA	IgG	IgA	IgG	IgA	IgG	IgA	IgG	IgA	IgG	IgA	IgG	IgA	IgG
HRV2x	2.5	3.5	0.3	34	1.5	0	4.5	2017	60	218	2	152	455	4080	0	3450
HRV3x	0.5	5.0	1.0	80	0	3	0	225	6	39	20	370	ND[b]	ND	ND	ND

Note:
[a]. Data summarized from Yuan et al. (1998).
[b]. Not determined.

In contrast to the negative outcome of the inactivated Wa HRV, another inactivated HRV (CDC-9, G1P[8]) vaccine demonstrated substantial protection against virus shedding after challenge with VirHRV in Gn pigs (Wang et al., 2010). Although the CDC-9 HRV has the same VP4 and VP7 sero-type as the Wa HRV, the inactivated CDC-9 vaccine candidate has two major differences from the inactivated Wa HRV. First, the method of inactivation was different. Instead of chemically inacti-vating the virus with BEI, a simple physical method, heat treatment for 6 h, was used to inactivate CDC-9 (Fix et al., 2020). Second, CDC-9 virion's triple-layered particle (TLP) is much more stable than other rotavirus strains and extremely resistant to experimental manipulations (Nair et al., 2017). In general, rotavirus TLP tends to uncoat during storage and experimental manipulation, leading to the partial loss of VP4 and VP7 and exposing the inner capsid VP6 as the major antigenic protein to be recognized by the host immune cells after injection. However, this is not the case for CDC-9. The CDC-9 strain grows in Vero cells at 7–8 logs with 90% of the particles maintaining their triple-layered structure after heat inactivation, as judged by electron micros-copy (Wang et al., 2010). The reason for the naturally occurring stability of CDC-9 TLP is not fully understood. The culture-adapted CDC-9 strain shared moderate to high amino acid sequence identities (range 81.3–97.9%) with the KU and other G1P[8] strains; however, the VP3 of CDC-9 is closely related to that of strain DS-1, not the KU or Wa HRV strains (Esona et al., 2010).

In this study, Gn pigs were inoculated IM three times on PID 0, 10, and 21 with 50 µg/dose purified CDC-9 TLP together with 600 µg/dose aluminum phosphate (Al[PO]$_4$) adjuvant. Control pigs received the alum adjuvant only. On PID 28, all pigs were challenged orally with VirHRV Wa strain at the dose of 10^5 FFU. Rectal swabs were collected from PCD 0–10 and virus shedding was measured by an antigen ELISA (Premium Rotaclone kit; Meridian Diagnostics, Cincinnati, OH). All five control pigs shed rotavirus antigen for three to five days; four of them had peak antigen titers of 8–32 on day four or five post-challenge. For six vaccinated animals, three shed no detectable antigen (50% protection rate) and the other three shed very low levels of antigen on day four only. Insignificant partial protection against diarrhea was found in this study and the data was not reported. Rotavirus-specific serum IgG and VN antibodies were induced by the vaccine and were further enhanced for an average of 57- and 11-fold post-challenge at PCD 15 compared to the titers at challenge (PID 28). This encouraging result from the Gn pig study provided supporting evidence for the filing of a US patent on the development of the CDC-9 inactivated vaccine in 2014 and the patent was granted in 2016 (Patent No. US 9,498,526 B2).

In a subsequent study in Gn pigs, CDC-9 TLP was administered intradermally (ID) or IM with a lower dose (Wang et al., 2016b). Two vaccination methods were tested. For ID vaccination, a hollow microneedle injection device MicronJet600® without adjuvant was used. Each dose of the ID vaccine contained 5 µg of CDC-9 TLP antigen. For IM vaccination, each vaccine dose con-tained 5 µg of CDC-9 TLP antigen formulated with 600 µg of Al(OH)3. Eleven three-day-old Gn pigs were randomly divided into three groups: diluent by ID device as placebo-control (n = 3); CDC-9 TLP vaccine by ID device (n = 4); or CDC-9 TLP vaccine with adjuvant by IM injection (n = 4). Pigs were inoculated on PID 0, 10, and 21 and were challenged with 5×10^5 FFU Wa VirHRV on PID 28. Both ID and IM regimens conferred a 75% protection rate against virus shedding and significantly reduced the duration and cumulative OD of the viral antigen shedding (Wang et al., 2016b). The ID immunization prevented diarrhea in all four pigs upon challenge. There was a trend in reduced diarrhea for both incidence (24% protection rate) and severity in the IM vaccinated pigs compared to the placebo controls, but no statistical significance was achieved. Only 66% (2/3) of pigs in the control group had diarrhea and the mean duration was only 1.3 days with a mean cumulative score of 2.7 post-challenge. Previous and more recent studies have shown that after challenge with 10^5 FFU Wa VirHRV at PID 28, control pigs develop diarrhea in 80–100% of the pigs with mean durations of 3.2 to 6 days and cumulative scores of 9.6 to 14 (Ramesh et al., 2019; Yuan et al., 1998). Thus, this pig challenge model did not perform well enough in this study for a valid evaluation of protection against rotavirus diarrhea by the can-didate vaccines. Both ID and IM immunized pigs developed higher or significantly higher titers of HRV-specific serum IgA and IgG antibody responses compared to control pigs at PID 21 and PID 28. The serum IgA titers at PID 21 and IgG titers at PID 21 and PID 28 were significantly higher in the IM immunized pigs than the ID pigs. The Wa HRV specific VN titers in the ID and IM immunized pigs were similar pre-challenge from PID 10 to PID 28 and increased significantly post-challenge with the titers in the IM group significantly higher than the ID group. The avidity index of serum HRV-specific IgG antibodies in the ID and IM groups were similar from PID 21 to PCD 14. There is no notable association between higher serum antibody responses induced by these vaccines and a higher degree of protection against rotavirus diarrhea.

Nonetheless, based on the studies discussed in this chapter, it is clear that in this animal model, the use of mucosal immunization routes and live replicating virus to induce an intestinal IgA antibody response is the more efficient regimen for rotavirus vaccination to induce protective immunity.

5

Virus-like Particles Given Intranasally or DNA Plasmids Given Intramuscularly Failed to Induce Any Protection in Gn Pigs

5.1 The 2/6-VLP vaccine with mLT or ISCOM mucosal adjuvant failed in Gn pigs

The success in producing different formulations of VLPs by coexpression of different combinations of rotaviral structural proteins in a baculovirus expression system has facilitated studies of potentially more cost-effective and safer, non-infectious rotavirus subunit vaccines (Crawford et al., 1994). The coexpression of VP2 and VP6, the major core and inner capsid proteins, results in their spontaneous assembly into double-layered 2/6-VLPs. VLPs are non-infectious because they lack nucleic acid, but they are morphologically and antigenically similar to the native virus. The advantages of VLP vaccines may include a lack of side effects seen after oral immunization with live rotavirus vaccines (diarrhea, fever, and possibly intussusception) and their stability during storage (Crawford et al., 1994). Double-layered 2/6-VLPs administered intranasally (IN) to adult mice with cholera toxin or mutant *Escherichia coli* heat-labile toxin (mLT) conferred protection against infection upon challenge with wild-type ECwt murine rotavirus (O'Neal et al., 1998; O'Neal et al., 1997). Therefore, VLPs were subsequently evaluated in Gn pigs (Yuan et al., 2000). In the study, 250 µg/dose of 2/6-VLPs were administered to Gn pigs IN with 5 µg/dose of LT-R192G (mLT) as a mucosal adjuvant. Pigs were challenged with Wa VirHRV at PID 21 (two-dose VLP regimen) or 28 (three-dose VLP regimen). In vivo antigen-activated ASC (effector B cells) and in vitro antigen-reactivated ASC (derived from memory B cells) from intestinal and systemic lymphoid tissues were quantitated by ELISPOT assays. Rotavirus-specific IgM, IgA, and IgG ASC and memory B-cell responses were detected at PID 21 or 28 in intestinal and systemic lymphoid tissues after IN inoculation with two or three doses of 2/6-VLPs with or without mLT. After challenge, anamnestic IgA and IgG ASC and memory B-cell responses were detected in intestinal lymphoid tissues of all VLP-inoculated groups, but serum virus-neutralizing antibody titers were not significantly enhanced compared to the challenged controls (because 2/6-VLPs do not content neutralizing antigens). Pigs inoculated with homologous Wa-RF 2/6-VLPs developed higher anamnestic IgA and IgG ASC responses in ileum after Wa HRV challenge compared to pigs inoculated with heterologous SA11 2/6-VLPs. Three doses of SA 11 2/6-VLP plus mLT induced the highest mean numbers of IgG memory B cells in MLN, spleen, and PBL among all groups post-challenge. However, no protection against diarrhea or virus shedding was evident in any of the 2/6-VLP (with or without mLT)-inoculated pigs after challenge with Wa VirHRV. These results indicate that 2/6-VLP vaccines are immunogenic in Gn pigs when inoculated IN and that the adjuvant mLT enhanced their immunogenicity. However, IN inoculation of Gn pigs with 2/6-VLPs did not confer protection against the HRV challenge (Yuan et al., 2000).

In another experiment, three IN immunizations of five Gn pigs with 2/6 VLP–ISCOM at 250 µg VLP and 1,250 µg ISCOM per dose conferred no protection against diarrhea and virus shedding upon challenge with Wa VirHRV (Gonzalez et al., 2004). Our findings demonstrated that the protective efficacy induced by 2/6-VLPs differ greatly between the adult mouse model of rotavirus shedding and the neonatal Gn pig model of HRV-induced diarrhea.

The development of rotavirus VLP vaccines did not move into human clinical trial until 2018 (NCT03507738) when a plant-derived VLP from Mitsubishi Tanabe Pharma Corporation, Japan, went under phase I clinical trial. The VLP vaccine candidate contains VP7 from RIX4414 (89-12 strain, G1P[8]), VP2/6 from Wa HRV, and using 0.2% aluminum hydroxide as adjuvant (Kurokawa et al., 2021). At the dose of 7 and 21 µg, the vaccine given IM 3 times was weakly immunogenic and induced only homotypic neutralizing antibody responses in 45.5–58.3% of infants (Kurokawa et al., 2021).

5.2 VP6 DNA vaccine failed in Gn pigs

DNA vaccines, including VP4, VP6, and VP7, given by gene gun or by IM injection or orally in microparticles have been shown to induce protection against rotavirus shedding in adult mice (Chen et al., 1999; Chen et al., 1998; Herrmann et al., 1999; Yang et al., 2001). Adult mice given bovine VP6 DNA vaccine IM were partially protected from virus infection upon challenge with murine rotavirus (Yang et al., 2001). DNA vaccines were shown to be effective in generating immunity in the presence of maternal antibodies in non-human primates (Premenko-Lanier et al., 2003). Because interference by the maternal antibody is considered a major cause of reduced rotavirus vaccine efficacy in low-income countries, this would be an important advantage of the rotavirus DNA vaccine if it is proven effective in generating immunity in the presence of maternal antibodies.

In this study, Gn pigs were inoculated IM with three doses of VP6 DNA (200 µg/dose) and were challenged orally with approximately 10^6 ID_{50} of Wa VirHRV at 42 days after the first inoculation to evaluate protection (Yuan et al., 2005). The VP6 DNA vaccine provided minimal to no protection against virus shedding (0%) and diarrhea (12.5%), contrary to results from the adult mouse model. The failure of the VP6 DNA vaccine to induce protection was associated with its inability to induce intestinal IgA antibody responses and VN antibodies (Yuan et al., 2005).

There are numerous reports of protective immunity against rotavirus infection in mice induced by various routes of inoculation using different forms of rotavirus antigen (e.g., live or inactivated, homologous or heterologous rotavirus; recombinant rotaviral proteins or VLPs; and DNA plasmids). However, the protective efficacy of the inactivated, 2/6-VLP, or VP6 DNA plasmid rotavirus vaccines in the adult mouse model (protection against infection) did not predict the lack of protective efficacy against rotavirus disease in the neonatal Gn pig model. To protect neonatal pigs against rotavirus diarrhea may require much higher levels of intestinal IgA antibodies or IgA antibodies with neutralizing specificity against VP4 and/or VP7 than to protect adult mice against rotavirus infection. On the other hand, both virulent Wa or live EDIM rotavirus-induced IgA antibodies in the intestinal lamina propria were correlated with the protection of mice and Gn pigs.

The pathogenesis of rotavirus disease in pigs and probably in human infants differs from that in mice; therefore, the mechanisms of protective immunity may differ in these models (Yuan et al., 2000). Rotavirus infection of young pigs and human infants induces watery diarrhea, anorexia, depression, and dehydration. Pronounced histologic changes (e.g., villous atrophy as a result of virus replication and epithelial cell lysis) occur in Gn pigs infected with Wa VirHRV and similar intestinal villous atrophy has been reported for human infants. In mice under 15 days of age, homologous rotavirus infection causes diarrhea and lethargy, but only mild intestinal lesions consisting of vacuolization of villous tip epithelial cells with little or no villous atrophy. In adult mice, rotavirus infection occurs without causing disease or histopathologic changes. Such differences are important to consider in extrapolating challenge results from the mouse model in which only protection against rotavirus shedding can be evaluated.

6

Prime-Boost Rotavirus Vaccine Regimens Are Highly Effective

Due to the nature of enteric virus infections (acute superficial infections of the intestinal epithelial cells), the presence of IgA antibodies in the intestinal mucosa at the time of exposure provides the most important first line of protection (Yuan and Saif, 2002). To stimulate intestinal IgA antibody responses, oral immunization is preferable because it mimics the route of natural infection. Combined vaccination regimens using the live AttHRV vaccine as oral priming dose and a non-replicating vaccine as parenteral boost were evaluated in Gn pigs and shown to be highly effective in inducing intestinal humoral immune responses and conferring high protection rates in Gn pigs. The non-replicating vaccines were either VLPs together with different adjuvants (Azevedo et al., 2010; Azevedo et al., 2004; Gonzalez et al., 2004; Iosef et al., 2002b; Nguyen et al., 2003; Yuan et al., 2001b) or DNA plasmids (Yuan et al., 2005).

Table 6.1 summarizes the IgA ASC responses in the intestinal lamina propria (mean numbers in duodenum and ileum), serum VN antibody titers at the time of challenge (PCD 0), and protection rates against diarrhea and virus shedding in Gn pigs immunized with prime-boost vaccine regimens and challenged with Wa VirHRV in comparison to various single regimens. These results clearly indicate that the combined immunization routes, oral priming followed by IN or IM boosting, were more effective in stimulating intestinal IgA ASC and serum VN antibody responses than each vaccine. The details of evaluating different types of combined prime-boost regimens in Gn pigs are reviewed in this chapter.

6.1 VLPs serve as effective booster doses in prime-boost regimens

The protection rates conferred by one of the combined regimens (AttHRV1x/VLP+ISCOM 2xIN) were superior over all the vaccine groups and only slightly lower than that conferred by a prior VirHRV infection (Azevedo et al., 2010; Gonzalez et al., 2004). In evaluating the AttHRV1x/VLP+ISCOM2xIN regimen, Gn pigs were given one oral immunization with Wa AttHRV (5×10^7 FFU) and two IN immunizations with 2/6 VLPs at a dose of 250 µg, 100 µg, or 25µg. The dose of ISCOM adjuvant was 1,250 µg per dose. Control pigs received one oral immunization with diluent and two IN immunizations with ISCOM matrix. Two regimens using only IN route were also tested in Gn pigs: 1) one IN dose of Wa AttHRV followed by two IN doses of 250 µg of 2/6VLP and 1,250 µg of ICOM (AttHRV1xIN/VLP+ISCOM2xIN), and 2) three IN doses of AttHRV (AttHRV3xIN). Because lymphocytes sensitized in the nasal lymphoid tissue (NALT) can relocate to distant effector sites through the common mucosal immune system, the respiratory tract was explored as an immunization route to improve rotavirus vaccine efficacy. In addition, AttHRV was found to replicate in nasopharyngeal epithelial cells in Gn pigs (Azevedo et al., 2005) and respiratory symptoms and rotavirus shedding in nasopharyngeal secretions of humans have been reported (Lewis et al., 1979; Nigro and Midulla, 1983; Thi Kha Tu et al., 2020).

The inoculations were performed at PID 0 (three to five days of age), PID 10, and PID 21. At PID 28, subsets of pigs were challenged with Wa VirHRV and observed daily for diarrhea and virus shedding. The AttHRV1x/VLP+ISCOM2xIN regimen conferred the highest protection rates against diarrhea (71%) and shedding (71%) after challenge compared to all other regimens (Table 6.1). Higher intestinal IgA ASC responses induced by the AttHRV1x/VLP+ISCOM2xIN regimen at PID 28 (PCD 0) and intestinal and systemic IgA and IgG ASC responses at PCD 7 were associated with the higher protection rate (Azevedo et al., 2010). The AttHRV1xIN/VLP+ISCOM2xIN regimen conferred a protection rate of 50% against virus shedding and

Table 6.1 Intestinal and Serum Humoral Immune Responses and Protection Rate in Gn Pigs

Vaccine group*	IgA ASC/5 ×10⁵ MNC		Serum VN		Protection rate (%) against		
	pig#	Mean No.	pig#	GMT	pig#	Diarrhea	Shedding
Combined regimens:							
AttHRV1x/VLP+mLT 2x IN	6	135	20	219	12	44	58
AttHRV1x/VLP+ISCOM 2x IN	5	83	7	500	7	71	71
AttHRV1x/VP6 DNA 2x IM	9	38	11	674	10	30	70
VLP+mLT2x IN/AttHRV1x	4	16	ND	ND	6	25	17
Single regimens:							
Inactivated HRV2x IM	6	3	12	4,000	12	6	0
2/6-VLP+mLT 3x IN	4	3	5	2	6	0	0
2/6-VLP+ISCOM 3x IN	3	11	5	2	5	0	0
VP6 DNA 3x IM	4	11	5	3	8	12.5	0

Note:
* Data are summarized from historical data (Azevedo et al., 2010; Gonzalez et al., 2004; Iosef et al., 2002b; Nguyen et al., 2003; Yuan et al., 2005; Yuan et al., 2001b; Yuan et al., 1998; Yuan et al., 1996).

17% against diarrhea. AttHRV3xIN regimen performed better with a protection rate of 67% against virus shedding and 50% against diarrhea (Azevedo et al., 2010). A VLP dose-effect was detected for the AttHRV/VLP+ISCOM2xIN regimen, with the protection rate against both virus shedding and diarrhea reduced to 33% and 33% in the 100 μg VLP dose group and further reduced to 33% and 0% in 25 μg VLP dose group. The lower protection rates were associated with reduced intestinal and systemic IgA and IgG ASC and serum IgM, IgA, and IgG antibody responses after immunization in the Gn pigs (Azevedo et al., 2010; Gonzalez et al., 2004).

6.2 DNA plasmids serve as effective booster doses in prime-boost regimens

VP6 DNA vaccine, although not effective when administered alone, boosted VN and VP4-specific antibody titers in pigs previously primed with AttHRV, possibly mediated by cross-reactive T helper cells (Yuan et al., 2005).

In this study, all pigs were inoculated at three to five days of age with the first dose of vaccine followed by booster vaccines at two and four weeks after the first inoculation. Vaccine groups included: 1) pigs inoculated orally with 5×10^7 FFU live Wa AttHRV, followed by two IM doses of 200 μg bovine rotavirus IND VP6 plasmid DNA vaccine (AttHRV1x/VP6 DNA2x); 2) pigs given the reverse inoculation sequence of the former group: two doses of VP6 DNA IM followed by one dose of AttHRV orally (VP6 DNA2x/AttHRV); 3) pigs inoculated IM with three doses of VP6 DNA (VP6DNA3x); 4) pigs inoculated with one dose of AttHRV orally followed by two doses of control plasmid IM (AttHRV1x/Control DNA2x); and 5) pigs mock-inoculated with one dose of diluent and two doses of control plasmid DNA (mock). A subset of pigs from each group was challenged orally with approximately 10^6 ID$_{50}$ of Wa VirHRV at 42 days after the first inoculation to evaluate protection.

The protective efficacy of the vaccine groups and the control group are summarized in Table 6.2. A protection rate of 70% against rotavirus shedding was induced in the AttHRV1x/VP6 DNA2x vaccinated pigs. The percent of pigs with virus shedding and the duration of days shed were significantly lower in this group compared to the control group. The duration of days shed in the AttHRV1x/VP6 DNA2x vaccine group was also significantly shorter compared to the AttHRV1x/Ctrl DNA2x group, indicating that boosting with VP6 DNA enhanced the protection against infection compared to boosting with control plasmid DNA. However, the protection rate against diarrhea was only 30% in the AttHRV1x/VP6 DNA2x vaccinated pigs.

Without the booster VP6 DNA vaccine, the protection rate of one dose of AttHRV (AttHRV1x/CtrlDNA2x group) was reduced to only 25% against shedding and 12.5% against diarrhea. Low or minimal protection was induced by the VP6 plasmid DNA priming and AttHRV booster approach (17% for virus shedding and 3% for diarrhea). Boosting, but not priming, with VP6 plasmid DNA enhanced the protective efficacy of the AttHRV oral vaccine. There were no significant differences in the percentage of pigs with diarrhea or duration of diarrhea among groups.

The finding that boosting with a DNA vaccine after priming with a replicating vaccine gave better protective efficacy against a mucosal pathogen is not unique to our system. Several other investigators have also shown that DNA vaccines were most effective when using replicating vaccines for priming and DNA vaccines for boosting (Doria-Rose et al., 2003; Eo et al., 2001; Patterson et al., 2004). Immunization of macaques with a vaccinia virus (expressing SHIV Gag-Pol or Env gp 160) priming and DNA plasmid (expressing all SHIV89.6 genes) boosting regimen conferred the highest protection against SHIV virus replication (lowest virus loads in plasma) compared to DNA priming/DNA boosting, DNA priming/inactivated virus boosting, and DNA priming/vaccinia virus boosting (Doria-Rose et al., 2003).

In this study, significantly higher protection against virus shedding in the AttHRV1x/VP6 DNA2x vaccinated pigs compared to the VP6 DNA3x vaccine and control groups was associated with the higher serum and intestinal IgA antibody titers, significantly higher serum VP4 and VP7 protein-specific IgA and IgG antibody titers, and significantly higher VN antibody titers at challenge (PID 42). Higher protection was also associated with the highest numbers of intestinal IgA and IgG ASC pre- and post-challenge. Substantial intestinal IgA antibody responses pre-challenge were detected only in pigs primed orally with the AttHRV. This result concurred with the findings

Table 6.2 Virus Shedding and Diarrhea in Gn Pigs Immunized with Bovine VP6 DNA Vaccine Alone or Combined with AttHRV and Challenged with Wa VirHRV

Groups	n	Virus shedding[a]				Diarrhea[a]		
		% shedding[b]	Mean duration days[c]	Avg. peak titer shed[c] (FFU/ml)	Protection (%) against shedding[b]	% diarrhea	Mean duration days[c]	Protection (%) against diarrhea
AttHRV1x/VP6 DNA 2x[d]	10	30%[b]	1.0[c]	4.1×10^{3ab}	70[a]	60%	1.5	30
VP6 DNA 2x/AttHRV	6	83%[ab]	1.6[bc]	8.7×10^{2b}	17[ab]	83%	3.2	3
AttHRV1x/control plasmid DNA 2x	4	75%[ab]	2.0[b]	8.4×10^{3ab}	25[ab]	75%	2.3	12.5
VP6 DNA 3x	8	100%[a]	1.6[bc]	2.2×10^{4ab}	0[b]	75%	1.7	12.5
diluent/control plasmid DNA 2x	7	100%[a]	3.1[a]	2.8×10^{4a}	0[b]	86%	2.2	0

Note:

[a] Means were calculated among the pigs that had virus shedding or diarrhea.

[b] Percent in the same column with different superscript letters differ significantly; shared letters or no letters indicate no significant difference (Fisher's exact test; $p < 0.05$).

[c] Duration and virus titers with different superscript letters differ significantly; shared letters or no letters indicate no significant difference (one-way ANOVA followed by Duncan's multiple rank test; $p < 0.05$).

This table is reproduced from Yuan et al. "Mucosal and systemic antibody responses and protection induced by a prime/boost rotavirus-DNA vaccine in a gnotobiotic pig model." *Vaccine* 2005. 23:3925–3936 with permission from Elsevier.

from the study of prime/boost DNA vaccine and replicating recombinant virus vector vaccines against HSV infection, suggesting that induction of mucosal immune responses requires priming at a mucosal site with replicating vaccines (Eo et al., 2001).

The use of the VP6 DNA plasmids as an IM boosting vaccine in pigs previously primed with AttHRV significantly boosted serum VN (4.2-fold), VP4 IgA (4.5-fold), and VP4 IgG (3.1-fold) antibody responses at challenge compared to pigs receiving only one dose of AttHRV. This finding is consistent with our previous observations that IN 2/6VLP vaccine boosted serum VN, VP4, and VP7 antibody responses in pigs previously primed with AttHRV (Gonzalez et al., 2004; Yuan et al., 2001b). The VP6DNA3x vaccine also enhanced the VN and IgA and IgG antibodies to VP4 post-challenge compared to the mock-vaccinated and challenged control pigs. This finding agreed with studies of mice that VP6 or 2/6VLP vaccination enhanced VN titers after challenge, although no VN antibodies were detected pre-challenge (Bertolotti-Ciarlet et al., 2003; Esquivel et al., 2000). The mechanism by which heterologous VP6 DNA plasmids or 2/6-VLP boosted VN (mostly contributable to VP4) antibody responses is speculated to be mediated by cross-reactive T helper cells. The T helper cells directed against VP6 may have provided cognate help to B cells specific for the surface viral proteins (VP4 and VP7) (Esquivel et al., 2000; Yuan et al., 2005).

The protection against virus shedding (70%) conferred by the AttHRV1x/VP6 DNA2x regimen was similar to that of AttHRV1x/VLP+ISCOM2x IN previously evaluated in Gn pigs (Table 6.1). However, the protection against diarrhea induced by the AttHRV1x/VP6 DNA2x regimen was only 30%, much lower than that induced by the AttHRV1x/VLP+ISCOM2x IN regimen (71%). The difference in the protection against diarrhea was associated with the differences in serum antibody titers at challenge between the two regimens. The geometric mean titers (GMT) of serum IgA antibodies in pigs receiving AttHRV1x/VLP+ISCOM2x IN were substantially higher than that in the pigs receiving the AttHRV1x/VP6 DNA2x regimen (GMT 920 versus 347) (Gonzalez et al., 2004). Intranasal boosting with 2/6VLP induced ~2.7-fold higher serum IgA antibody GMT at challenge than the VP6 DNA IM boosting of pigs previously primed with AttHRV. Therefore, mucosal boosters appear to be more effective than parenteral boosters in boosting serum IgA antibody responses. However, at challenge, the IgA antibody GMTs in the small intestinal contents (SIC) of pigs receiving the AttHRV1x/VLP+ISCOM2x IN and the AttHRV1x/VP6 DNA2x regimen were similar (GMT 813 and 1,024, respectively) (Gonzalez et al., 2004); therefore, intestinal IgA antibody levels, known to correlate with protection in previous studies, may not explain the difference in the protection against diarrhea. T cell–mediated immunity might have played an important role in the significantly reduced virus shedding seen in the AttHRV1x/VP6 DNA2x group (Yuan et al., 2005).

Since our study, there has been no meaningful development of rotavirus-DNA vaccines. Overall, the potential of DNA vaccines against infectious diseases has not been realized.

7

Rotavirus P2-VP8* and P24-VP8* Intramuscular Vaccines Evaluated in Gn Pigs

Rotavirus outer capsid spike protein VP4, in the presence of trypsin, is cleaved to produce VP8* (~28 kDa) and VP5* (~60 kDa). Bacterially expressed ΔVP8* subunit protein containing amino acid (aa) residues 64 (or 65)-223 with P[8], P[4], or P[6] specificity elicited high levels of homo-typic and varying levels of heterotypic VN antibodies in small animals immunized IM, thus dem-onstrating a solid vaccine potential (Wen et al., 2012d).

7.1 Highly immunogenic P2-VP8* subunit rotavirus vaccine demonstrated protective effects against diarrhea and virus shedding in Gn pigs

In general, the magnitude of protection provided by a parenteral vaccine against orally trans-mitted diseases varies depending upon the levels of protective antibodies induced by the vac-cine (Offit and Clark, 1985). ΔVP8* subunit vaccines are reasonably immunogenic; to further increase the immunogenicity, Dr. Yasutaka Hoshino's team, the Rotavirus Vaccine Development Section in the Laboratory of Infectious Diseases, National Institute of Allergy and Infectious Diseases, NIH, incorporated a universal T cell epitope in the construct creating the P2-VP8* vaccine candidate (Wen et al., 2014b). Selected T cell epitopes from tetanus toxoid (TT) have been extensively studied. They are shown to be widely recognized in association with a large number of MHC class II molecules and to be universally immunogenic in humans and mice (Kaumaya et al., 1993). The P2 universal CD4+ T cell epitope (aa 830–844) was reported to enhance the immunogenicity of a synthetic peptide vaccine for malaria (Franke et al., 1997) as well as tandem copies of a bovine VP8 epitope (Kovacs-Nolan and Mine, 2006). In addition, the P2 has been proven to be as efficient as the whole TT molecule as a carrier without inducing epitope suppression (Kumar et al., 1992). The immunity enhancing effect of the TT P2 universal T cell epitope is thought to be due to an activation of specific T cells for the synthesis of certain cytokines that are necessary for T and B cell activation and interaction as well as proliferation (Zegers and Boersma in Zegers, 1995).

The P2-P[8]VP8* and P2-P[6]VP8* constructs were expressed in *Escherichia coli* and the immu-nogenicity and protective efficacy of these recombinant fusion proteins were characterized in guinea pigs (in Dr. Hoshino's lab at NIH) and Gn pigs (in Dr. Yuan's Lab at Virginia Tech), respec-tively (Wen et al., 2014b). The encouraging protective efficacy demonstrated in Gn pigs led to further development of the P2-VP8* vaccine by PATH for optimized formulations in the presence of aluminum adjuvants (Agarwal et al., 2020a; Agarwal et al., 2020b; Agarwal et al., 2020c; Lakatos et al., 2020). The trivalent P[8], P[4], and P[6] P2-VP8* vaccine has undergone phase 1 human clinical trials in both healthy adults (Fix et al., 2015), infants, and young children (Groome et al., 2017), and a phase 1/2 clinical trial in South Africa with the financial support of the Bill and Melinda Gates Foundation.

In the Gn pig study, pigs were immunized IM three times (50 μg/dose) with the P2-P[8]ΔVP8 vaccine with aluminum phosphate (AP) adjuvant (ADJU-PHOS™, Brenntag) (600 μg/dose, in 0.5 ml total volume) at PID 0 (five days of age), PID 10, and PID 21. At PID 28, the vaccinated and mock-vaccinated control pigs were challenged orally with 10^5 FFU of the Wa VirHRV to assess the protection against G1P[8] HRV-induced diarrhea and fecal virus shedding. Pigs were

DOI: 10.1201/b22816-7

monitored and fecal swabs were collected daily from PCD 0 to 7 for detection of virus shedding. The pigs vaccinated with P2-P[8]ΔVP8* had significantly delayed onset of diarrhea (mean onset days from 1.4 to 4.2) and significantly reduced duration of diarrhea (mean duration from 5.6 to 1.6 days) compared to the controls, although all vaccinated pigs had diarrhea for one to three days (Figure 7.1). Their diarrhea scores were also significantly reduced compared to the controls (from 14.4 to 8.8).

P2-P[8]ΔVP8* vaccine induced high titers of serum VN antibodies and VP4-specific IgG antibody response, and primed the Gn pigs for IgA antibody response post-challenge (Figure 7.2). Significantly higher titers of VN antibody titers were detected in the sera of vaccinated pigs at challenge (PID 27) and post-challenge (PCD 2 and 10) compared to the control pigs ($p <$ 0.0001). HRV-specific IgA antibody was not detectable pre-challenge. However, significantly higher IgA antibody titers at PCD 10 in the vaccinated pigs compared to the control pigs suggest that the P2-VP8* IM vaccine primed the pigs for serum IgA antibody production post-challenge. The P[8] VP4-specific IgG antibody titers measured by an Sf-9 cell immunocytochemical staining assay (Yuan et al., 2004) increased significantly after each vaccination and after challenge. However, VP4-specific IgA antibody responses in serum were not detected pre- or post-challenge, which is consistent with the lack of detectable serum virus-specific IgA antibodies until PCD 10.

P2-P[8]ΔVP8* vaccine-induced rotavirus-specific IFN-γ producing T cell responses in intestinal and systemic lymphoid tissues (Figure 7.2). Pigs vaccinated IM with P2-P[8]ΔVP8* and orally challenged with the VirHRV had significantly higher frequencies of virus-specific IFN-γ producing CD8+ T cells in ileum compared to the challenged control pigs at PCD 10. Vaccinated pigs also had significantly higher frequencies of IFN-γ producing CD4+ T cells in the ileum, spleen, and blood, suggesting that IM immunization with the P2-VP8* effectively primed for not only systemic but also intestinal virus-specific memory T cell responses (Wen et al., 2014b).

The results of the immunogenicity of P2-P[8]ΔVP8* vaccine in Gn pigs successfully predicted the immunogenicity of this vaccine in human clinical trials. In a multisite, randomized, double-blind, placebo-controlled trial, infants received an injection of vaccine or placebo on days 0, 28, and 56, at approximately 6, 10, and 14 weeks of age (Groome et al., 2020). Among 528 infants, adjusted anti-P2-VP8* IgG seroresponse rates (≥4-fold increase) to P[4], P[6], and P[8] antigens were significantly higher (99–100%) in all three dose groups (15 µg, 30 µg, and 90 µg) than in the placebo group (10–29%; $p <$ 0.0001) measured four weeks after the injection of the final dose. Adjusted VN antibody seroresponse rates (≥2.7-fold increase) to P[4] (DS-1), P[6] (1076), and P[8] (Wa) HRV were also significantly higher in vaccine recipients than in placebo recipients with $p <$ 0.0001 for all comparisons. Anti-P2-VP8* IgA seroresponse rate to each individual antigen was modest (20–34%) across the three dose groups, but still significantly higher than in

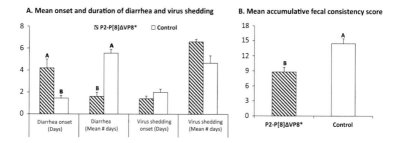

Figure 7.1   Clinical signs and virus shedding in P2-P[8]ΔVP8* vaccinated Gn pigs post-challenge. After the challenge, the mean onset and duration of diarrhea and virus shedding (panel A) and the mean accumulative fecal consistency scores (panel B) were calculated among all pigs in each group from PCD 0–7. The error bars represent the standard errors of the mean. Different letters on top of bars indicate a significant difference between the pig groups (ANOVA-GLM; $p <$ 0.05), while shared letters indicate no significant difference. (Source: Figure 7.1 is reproduced from Wen et al. "Inclusion of a universal tetanus toxoid CD4+ T cell epitope P2 significantly enhanced the immunogenicity of recombinant rotavirus ΔVP8* subunit parenteral vaccines." *Vaccine* 2014. 32:4420–4427 with permission from Elsevier.)

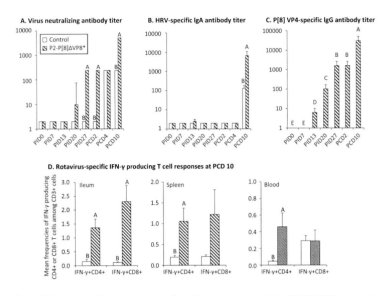

Figure 7.2   Serum antibodies and virus-specific T cell responses induced by the P2-P[8]ΔVP8* vaccine in Gn pigs. Gn pigs were immunized IM with 50 μg (n = 5) of the purified P2-P[8]ΔVP8* protein with 600 μg/dose of aluminum phosphate or mock vaccinated (n = 9) at PID 0, 10, and 21 and challenged with the virulent HRV Wa strain at PID 28. Serum samples were collected weekly from vaccinated and control pigs. Virus neutralizing antibody titers and Wa HRV-specific IgA antibody titers were compared between vaccinated and control pig groups on each time point (panels A and B); P[8] VP4-specific IgG antibody titers of the vaccinated pigs were compared among all the time points (panel C). Representative dot plots for detecting IFN-γ producing CD4+ and CD8+ T cells in the vaccinated and control pigs by flow cytometry are shown on the left of panel D. Frequencies of virus-specific IFN-γ producing T cells were compared between vaccinated and control pig groups at PCD 10 for the same cell type from the same tissue. The error bars represent the standard errors of the mean. Different letters on top of bars indicate a significant difference between the pig groups (Kruskal-Wallis rank-sum test; p < 0.05, n = 3–5); while shared letters or no letters indicate no significant difference. (Source: Figure 7.2 is reproduced from Wen et al. "Inclusion of a universal tetanus toxoid CD4+ T cell epitope P2 significantly enhanced the immunogenicity of recombinant rotavirus ΔVP8* subunit parenteral vaccines." *Vaccine* 2014. 32:4420–4427 with permission from Elsevier.)

placebo recipients (9–10%) (Groome et al., 2020). These results support further efficacy testing of this trivalent P2-VP8* vaccine. A phase 3 trial of the vaccine at 90 μg dose started recruiting in July 2019 with the estimated completion date of December 2024 (NCT04010448).

7.2 P24-VP8* nanoparticle vaccine conferred strong protection against rotavirus diarrhea and virus shedding in Gn pigs

The norovirus (NoV) P particle, referred to as the P24 particle, is an octahedral nanoparticle (≈840 kDa) composed of 24 copies of the protrusion (P) domain of the NoV capsid protein. It can be easily produced in large quantities using an *E. coli* expression system. The distal surface of each P domain, corresponding to the outermost surface of the P particle, contains three surface loops, which can tolerate large sequence insertions. Based on this concept, a nanoparticle vaccine was developed by inserting the VP8* of HRV into the loop sections of the P domains. The P24-VP8* nanoparticle consists of a 24-valent core of NoV P particle and 24 surface-displayed VP8*. The P24-VP8* nanoparticle shares the features of the P24 particle in self-formation, easy production, and high stability over a wide range of temperatures (Tan et al., 2011). Efficacy studies in mice revealed that the P24-VP8* nanoparticle vaccine is highly immunogenic and capable of inducing a significantly higher VP8* specific antibody response as compared with free VP8* particles even without adjuvant (Tan et al., 2011).

The immunogenicity and protective efficacy of the P24-VP8* nanoparticle vaccine were evaluated in the Gn pig model of HRV infection and disease (Ramesh et al., 2019). This study was led by my Ph.D. student Ashwin Ramesh and was funded by a contract award titled "Evaluation of P24-VP8* Vaccine in a Gnotobiotic Pig Model of Human Rotavirus Infection and Diarrhea" from Anhui Zhifei Longcom Biopharmaceutical, China (PI: Lijuan Yuan. 12/31/2016–13/31/2018).

The P24-VP8* vaccine was comprised of 200 µg of P24-VP8* proteins and 600 µg aluminum hydrogel (Al[OH]$_3$) adjuvant. The dosage of the P24-VP8* vaccine was selected based on a similar VP8* molar amount of the P2-VP8 subunit vaccine used in the clinical trial (Fix et al., 2015; Groome et al., 2017; Wen et al., 2014b). The VP8* region used in this vaccine was designed based on the sequence of Wa HRV. As the negative control, the Al(OH)$_3$ adjuvant was diluted in sterile PBS to form a final concentration of 600 µg/mL. The purified P24-VP8* proteins were used as the detector antigen in ELISA for the detection of serum IgA and IgG antibody responses and as stimulating antigens in the intracellular IFN-γ staining assay (Wen et al., 2014a; Yuan et al., 2008).

Pigs were administered IM with an equal volume (1 mL) of either P24-VP8* vaccine formulated with adjuvant or adjuvant alone at five days of age (PID 0), followed by two booster doses at PID 10 and PID 21. The human clinical trials carried out to evaluate the effects of the P2-VP8* vaccine demonstrated that participants who received a three-dose vaccination regime shed fewer attenuated rotavirus in feces as compared to trial participants who received two doses (Fix et al., 2015; Groome et al., 2017). Based on this rationale, we opted to use the three-dose regimen in this study. Serum was collected at PID 0, PID 10, PID 21, PID 28, and PCD 7 for the detection of VP8*-specific IgA, IgG, and Wa HRV-specific neutralizing antibody responses. One subset of pigs ($n = 3$–7) from each group was euthanized before the challenge at PID 28. Another subset of pigs ($n = 7$–8) was orally challenged with 1×10^5 FFU of VirHRV Wa strain and monitored from PCD 1 to PCD 7 to assess the protection against virus shedding and diarrhea conferred by the vaccine before euthanasia on PCD 7.

Vaccinated and control Gn pigs were challenged with VirHRV at PID 28 and were monitored daily for diarrhea and virus shedding from PCD 1 to PCD 7. Gn pigs that were administered with P24-VP8* vaccine had a significantly delayed onset of diarrhea (from 1.6 to 4.4 days), a significantly reduced duration of diarrhea (from 6.0 to 3.3 days), significantly lower mean diarrhea scores on PID 1 and 2, and a significantly lower cumulative fecal consistency score (from 14.3 to 9.1) as compared to the mock-vaccinated control group (Table 7.1 and Figure 7.3). Delayed onset of virus shedding, a reduced peak titer, a reduced cumulative virus titer (presented as the area under the curve, AUC), and a significantly reduced duration (from 5.9 to 2.5 days) of virus shedding were observed in P24-VP8* vaccinated pigs when compared to the control group. In addition, the mean daily virus shedding titer in the vaccinated pigs was significantly reduced at PCD 2 (Figure 7.3), and the reduction of total virus shed (AUC) was 2.27-fold compared to the control pigs (Table 7.1). However, the vaccine did not significantly reduce the incidence (%) of diarrhea and virus shedding.

P24-VP8* vaccine induced strong antigen-specific IgG and Wa HRV-specific VN antibody responses in serum. P24-VP8*-specific IgG antibody titers in serum were significantly higher ($p < 0.001$) in vaccinated pigs at PID 10, PID 21, PID 28, and PCD 7 when compared to pigs in the control group. However, serum IgA titers were only detectable after challenge (PCD 7) with VirHRV (Figure 7.4). VN antibodies were detected in the serum of P24-VP8* vaccinated pigs starting from PID 21 and were observed to increase similarly with P24-VP8*-specific IgG titers until euthanasia at PCD 7. In control pigs, VN antibodies were only detectable after challenge with VirHRV and were at significantly lower levels compared to the vaccinated pigs. P24-VP8* vaccine did not induce strong P24-VP8*-specific effector T cell responses in intestinal or systemic lymphoid tissues (Ramesh et al., 2019).

The observed protection and immune responses data together suggest that the protection conferred by the P24-VP8* vaccine against diarrhea and virus shedding upon challenge with the VirHRV was mediated by the vaccine-induced antibodies in the serum. Although there were no antibodies present at the lumen of the small intestine, the site of HRV infection, at the time of challenge to prevent the initiation of RV infection, the viruses disseminated into the blood from the infected small intestinal epithelial cells could have been neutralized by the high

Table 7.1 Diarrhea and Rotavirus Fecal Virus Shedding in P24-VP8* Vaccinated and Control Gn Pigs after the VirHRV Challenge

Treatment	n	Diarrhea				Fecal virus shedding				AUC
		% with diarrhea	Mean days to onset	Mean duration days	Mean cumulative fecal score	% shedding virus	Mean days to onset	Mean duration days	Mean peak titer (FFU/mL)	
P24-VP8	8	87.5	4.4 (0.5) [a],*	3.3 (0.75) *	9.1 (1.23) *	75	4.8 (1.0)	2.5 (0.89) *	8,500 (2,196) *	11,750 (3,172)
Control	7	100	1.6 (0.3)	6.0 (0)	14.3 (0.44)	85.7	1.9 (0.14)	5.9 (0.14)	11,492 (4,300)	26,664 (10,489)

Note:
[a] Standard error of the mean. Asterisk indicates statistical significance between the groups ($p < 0.05$). Student's t-test was used for comparison between vaccine and control groups. This table is adapted from Ramesh et al. (2019).

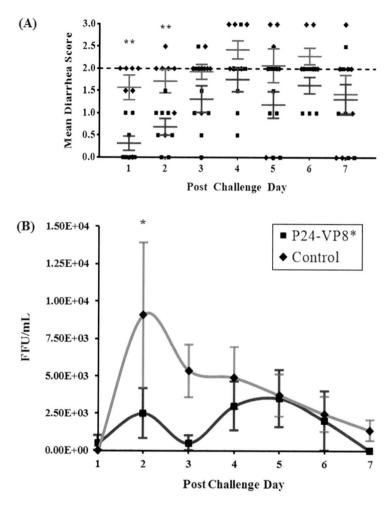

Figure 7.3 P24-VP8* vaccine protected against VirHRV diarrhea and reduced overall virus shed among vaccinated pigs. Fecal consistency (A) and virus shedding (B) were monitored daily from PCD 1 to PCD 7 after the challenge with VirHRV. Statistical significance between vaccinated and control groups, determined by multiple t-tests, are indicated by asterisks (*, $p < 0.05$; **, $p < 0.01$). (Source: this figure is adapted from Ramesh et al. (2019).)

titers of VP8*-specific IgG and VN antibodies during the phase of viremia. Such mechanisms can reduce the chance of infection of more epithelial cells by the virus from the basolateral side (Azevedo et al., 2005). Studies show that passively transferred serum antibodies can suppress or delay viral infection in rotavirus-challenged pigtailed macaques (Westerman et al., 2005), and the inactivated CDC-9 HRV IM vaccine reduced virus shedding in Gn pigs upon challenge with Wa VirHRV (Wang et al., 2010), likely sharing the same protection mechanism with the P24-VP8* vaccine. The serum IgG and VN antibody responses induced by the P24-VP8* IM nanoparticle vaccine had similar dynamics and magnitude as the aluminum phosphate adjuvanted inactivated CDC-9 and P2-VP8* IM vaccines in Gn pigs (Wang et al., 2010; Wen et al., 2014b). The P24-VP8* vaccine demonstrated a similar degree of protection against diarrhea

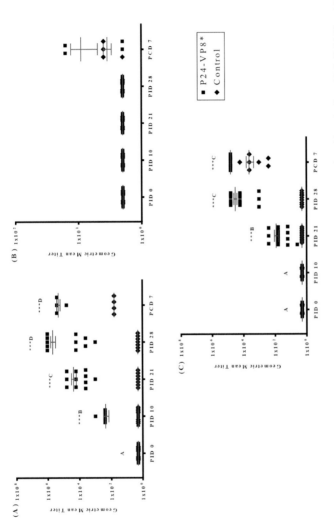

Figure 7.4 Geometric mean P24-VP8*-specific IgG (A) and IgA (B) and Wa HRV neutralizing (C) antibody titers in serum collected from Gn pigs at PID 0, 10, 21, 28, and PCD 7. Comparisons between groups at the same time points were carried out using student's t-test and significant differences are identified by *** (n = 10–15; $p < 0.001$). Tukey–Kramer HSD was used for the comparison of different time points within the same group, where different capital letters (A, B, C, D) indicate a significant difference, $p < 0.01$, and shared letters indicate no significant difference. (Source: this figure is adapted from Ramesh et al. (2019).)

but stronger protection against virus shedding in Gn pigs as compared to the P2-VP8* vaccine (Wen et al., 2014b).

There was a trend of inverse correlation between serum VP8*-specific IgG titers at PID 28 and cumulative diarrhea scores post-challenge in the vaccinated pigs (Pearson's rank correlation, r = −0.6699 and p = 0.0691), suggesting that vaccinated pigs with higher serum VP8*-specific IgG responses are more likely to be protected against severe diarrhea, which is in agreement with the study of serum IgG antibody in human adults showing that VP4-specific IgG titer was correlated with resistance to HRV infection (Yuan et al., 2009). As reviewed by Jiang et al., serum antibodies, if present at critical levels, are either protective themselves or are an important and powerful correlate of protection against rotavirus disease (Jiang et al., 2002).

8

Norovirus P Particle and VLP Vaccines Evaluated in Gn Pig Model of GII.4 HuNoV Infection and Diarrhea

This norovirus vaccine study was led by my Ph.D. student Jacob Kocher (Kocher et al., 2014). It was funded by an R01 sub-award from NIAID, NIH titled "Novel Vaccine against Norovirus" (R01AI089634, 2010-2015, Xi Jiang PI; Lijuan Yuan subcontract PI).

8.1 Introduction

Due to the lack of a cell culture system to isolate and propagate HuNoV, the development of live-attenuated or inactivated vaccines is impossible. HuNoV vaccine development has relied upon recombinant NoV capsid proteins, such as VLPs and P particles. The capsid of NoV is composed of a single major structural protein of 55–60 kDa (VP1) that is divided into the shell (S) and protruding (P) domains linked by a short hinge (Jiang et al., 1993). VLP forms through the expression of VP1 in prokaryotic and eukaryotic expression systems and retains the antigenic structure and HBGA receptor binding function (Harrington et al., 2004; Huo et al., 2018; Marionneau et al., 2002). P particle forms when the P domain is expressed in *E. coli* (Tan and Jiang, 2005b). Each P particle contains 24 copies of the P domain with a total molecular weight of ~840 kDa and a diameter of ~20 nm, which is an ideal size for an immunogen (Tan and Jiang, 2005b). P particles contain the P2 binding domain in the outer layer and the P1 domain in the inner core (Tan and Jiang, 2005b). Thus, P domain complexes retain the P2 HBGA binding domain (Tan et al., 2003) and have the same binding profile as the native capsid and VLPs (Tan and Jiang, 2005b).

P particles and VLPs elicit similar innate, humoral, and cellular immune responses as VLPs in mice (Fang et al., 2013). A study by Tamminen et al. (2012) comparing the immunogenicity of VLPs and P particles in mice suggested that VLPs induce a superior immune response than P particles. However, mice are resistant to HuNoV infection so the protective efficacy cannot be evaluated to address the disparities in these studies. In addition, Tan and Jiang (2012) raised concerns that this study utilized P dimers instead of P particles. Another study indicated that P dimers induce weaker immune responses than P particles (Fang et al., 2013), which may have impacted Tamminen et al.'s (2012) results.

8.2 Intranasal P particle and VLP vaccine provided partial protection against GII.4 HuNoV diarrhea and shedding in Gn pigs

In our study (Kocher et al., 2014), P particles and VLPs derived from GII.4 VA387 (1997 Farmington Hills variant) were evaluated in the Gn pig model of GII.4/2006b (092895) challenge. The amino acid sequences for VA387 VP1 (GenBank Accession no: AY038600 and AAK84679) and 092895 VP1 (GenBank Accession no: KC990829) have 93.5% sequence homology. The amino acid alignment showed that 33 substitutions were present, including 21 substitutions in the P2 domain. There are four substitutions in epitope A, one substitution in epitope B, one

DOI: 10.1201/b22816-8

substitution in epitope C, three substitutions in epitope D (including one deletion), and three substitutions in epitope E (Kocher et al., 2014).

Synthetic MPL and chitosan were used as mucosal adjuvants. Vaccines contained 100 or 250 µg of P particles or 100 µg of VLPs, 5 mg chitosan, 50 µg MPL, and TNC buffer to a final volume of 1 ml. Endotoxin levels of the P particles were determined with the ToxinSensor™ Chromogenic LAL Endotoxin Assay (GenScript) to be 0.8 EU/ml, which is below the recommended level for a recombinant subunit vaccine (Brito and Singh, 2011).

Pigs were confirmed to be A+ or H+ prior to inoculation. Each group is composed of at least six pigs that were derived from at least three different litters—three experimental replicates. Pigs in P particle or VLP groups were inoculated IN with three doses of the vaccines using mucosal atomization devices (MADs, LMA North America) at PID 0 (five days of age), PID 10, and PID 21. Pigs in the HuNoV infection group (to mimic nature infection) were orally inoculated with 10 ID_{50} HuNoV (2.74×10^4 viral RNA copies) at five days of age. Control pigs received diluent or adjuvants only. Pigs were given 4 ml of 200 mM sodium bicarbonate ten minutes prior to oral inoculation to reduce gastric acidity. A subset of pigs in each group was orally challenged with 10 ID_{50} HuNoV (6.43×10^5 viral RNA copies) at PID 28 (PCD 0) and monitored daily for clinical signs and virus shedding until PCD 7. All pigs were euthanized at PID 28 or PID 35 (PCD 7) to isolate MNC from duodenum, ileum, spleen, and blood as previously described (Yuan et al., 1996).

The high dose P particle vaccine and VLP vaccine provided the same protection rate (60%), which were lower than from a primary HuNoV infection, against diarrhea after a homotypic (but different variant) GII.4/2006b HuNoV challenge (Table 8.1). Primary HuNoV infection (Live HuNoV) significantly reduced the occurrence of diarrhea (protection rate 82.9%). The P particle and VLP vaccines reduced diarrhea incidence in the pigs at a similar rate (46.7%, 60.0%, 60.0%). Compared to control pigs, HuNoV, 250 µg of P partials, and 100 µg of VLPs slightly shortened the mean duration of diarrhea, while P particles reduced the mean area under the curve (AUC) of diarrhea, but these differences were not statistically significant.

Primary HuNoV infection and high dose P particles (250 µg) not only provided substantial protection against diarrhea but also from fecal virus shedding (protection rate 48.6% and 60%, respectively). HuNoV primary infection and P particles also decreased the mean AUC of virus shedding, especially the high dose P particles (53-fold).

8.3 Higher dose of P particle vaccine induced stronger T cell responses

To assess the qualitative relationship between the numbers of immune cells and the differences in functional protective immunity, we analyzed absolute numbers of T cell subsets in intestinal and systemic lymphoid tissues of Gn pigs in different vaccination groups. We calculated the correlation (with Spearman's rank correlation coefficient) between the T cell expansion (change in absolute numbers from pre- to post-challenge) with the protection rate against diarrhea.

8.3.1 High dose P particle (HiPP) primed for increased Th and CTL in intestine and blood

The pre- and post-challenge total Th (CD3+CD4+) and CTL (CD3+CD8+) are depicted in Figure 8.1A. High dose (250 µg) P particle vaccinated pigs (HiPP) had significantly higher numbers of Th and CTL in duodenum and ileum, respectively, compared to low dose (100 µg) P particle vaccinated pigs (LoPP) post-challenge. There were significant increases in numbers of Th and CTL in duodenum, ileum, and PBL of HiPP pigs post-challenge. The fold expansions of Th and CTL following challenge are shown in Figure 8.1B. HiPP induced the most expansion of Th and CTL in each tissue examined, while LoPP induced the lowest expansion, though these differences were not statistically significant.

8.3.2 HiPP primed for increased activated T cells post-challenge

The activated non-regulatory CD4+ and CD8+ T cells are shown in Figure 8.2A. Following challenge, HiPP pigs had significantly higher numbers of activated CD8+ T cells in the ileum

Table 8.1   Clinical Signs and Protective Efficacy in Previously Infected or Vaccinated Gn Pigs after Challenge with GII.4 2006b HuNoV[a]

Group	n	Diarrhea Mean percent diarrhea*	Mean number of days with diarrhea[bc]**	Mean AUC of diarrhea[cd]**	Protection rate against diarrhea (%)[e]	Virus shedding Mean percent shedding*	Mean number of days with shedding[b]**	Mean AUC of virus shedding[d]**	Protection rate against shedding (%)[e]
Live HuNoV (oral)	7	14% (1/7)[b]	0.3 (0.3)	6.7 (0.6)	82.9	43% (3/7)	2.9 (1.4)	20,775	48.6
250 µg P particle + MPL/chitosan (IN)	6	33% (2/6)[ab]	0.8 (0.7)	5.1 (1.1)	60.0	33% (2/6)	0.5 (0.3)	2,191	60.0
100 µg P particle + MPL/chitosan (IN)	9	44% (4/9)[ab]	1.3 (0.6)	5.4 (0.8)	46.7	89% (8/9)	1.8 (0.4)	38,216	0.0
100 µg VLP + MPL/chitosan (IN)	6	33% (2/6)[ab]	0.7 (0.4)	6.2 (1.0)	60.0	100% (6/6)	2.0 (0.4)	96,428	0.0
Diluent #5 (oral) or MPL/chitosan (IN)	6	83% (5/6)[a]	1.8 (0.6)	6.7 (1.0)	NA	83% (5/6)	2.0 (0.7)	116,960	NA

Note:

a   Gn pigs were challenged with a human NoV GII.4 2006b variant 092895 at 33–34 days of age. Rectal swabs were collected daily after challenge to determine diarrhea and virus shedding by conventional and real-time RT-PCR. Virus shedding was also detected in intestinal contents.

b   Calculated as the mean number of days with diarrhea or shedding from PCD 0 to PCD 7.

c   Fecal scoring system: 0) solid, 1) pasty, 2) semi-liquid, 3) liquid. Scores of 2 or greater were considered diarrheic.

d   AUC, area under the curve of viral diarrhea or shedding from PCD 1–7.

e   Protection rate = (1 − [percentage of immunized pigs in each group with diarrhea or shedding/percentage of control pigs with diarrhea or shedding]) × 100.

*   Proportions in the same column followed by different letters (A, B) differ significantly (Fisher's exact test, $p < 0.05$); while shared letters indicate no significant difference.

The values in parentheses are pig numbers with diarrhea/shedding over total numbers in the group or the standard error of the mean.

**   Means in the same column followed by different letters (A, B) differ significantly (one-way ANOVA, $p < 0.05$); while shared letters indicate no significant difference.

This table is modified from Kocher (2014) Ph.D. dissertation with permission from the author.

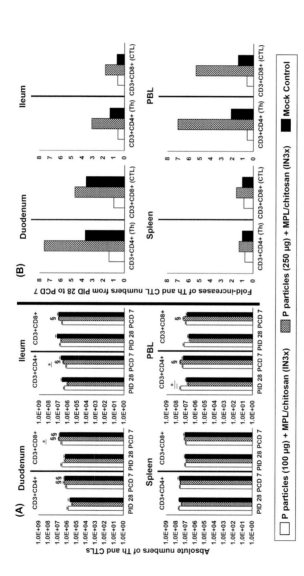

Figure 8.1   Th and CTL responses induced by low dose and high dose P particle vaccination pre- and post-challenge. Mean total numbers + SEM (n = 6–10) of CD3+CD4+ (Th) and CD3+CD8+ (CTL) T cells in intestinal (duodenum, ileum) and systemic (spleen, PBL) tissues pre- and post-challenge (A). Frequencies of Th and CTL were determined using intracellular staining and multicolor flow cytometry. Total numbers were calculated using the volume and concentration of isolated MNCs. The sign "*" on top of error bars indicates significant differences among groups for the same cell type and tissue at the same time point (Kruskal-Wallis one-way ANOVA, $p < 0.05$). "§" indicates that numbers increased significantly following challenge among the same group and "#" indicates that numbers decreased significantly following challenge among the same group. Total expansion of Th or CTL following HuNoV challenge (B). Expansion of Th and CTL were calculated by dividing post-challenge numbers by pre-challenge numbers. (Source: Figures 8.1–8.5 are reproduced from Kocher (2014) Ph.D. dissertation with permission from the author.)

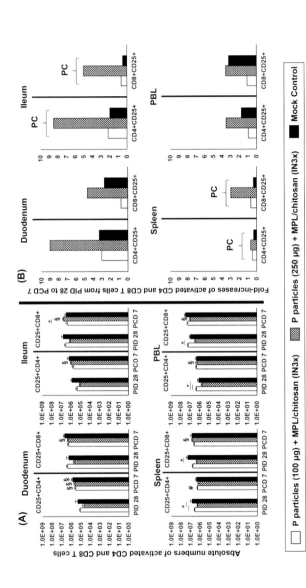

Figure 8.2 Activated non-regulatory CD4+ and CD8+ T cell responses induced by low dose and high dose P particle vaccination pre- and post-challenge. Mean total numbers + SEM (n = 6–10) of CD4+CD25+FoxP3– and CD8+CD25+FoxP3– T cells in intestinal (duodenum, ileum) and systemic (spleen, PBL) tissues pre- and post-challenge (A). Frequencies of activated non-regulatory T cells were determined using intracellular staining and multicolor flow cytometry. Total numbers were calculated using the volume and concentration of isolated MNCs. Total expansion of activated CD4+ or CD8+ T cells following HuNoV challenge (B). Expansion of activated CD4+ and CD8+ T cells was calculated by dividing post-challenge numbers by pre-challenge numbers. Correlations between protection rate against diarrhea and T cell subset expansion were determined using Spearman's rank correlation coefficient. A "(PC)" indicates a positive correlation between protection rate against diarrhea and T cell subset expansion (R = 1, p < 0.0001), while a "(NC)" indicates a negative correlation (R = −1, p < 0.0001). See Figure 8.1 legend for a description of statistical analysis.

compared to LoPP pigs. HiPP pigs had significant increases in both activated T cell subsets in all tissues except for CD4+ T cells in the spleen post-challenge. The HuNoV challenge-induced fold expansions of activated T cells are shown in Figure 8.2B. HiPP pigs had the biggest fold expansion of activated T cells in all tissues examined. Activated CD4+ and CD8+ T cells in the ileum and spleen were positively correlated with the protection rate against diarrhea.

8.3.3 HiPP primed for increased IFN-γ producing effector T cells

The fold expansions of IFN-γ+ T cells following challenge are shown in Figure 8.3. HiPP pigs had high levels of expansion, ranging from 20-fold to 975-fold, following HuNoV challenge.

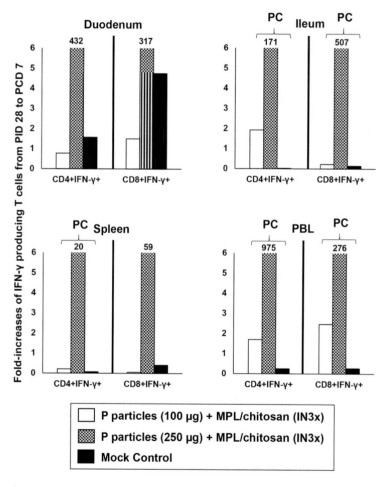

Figure 8.3   Expansion of HuNoV specific IFN-γ producing CD4+ and CD8+ T cell responses induced by low dose and high dose P particle vaccination. Expansion of IFN-γ producing CD4+ and CD8+ T cells was calculated by dividing post-challenge numbers by pre-challenge numbers. Correlations between protection rate against diarrhea and T cell subset expansion were determined using Spearman's rank correlation coefficient. A "(PC)" indicates a positive correlation between protection rate against diarrhea and T cell subset expansion (R = 1, $p < 0.0001$), while a "(NC)" indicates a negative correlation (R = −1, $p < 0.0001$).

Additionally, there were positive correlations between protection rates from diarrhea and expansion of ileal CD4+IFN-γ+ and CD8+IFN-γ+ T cells, splenic CD4+IFN-γ+ T cells, and PBL CD4+IFN-γ+ and CD8+IFN-γ+ T cells.

8.3.4 HiPP down-regulated systemic regulator T cell (Treg) responses

Treg responses of vaccinated and control pigs are shown in Figure 8.4A. Following challenge, HiPP pigs had significantly lower numbers of CD25-FoxP3+ Tregs in the duodenum and spleen and CD25+FoxP3+ Tregs in the spleen compared to both LoPP and control pigs. Challenge significantly increased numbers of both Treg subsets in the duodena of all groups compared to pre-challenge numbers. HiPP pigs also had significantly increased CD25+FoxP3+ Tregs in the ileum and both subsets in PBL post-challenge. The fold expansions of Tregs in each group following challenge are shown in Figure 8.4B. The fold expansions of CD25+FoxP3+ Tregs in the ileum and CD25-FoxP3+ Tregs in PBL were positively correlated with protection rate from diarrhea, while expansion of CD25-FoxP3+ Tregs in the duodenum was negatively correlated.

8.3.5 Correlations between protection rate against diarrhea and Treg expansion

The TGF-β producing Treg responses in intestinal and systemic tissues are shown in Figure 8.5A. HiPP pigs had significantly lower numbers of TGF-β+CD25+ Tregs in the spleen compared to LoPP pigs pre-challenge. It is worth noting that HiPP pigs had no detectable TGF-β+CD25– Tregs in PBL at PID 28. Following challenge, HiPP pigs had significantly higher numbers of TGF-β+CD25– Tregs in the ileum and PBL and TGF-β+CD25+ Tregs in the spleen compared to control pigs. HiPP pigs had significantly lower numbers of TGF-β+CD25– Tregs in the spleen compared to LoPP pigs. HiPP pigs had significantly increased TGF-β+CD25– Tregs in PBL and TGF-β+CD25+ Tregs in the spleen compared to pre-challenge. The fold expansions of TGF-β+ Tregs following challenge are shown in Figure 8.5B. We identified a positive correlation between the protection rate against diarrhea and expansion of TGF-β+CD25+ Tregs in the spleen and PBL and TGF-β+CD25– Tregs in the ileum.

8.4 Summary

In this study, we demonstrated the dose-dependent protective efficacy and immunogenicity of P particle vaccines. Studies in human clinical trials with HuNoV VLPs have indicated the proper dose for a HuNoV vaccine likely ranges from 100 μg to 1000 μg per vaccine course (two doses at 50–500 μg/dose) based on protective and immunological parameters (Atmar et al., 2011; El-Kamary et al., 2010; Ramirez et al., 2012; Treanor et al., 2014). The present study utilizes two vaccine regimens (three doses at 100 and 250 μg/dose) that fall within this range (total 300 μg and 750 μg per vaccine course). P particle vaccine at high dose (250 μg) increased protective efficacy against diarrhea and shedding and primed for increased activated T cells and IFN-γ producing T cells post-challenge compared to the low dose (100 μg).

P particles and VLPs provided similar protection against HuNoV diarrhea. Prior HuNoV infection provided the highest protection rate against diarrhea as well as partial protection against HuNoV shedding whereas only the high dose P particles provided partial protection against virus shedding. HiPP group had slightly increased protection against shedding compared to prior HuNoV infection. The protection rates against diarrhea conferred by the higher dose P particle vaccine at three doses in Gn pigs are slightly higher than the protection rate of a two-dose intranasal Norwalk-derived VLP vaccine in humans (47%) (Atmar et al., 2011). However, the VLP vaccine only provided 26% protection against infection in humans (Atmar et al., 2011) while HiPP provided 60% protection in Gn pigs. Since VLPs provided similar protection against diarrhea as HiPP but did not provide any protection against infection in Gn pigs, these results suggest that a HiPP regimen may be more protective and immunogenic than VLPs.

Previous studies have indicated that CD4+ helper T cells are required for protection (Zhu et al., 2013), while CD8+ effector T cells play a role in the clearance of primary NoV infection (Lindesmith et al., 2005; Tacket et al., 2003; Tomov et al., 2013). Our study identified several

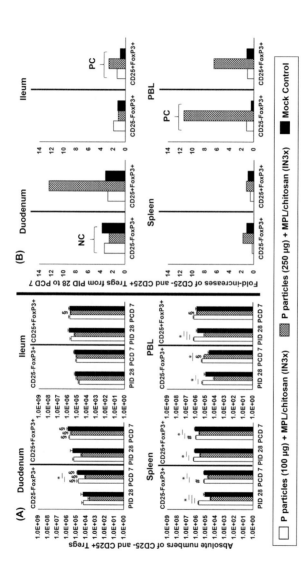

Figure 8.4 Treg responses induced by low dose and high dose P particle vaccination pre- and post-challenge. Mean total numbers + SEM (n = 6–10) of CD4+CD25-FoxP3+ and CD4+CD25+FoxP3+ Tregs in intestinal (duodenum, ileum) and systemic (spleen, PBL) tissues pre- and post-challenge (A). Frequencies of Tregs were determined using intracellular staining and multicolor flow cytometry. Total numbers were calculated using the volume and concentration of isolated MNCs. Fold expansion of Tregs following NoV challenge (B). Expansion of Tregs was calculated by dividing post-challenge numbers by pre-challenge numbers. Correlations between protection rate against diarrhea and T cell subset expansion were determined using Spearman's rank correlation coefficient. A "(PC)" indicates a positive correlation between protection rate against diarrhea and T cell subset expansion (R = 1, p < 0.0001), while a "(NC)" indicates a negative correlation (R = −1, p < 0.0001). See Figure 8.1 legend for the description of statistical analysis.

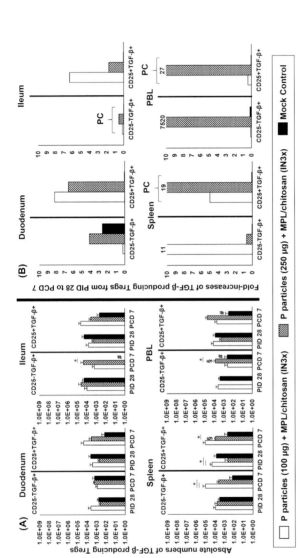

Figure 8.5   TGF-β producing Treg responses induced by low dose and high dose P particle vaccination pre- and post-challenge. Mean total numbers + SEM (n = 6–9) of CD4+CD25-FoxP3+TGF-β+ and CD4+CD25+FoxP3+TGF-β+ Tregs in intestinal (duodenum, ileum) and systemic (spleen, PBL) tissues pre- and post-challenge (A). Frequencies of TGF-β producing Tregs were determined using intracellular staining and multicolor flow cytometry. Total numbers were calculated using the volume and concentration of isolated MNCs. Total expansion of TGF-β producing Tregs following HuNoV challenge (B). Expansion of TGF-β+ Tregs was calculated by dividing post-challenge numbers by pre-challenge numbers. Correlations between protection rate against diarrhea and T cell subset expansion were determined using Spearman's rank correlation coefficient. A "(PC)" indicates a positive correlation between protection rate against diarrhea and T cell subset expansion (R = 1, $p < 0.0001$), while a "(NC)" indicates a negative correlation (R = −1, $p < 0.0001$). See Figure 8.1 legend for the description of statistical analysis.

positive correlations in ileum (activated CD4+ and CD8+ T cells, CD4+IFN-γ+ and CD8+IFN-γ+ T cells, CD25+ Tregs, and CD25-TGF-β+ Tregs), spleen (activated CD4+ and CD8+ T cells, CD4+IFN-γ+ T cells, and CD25+TGF-β+ Tregs), and PBL (CD4+IFN-γ+ and CD8+IFN-γ+ T cells, CD25- Tregs, and CD25+TGF-β+ Tregs). We also identified a negative correlation between the expansion of CD25- Tregs in the duodenum and the protection rate. We found an overall positive correlation between activated CD4+ and CD8+ T cells in the spleen and protection rate against diarrhea, as well as a negative correlation with CD25- Treg expansion in duodenum and protection rate against diarrhea. These findings illustrate the importance of vaccines down-regulating CD25- Tregs in the duodenum following expansion and could have implications in evaluating host susceptibility to HuNoV disease.

Overall, our data on T cell responses indicate that HiPP vaccination promotes the development of activated T cells in secondary lymphoid tissues and down-regulates expansion of Tregs in the primary effector site following challenge. However, it is worth noting that all groups in this study displayed significant increases in total numbers of duodenal Tregs from pre-challenge to post-challenge, indicating that HuNoV infection primarily induces a Treg response. HuNoV vaccines should aim to down-regulate this expansion.

In conclusion, three-dose (250 µg/dose) intranasal P particle vaccine (derived from VA387 GII.4) conferred higher protection rates against diarrhea and infection (60% and 60%, respectively) in Gn pigs compared to the protection rates (47% and 26%, respectively) conferred by the two-dose (100 µg/dose) intranasal VLP vaccine (derived from Norwalk GI.1) in humans (Atmar et al., 2011). Our results in the Gn pig study demonstrate that P particles are a viable vaccine alternative to VLPs.

9

Simvastatin Reduces Protection and Intestinal T cell Responses Induced by a Norovirus P Particle Vaccine in Gnotobiotic Pigs

Cholesterol pathways play an important role in viral infection and immunity. The effects of cholesterol-reducing drugs on the immunogenicity and protective efficacy of viral vaccines have not been well studied. It is an important question that needs to be addressed considering the high percentage of the human population taking cholesterol-reducing drugs. In this chapter, we described our study of the impacts of simvastatin on the protective efficacy of a candidate HuNoV vaccine and the T cell immune responses in Gn pigs (Kocher et al., 2021). This study was also led by my Ph.D. student Jacob Kocher. It was funded by an R01 sub-award I received from NIAID, NIH titled "Novel Vaccine against Norovirus" (R01AI089634, 2010–2015, Xi Jiang PI; Lijuan Yuan subcontract PI).

9.1 Introduction

Simvastatin is a cholesterol-reducing drug that inhibits HMG-CoA reductase, an enzyme in the cholesterol biosynthesis pathway, resulting in the reduction of low-density lipoprotein (LDL) cholesterol levels (Goldstein and Brown, 2009). Forty mg of simvastatin decreases LDL cholesterol and the risk of cardiovascular events by 23% over five years (Heart Protection Study Collaborative et al., 2011) with similar effects witnessed in low-risk populations (Cholesterol Treatment Trialists et al., 2012). Based on the report by the National Center for Health Statistics, 50% of men and 36% of women who are 65–74 years old took statin-type drugs in 2010. In 2013, the American Heart Association and American College of Cardiology released new guidelines which expand the recommendation for the use of simvastatin in the prevention of heart diseases even to people without high LDL levels.

Simvastatin has been shown to reduce the severity of several other diseases, including rheumatoid arthritis (Cojocaru et al., 2013; Kanda et al., 2007; Leung et al., 2003), multiple sclerosis (Zhang et al., 2013b), and periodontitis (Dalcico et al., 2013; Nassar et al., 2014). These pleiotropic effects can be explained by simvastatin's role in the down-regulation of IFN-induced MHC II expression by inhibition of the *CIITA* gene (Kwak et al., 2000; Kwak et al., 2001). This down-regulation of induced MHC II has many downstream effects, including reduced NK cell cytotoxicity (Hillyard et al., 2004; Hillyard et al., 2007), reduced *in vitro* T cell proliferation (Hillyard et al., 2004), reduced production of IL-2 and IFN-γ (Okopien et al., 2004), reduced CD4/CD8 and Th1/Th2 ratios (Kanda et al., 2007), increased Tregs (Mausner-Fainberg et al., 2008; Meng et al., 2012), and impaired lymphocyte homing to secondary lymphoid organs (Ulivieri et al., 2008).

Cholesterol pathways have been shown to influence murine NoV and bovine NoV VLP cellular entry (Gerondopoulos et al., 2010; Mauroy et al., 2011; Perry and Wobus, 2010) and Norwalk replication (Chang, 2009). Norwalk replication was directly associated with increased expression of LDL receptor (LDLR) mRNA in cells bearing the Norwalk replicon (Chang, 2009). Simvastatin has been shown to increase the expression of LDLR (Goldstein and Brown, 2009) and the production of Norwalk proteins and RNA in replicon-bearing cells (Chang, 2009). Additionally, 25-hydroxycholesterol has been shown to reduce murine NoV replication in RAW264.7 cells (Shawli et al., 2019). During hepatitis E virus (HEV) infection, simvastatin treatment led to a

DOI: 10.1201/b22816-9

significantly increased viral release *in vitro* and elevated viral loads in the patient sera (Glitscher et al., 2021). On the contrary to HEV, statins significantly repressed rotavirus replication through down-regulation of cholesterol synthesis (Ding et al., 2021)

The roles of simvastatin on HuNoV infection and disease have also been investigated *in vivo*. Gn pigs treated with simvastatin had an earlier onset and longer duration of fecal HuNoV shedding, as well as increased viral titers compared to pigs that were not fed simvastatin (Jung et al., 2012). Oral inoculation of IFN-α reduced the effects of simvastatin on HuNoV infectivity, indicating simvastatin down-regulates innate immunity (Jung et al., 2012). Previously, our study showed that simvastatin feeding increased the susceptibility of Gn pigs to infection by a HuNoV GII.4 variant, increased incidence of diarrhea compared to non-simvastatin-fed Gn pigs, and reduced the ID_{50} of the GII.4 variant in Gn pigs (Bui et al., 2013).

In this study, we used the well-established Gn pig model to evaluate the effects of simvastatin on the protective efficacy of the P particles vaccine following cross-variant GII.4 HuNoV challenge. Since simvastatin's immunomodulatory effects have been previously shown to mainly impact T cell responses (Kanda et al., 2007; Kim et al., 2010; Lee et al., 2010), we examined simvastatin's effects on the T cell profile induced by the P particles in the intestinal and systemic lymphoid tissues in Gn pigs. Simvastatin is primarily consumed by the elderly and aging, one of the target populations for NoV vaccines (Aliabadi et al., 2015), but its effects on HuNoV vaccine-induced immunity have not been investigated.

Near-term Large White cross pigs were derived via hysterectomy and maintained in germ-free isolator units. Pigs (both male and female) were randomly assigned into four groups: 1) P particles without simvastatin (P+S–), 2) P particles with simvastatin (P+S+), 3) control without simvastatin (CS–), and 4) control with simvastatin (CS+). Each group was composed of six to ten pigs; simvastatin groups were conducted in sequential experiments after the non-simvastatin-fed groups (Kocher et al., 2014) and were vaccinated and challenged with the same vaccine and virus preparations, fed the same diet, and kept under the same housing conditions in the same facility. The challenge virus inoculums were prepared from the single clinical isolate to avoid variabilities. All pigs in the P+ groups were intranasally inoculated with three doses of the vaccine using mucosal atomization devices (MADs, LMA North America, Inc. San Diego, CA, USA) at post-partum day (PPD) five (post-inoculation day [PID] 0), PID 10, and PID 21. Control pigs received adjuvants alone at the same time points. Simvastatin-treated pigs were orally inoculated with 8 mg/day/pig of simvastatin in a diluent for 11 days before challenge (PID 17–27, approximately 1.5–2.0 kg of body weight). At PID 28, a subset of pigs from each group was challenged with 10 ID_{50} of the GII.4/2006b variant 092895 at PID 28 (post-challenge day [PCD] 0). Four ml of 200 mM sodium bicarbonate were given to pigs 10 minutes before HuNoV oral inoculation to reduce gastric acidity. Challenged pigs were monitored daily for diarrhea and virus shedding until PCD 7. All pigs were euthanized at PID 28 (pre-challenge) or PCD 7 (post-challenge). Mononuclear cells (MNCs) from the duodenum (20 cm in length), ileum (20 cm in length), spleen (whole organ), and blood (70 ml) were isolated as previously described (Yuan et al., 1996) for analysis of MNC and T cell subsets with flow cytometry. The total numbers of MNCs were calculated by multiplying the volume and concentration of cells isolated from each tissue.

9.2 Simvastatin feeding significantly reduced serum cholesterol in Gn pigs

Simvastatin (Dr. Reddy's Laboratories, Ltd, Hyderabad, Telangana, India) was prepared as previously described (Bui et al., 2013). Tablets (80 mg) of simvastatin were dissolved in 100% ethanol for a final concentration of 8 mg/ml and filter sterilized. Serum cholesterol concentrations in Gn pigs pre- and post-feeding were evaluated by the Virginia-Maryland College of Veterinary Medicine hospital laboratory to verify simvastatin's effects. Data were presented as the mean serum cholesterol levels in mg/dL.

Serum cholesterol was monitored before and after simvastatin feeding to verify its effects in Gn pigs. Simvastatin feeding significantly reduced serum cholesterol after 11 days of feeding. Mean serum cholesterol levels decreased from 133 mg/dL at post-inoculation day (PID) 17 to 72 mg/dL at PID 27 for a 44% reduction ($p < 0.0001$). Mean serum cholesterol levels in age-matched non-simvastatin-fed pigs were 110 mg/dL at PID 17 and 99 mg/dL at PID 27 for an 8% reduction ($p > 0.05$).

9.3 Protective efficacy conferred by the P particle vaccine in simvastatin-fed pigs

The protective efficacy of P particle vaccines against HuNoV diarrhea and shedding following GII.4/2006b HuNoV challenge was evaluated in simvastatin-fed pigs in comparison to non-simvastatin-fed pigs. Diarrhea and fecal HuNoV shedding were monitored from PCD 1 to PCD 7 (Table 9.1). All P+S+ and CS+ pigs had a similar occurrence of diarrhea (100% and 83%). P+S+ and CS+ pigs had significantly higher diarrhea AUCs compared to P+S− pigs (8.8 vs. 5.4). Simvastatin did not affect the occurrence of shedding; however, it reduced the fold-reduction of AUC in vaccinated pigs from the corresponding controls (P+S− 3.1-fold vs. P+S+ 2.3-fold). Additionally, the P particle vaccine significantly shortened the duration of virus shedding (4.2 days vs. 1.7 days) and reduced the mean AUC (2.3 fold) of virus shedding compared to control pigs following the HuNoV challenge (Table 9.1). The protective effects conferred by the same P particle vaccine against the same GII.4/2006b HuNoV challenge in non-simvastatin-fed pigs were presented previously in Chapter 8 (Kocher et al. 2014).

9.4 Simvastatin decreases frequencies of proliferating intestinal CD8+ T cells after infection

To determine how simvastatin affects proliferating T cells, MNCs were isolated from HuNoV-vaccinated pigs post-challenge at PCD 7. MNCs were stimulated with P particles (virus-specific) or phytohemagglutinin (PHA) (non-specific), with or without simvastatin, and cultured in the presence of BrdU. The frequencies of virus-specific and non-specific proliferating T cells are shown in Figure 9.1. Simvastatin treatment significantly reduced proliferation of virus-specific CD4+ and CD8+ T cells isolated from spleen and CD8+ T cells from the duodenum (Figure 9.1A and 9.1C). Curiously, simvastatin significantly increased virus-specific proliferating CD8+ T cells isolated from blood. However, simvastatin significantly reduced the non-specific proliferating CD4+ and CD8+ T cells in PBL and CD8+ T cells in the duodenum (Figure 9.1B and 9.1D). Taken together, these results indicate that simvastatin impacts the proliferation of both HuNoV-specific and non-specific T cells.

9.5 Simvastatin feeding decreased total MNCs isolated from duodenum pre-challenge and PBL post-challenge

To evaluate how simvastatin feeding affected the development of the neonatal immune system, we calculated the total numbers of MNCs isolated from each tissue pre- and post-challenge (Figure 9.2). There are four treatment groups designated as 1) P particles without simvastatin (P+S−), 2) P particles with simvastatin (P+S+), 3) control without simvastatin (CS−), and 4) control with simvastatin (CS+). The effects of simvastatin in control pigs or vaccinated pigs were revealed by comparisons between CS+ and CS− groups or between P+S+ and P+S− groups, respectively. To assess the immunogenicity of P particles in the presence of simvastatin, the simvastatin-fed vaccinated pigs were compared to simvastatin-fed control pigs (P+S+ vs. CS+).

At PID 28, simvastatin feeding significantly reduced the total number of duodenal MNCs in vaccinated pigs (P+S− vs. P+S+) and also reduced the numbers in control pigs (CS− vs. CS+). Interestingly, CS+ pigs had significantly higher numbers of MNCs compared to P+S+ pigs in the duodenum. At PCD 7, simvastatin-fed pigs had significantly lower numbers of MNCs in PBL compared to non-simvastatin-fed pigs with or without vaccination. Total MNCs significantly increased in the duodenum of all groups post-challenge except P+S− pigs. Total MNCs also increased significantly in the ileum of CS+ pigs post-challenge (Figure 9.2).

Table 9.1 Clinical Signs and Protective Efficacy in P Particle-Vaccinated Gn Pigs after Challenge with GII.4 2006b HuNoV[a]

Group	n	Diarrhea					Virus shedding				
		Percent of pigs with diarrhea (no. of pigs with diarrhea/total no.)	Mean no. of days with diarrhea (SEM)	Mean AUC (SEM)**	Fold-reduction in AUC	Rate of protection against diarrhea	Percent of pigs shed virus (no. of pigs with shedding/total no.)	Mean no. of days with shedding (SEM)**	Mean AUC	Fold-reduction in AUC	Rate of protection against shedding
P+S−	9	44% (4/9)	1.3 (0.6)	5.4 (0.8)[b]	−1.6	46.7%	89% (8/9)	1.8 (0.4)[b]	3.82×10^4	−3.1	0%
P+S+	6	100% (6/6)	3.0 (0.7)	8.8 (1.1)[a]	0.0	0%	83% (5/6)	1.7 (0.4)[b]	1.87×10^4	−2.3	0%
CS−	6	83% (5/6)	1.8 (0.6)	6.7 (1.0)[ab]	NA[b]	NA	83% (5/6)	2.0 (0.7)[ab]	1.17×10^5	NA	NA
CS+	6	83% (5/6)	2.5 (0.6)	8.8 (0.7)[a]	NA	NA	83% (5/6)	4.2 (1.3)[a]	4.22×10^4	NA	NA

Note:

[a] All pigs in the P+ groups were intranasally inoculated with three doses of the P particle vaccine using mucosal atomization devices (MADs, LMA North America) at post-partum day (PPD) 5 (PID 0), PID 10, and PID 21. Control pigs (CS− and CS+) received adjuvants alone at the same time points. Simvastatin-treated pigs (P+S+ and CS+) were orally inoculated with 8 mg/day of simvastatin in diluent #5 for 11 days before challenge (PID 17–27). Gn pigs were challenged with 10 ID50 of a HuNoV GII.4 2006b variant 092895 at PID 28. Rectal swabs were collected daily after challenge to determine diarrhea and virus shedding by real-time RT-qPCR. Virus shedding was also detected in intestinal contents.

[b] NA, not applicable

* Abbreviated group names: P+S−, non-simvastatin-fed P particle-vaccinated; P+S+, simvastatin-fed P particle-vaccinated; CS−, non-simvastatin-fed placebo control; CS+, simvastatin-fed placebo control. Pigs in all groups received MPL/chitosan adjuvants.

** Means in the same column followed by different letters (A, B) differ significantly (One way ANOVA, $p < 0.05$); while shared letters indicate no significant difference. This table is modified from Kocher et al. (2021).

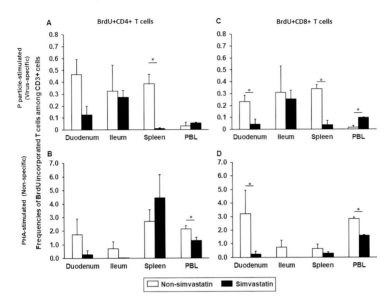

Figure 9.1   Virus-specific (A and C) and non-specific (B and D) proliferating T cells in the presence or absence of simvastatin. MNCs were stimulated with P particles or PHA and cultured with BrdU. Mean frequencies of CD3+CD4+BrdU+ and CD3+CD8+BrdU+ T cells in intestinal (duodenum, ileum) and systemic (spleen, PBL) tissues were analyzed by flow cytometry. Note the difference in the Y-axis scale between the upper (virus-specific) and lower (non-specific) panels. An asterisk above the bars indicates a significant difference between groups for the same cell type and tissue ($p < 0.05$ by Kruskal-Wallis rank-sum test).

9.6 Simvastatin decreased total number and frequency of CTLs in duodenum of vaccinated pigs pre-challenge but increased Th cells in PBL post-challenge

To understand the effects of simvastatin on P particle vaccine-induced T cell responses, we evaluated total Th, CTLs, activated non-regulatory CD25+FoxP3− T cells, IFN-γ producing effector/memory T cells, and CD4+CD25-FoxP3+ and CD4+CD25+FoxP3+ Tregs in intestinal (duodenum, ileum) and systemic (spleen, blood) lymphoid tissues. The total numbers and frequencies of Th and CTLs were compared among P particle-vaccinated or control pigs with or without simvastatin pre- and post-challenge (Figure 9.3A and 9.3B).

At PID 28, P+S+ pigs had significantly lower numbers of Th and CTLs in duodenum compared to P+S− pigs. Following NoV challenge, duodenal Th and CTLs increased significantly from PID 28 to PCD 7 in all groups except P+S− pigs. In the duodenum, CS+ pigs had significantly lower numbers of CTLs compared to CS− pigs, but P+S+ pigs had significantly higher numbers of CTLs compared to P+S− and CS+ pigs. In the ileum, CS+ pigs had significant increases in Th compared to pre-challenge and significantly higher Th compared to P+S+ pigs. In PBL, P+S+ pigs had significantly higher Th numbers compared to P+S− pigs post-challenge (Figure 9.3A). There were no significant differences in the numbers of splenic Th and CTLs among the treatment groups at either timepoint.

When comparing the frequencies of Th cells and CTLs among P particle-vaccinated or control pigs with or without simvastatin pre- and post-challenge (Figure 9.3B), the overall trends are similar to comparing the total numbers of Th cell and CTL in each tissue (Figure 9.3A). There are several statistically significant differences observed only in either the number or the frequency

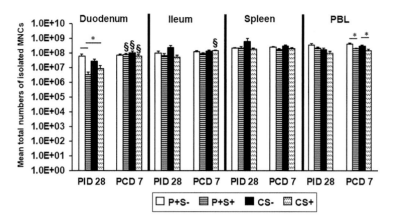

Figure 9.2 Total numbers of MNCs in intestinal and systemic lymphoid tissues of Gn pigs vaccinated with P particles with or without simvastatin feeding pre- and post-challenge. Total MNC numbers were calculated based on the concentration and total volume of MNCs isolated from the tissues. Total MNC numbers plus standard errors of the means (n = 6 to 10) pre-challenge and post-challenge in intestinal (duodenum, ileum) and systemic (spleen, PBL) tissues are presented. An asterisk * above the bars indicates a significant difference among groups for the same cell type and tissue at the same time point; a section sign § indicates that the numbers increased significantly following challenge in the same group; a number sign # indicates that the numbers decreased significantly following challenge in the same group (*p* < 0.05 by Kruskal-Wallis rank-sum test). Abbreviations of treatment groups in Figures 9.2–9.6: P particles without simvastatin (P+S–), P particles with simvastatin (P+S+), control without simvastatin (CS–), control with simvastatin (CS+).

data. The most noteworthy difference is that in blood, the frequencies (not the total numbers) of Th cell and CTL in the simvastatin-fed pigs (P+S+ and CS+) were significantly higher than the non-simvastatin-fed vaccinated and control pigs (P+S– and CS–) at PID 28 (Figure 9.3B).

9.7 Simvastatin reduced numbers and/or frequencies of activated CD4+ and CD8+ T cells in the intestinal tissues and blood pre-challenge and CD8+ T cells in spleen and blood post-challenge

To evaluate the effect of simvastatin on T cell activation, activated non-regulatory T cells pre- and post-challenge were enumerated. MNCs were stained freshly on the day of isolation and gated for FoxP3-CD25+CD4+ or FoxP3-CD25+CD8+ T cells. The mean numbers and frequencies of activated CD4+ and CD8+ T cells are shown (Figure 9.4A and 9.4B). Pre-challenge, P+S+ pigs had significantly lower numbers of activated CD4+ and CD8+ T cells in the duodenum compared to P+S– pigs. Following challenge, the activated CD4+ T cells in the duodenum increased significantly from pre-challenge in all groups. P+S+ pigs had significant increases in duodenal activated CD8+ T cells but significant decreases in splenic activated CD8+ T cells. In the ileum, CS+ pigs had significant increases in activated CD4+ T cell numbers compared to pre-challenge. Similar to the trend for Th cells, CS+ pigs had significantly higher numbers of activated CD4+ T cells compared to P+S+ pigs in the ileum. In spleen and PBL, P+S+ pigs had significantly lower numbers of activated CD8+ T cells compared to P+S– pigs (Figure 9.4A).

When comparing the frequencies of activated CD4 and CD8 T cells among P particle-vaccinated or control pigs with or without simvastatin pre- and post-challenge (Figure 9.4B), the overall trends are similar to comparing the numbers of activated CD4 and CD8 in most tissues (Figure 9.4A), but the statistically significant differences in the number data were not observed in the frequency data. Two distinct significant differences were detected from the frequency data only. Frequencies of activated CD4+ T cells in ileum and activated CD8+ T cells in the blood

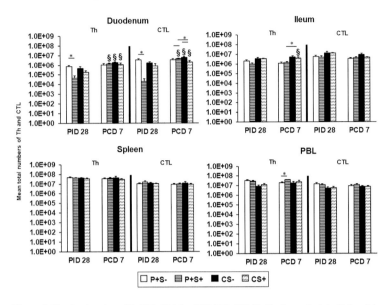

Figure 9.3A   Total numbers of Th (CD3+CD4+) and CTL (CD3+CD8+) in Gn pigs vaccinated with P particles with or without simvastatin feeding pre- and post-challenge. MNCs were gated as previously described (Kocher et al., 2014) following *in vitro* stimulation. Total numbers plus standard errors of the mean (n = 6 to 10) of CD3+CD4+ Th and CD3+CD8+ CTL in intestinal (duodenum, ileum) and systemic (spleen, PBL) tissues pre- and post-challenge are presented. See Figure 9.2 legend for statistical analysis and an explanation of the symbols indicating statistical significance.

of simvastatin-fed control pigs (CS+) were significantly lower than the non- simvastatin-fed control pigs (CS–) at PID 28 (Figure 9.4B).

9.8 Simvastatin feeding reduced numbers of CD8+IFN-γ+ T cells in duodenum and PBL at PID 28

We previously showed that HuNoV challenge induced a 37-fold expansion in the total numbers of duodenal CD4+IFN-γ+ T cells in previously HuNoV-infected pigs (Kocher et al., 2014). The results indicate the importance of an intestinal Th1 response in HuNoV immunity. In the present study, we identified IFN-γ+ producing CD4+ and CD8+ T cells in intestinal and systemic lymphoid tissues at PID 28 and PCD 7 by flow cytometry following *in vitro* stimulation with P particles. IFN-γ+ T cell numbers were determined after subtraction of isotype controls and mock-stimulated MNCs and are presented as mean total numbers (Figure 9.5). Pre-challenge at PID 28, simvastatin significantly reduced IFN-γ producing CD4+ and CD8+ T cell numbers in duodenum and CD8+IFN-γ+ T cell numbers in PBL compared to non-simvastatin pigs in both vaccinated and control pigs. In vaccinated pigs, simvastatin also significantly reduced IFN-γ producing CD4+ and CD8+ T cells in the spleen.

Post-challenge at PCD 7, P+S+ pigs had significantly lower numbers of CD4+IFN-γ+ and CD8+IFN-γ+ T cells in the spleen compared to P+S– pigs. In fact, P+S+ pigs had no detectable CD4+IFN-γ+ and CD8+IFN-γ+ T cells in the spleen post-challenge. Similarly, CS+ pigs had a complete lack of CD8+IFN-γ+ T cells in PBL, which were significantly lower than CS– pigs. Interestingly, CS+ pigs had significantly higher CD4+IFN-γ+ T cells in the spleen compared to CS– pigs (Figure 9.5). Frequencies of the CD4+IFN-γ+ and CD8+IFN-γ+ T cells did not show any statistically significant difference among any treatment groups (data not shown).

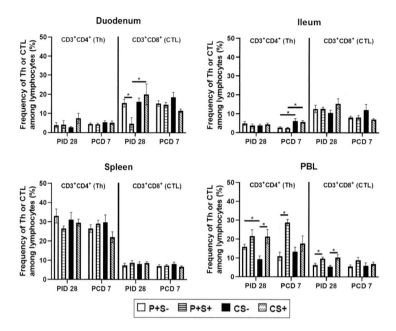

Figure 9.3B Frequencies of Th (CD3+CD4+) and CTL (CD3+CD8+) in Gn pigs vaccinated with P particles with or without simvastatin feeding pre- and post-challenge. Data presented are mean frequencies +/− SEM (n = 6–10) of Th cells or CTLs among total lymphocytes in intestinal (duodenum, ileum) and systemic (spleen, PBL) tissue at PID 28 or PCD 7. An asterisk indicates statistical significance among groups for the same cell type at the same time point (*p* ≤ 0.05, Kruskal-Wallis rank-sum test).

9.9 Simvastatin-fed pigs had reduced CD25-FoxP3+ and CD25+FoxP3+ Tregs in duodenum

Previously, we showed the importance of reducing Tregs in the duodenum in HuNoV protective immunity (Kocher et al., 2014), though simvastatin has been shown to increase FoxP3 expression in murine CD4 T cells *in vitro* and in tumor cells in mice (Kim et al., 2010; Lee et al., 2010). In the present study, we evaluated simvastatin's effects on P particle-vaccine induced Tregs pre- and post-challenge. The mean total numbers and frequencies of CD25-FoxP3+ and CD25+FoxP3+ Tregs are shown in Figures 9.6A and 9.6B.

At PID 28, simvastatin feeding significantly reduced the numbers of CD25-FoxP3+ Tregs in the duodenum and spleen of both vaccinated and control pigs. P+S+ pigs had significantly lower numbers of CD25-FoxP3+ Tregs in the ileum and PBL and CD25+FoxP3+ Tregs in duodenum and PBL compared to P+S− pigs. Following challenge, simvastatin feeding significantly reduced the numbers of Tregs in PBL compared to non-simvastatin-fed pigs. P+S+ pigs had significantly lower numbers of both Treg subsets in duodenum and spleen compared to P+S− pigs; P+S+ pigs also had significantly lower numbers of CD25+FoxP3+ Tregs in ileum compared to P+S− pigs. However, P+S+ pigs still had significantly reduced numbers of CD25-FoxP3+ Tregs in the duodenum and spleen and CD25+FoxP3+ Tregs in ileum and spleen compared to CS+ pigs. CS+ pigs had significant increases in both Treg subsets in ileum compared to pre-challenge. Interestingly, all groups had increases or significant increases in CD25-FoxP3+ and CD25+FoxP3+ Tregs in the duodenum following challenge. In the spleen, both P+S+ and P+S− groups had significant decreases in both Treg subsets following challenge compared to pre-challenge numbers. In PBL, simvastatin-fed pigs (P+S+ and CS+) had significant decreases in CD25+FoxP3+ Tregs compared to pre-challenge.

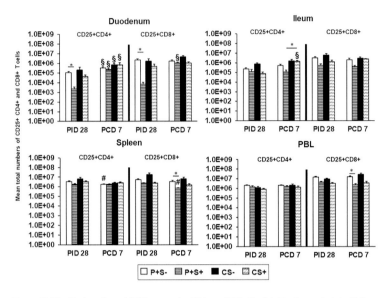

Figure 9.4A   Total numbers of CD25+ expressing CD4+ and CD8+ T cells in Gn pigs vaccinated with P particles with or without simvastatin feeding pre- and post-challenge. MNCs were gated as previously described (Kocher et al., 2014) and analyzed freshly on the day of cell isolation by flow cytometry. Total numbers plus standard errors of the means (n = 6 to 10) of CD4+CD25+FoxP3– and CD8+CD25+FoxP3– activated T cells pre-challenge and post-challenge in intestinal (duodenum, ileum) and systemic (spleen, PBL) tissues are presented. See Figure 9.2 legend for statistical analysis and an explanation of the symbols indicating statistical significance.

When comparing frequencies of CD25- and CD25+ Tregs among P particle-vaccinated or control pigs with or without simvastatin pre- and post-challenge (Figure 9.6B), the overall trends and statistical significances are very similar to comparing the numbers of CD25– and CD25+ Tregs in all tissues, except that in the duodenum the frequencies, not numbers, of CD25+FoxP3+ Treg of CS+ pigs were significantly higher than the CS– pigs at PCD 7 (Figure 9.6A and 9.6B).

9.10 Summary

In this study, we identified simvastatin's effects on HuNoV vaccine-induced protection and T cell immune responses. First, we found that simvastatin impaired the P particle vaccine-induced protection against HuNoV shedding and abolished the protection against HuNoV diarrhea when compared to non-simvastatin-treated Gn pigs in our previous study (Kocher et al., 2014). Secondly, we demonstrated that simvastatin impaired the P particle vaccine-induced T cell responses, especially those at the effector site of the gut-associated lymphoid tissues (duodenum). We showed that simvastatin impaired total MNC development *in vivo* and proliferating T cells *in vitro*, suggesting simvastatin can affect vaccine-induced immunity against HuNoV.

Despite recent efforts, HuNoV T cell immunity remains understudied relative to B cell and antibody immunity (Pattekar et al., 2021). Murine models have indicated that both CD4 and CD8 T cells are required for protection from murine norovirus (MNV) infection (Chachu et al., 2008), CD8 T cells are required for MNV clearance (Tomov et al., 2013), and CD4 T cells are a correlate of protection from MNV (Zhu et al., 2013). Previous studies have also indicated human NoVs induce a strong Th1 response in humans (Lindesmith et al., 2005) and Gn pigs (Kocher et al., 2014). The present study builds upon our previous findings in the role of T cells in HuNoV immunity. Among all the treatment groups, P particles without simvastatin (P+S–) had significantly higher numbers of Th cells, CTLs, activated CD4+ and

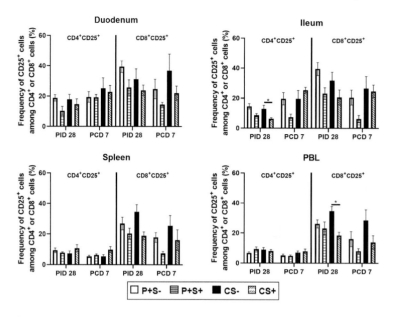

Figure 9.4B   Frequencies of activated CD4+ or CD8+ T cells in Gn pigs vaccinated with P particles with or without simvastatin feeding pre- and post-challenge. Data presented are mean frequencies +/– SEM (n = 6 to 10) of FoxP3-CD25+ cells among CD4+ or CD8+ cells in intestinal (duodenum, ileum) and systemic (spleen, PBL) tissue at PID 28 or PCD 7. An asterisk indicates statistical significance among groups for the same cell type at the same time point (p ≤0.05, Kruskal-Wallis rank-sum test).

CD8 T cells, and IFN-γ producing CD4+ and CD8+ T cells in the duodenum compared to P particle-vaccinated simvastatin-fed pigs (P+S+) at challenge (PID 28); additionally, it was the only group to have partial protection against HuNoV diarrhea (Kocher et al., 2014). This association between T cell responses and protection is similar to the findings in mice (Tomov et al., 2013; Zhu et al., 2013).

One of the goals of this study was to determine how simvastatin impacts the partial cross-variant protection (46.7% protection rate) against HuNoV diarrhea conferred by the P particle vaccine (Kocher et al., 2014). Simvastatin feeding totally abolished the previously observed protection as all the pigs in the P particles with simvastatin (P+S+) group had diarrhea. Simvastatin also resulted in a significantly higher AUC of diarrhea compared to non-simvastatin-fed pigs (8.8 vs. 5.4), eliminating the slight reduction in AUC of diarrhea among vaccinated pigs. It is important to note that although simvastatin alone can cause diarrhea in Gn pigs (Jung et al., 2012), the non-vaccinated control pigs with or without simvastatin did not differ significantly in the percent incidence (83%, 83%), duration (2.5, 1.8), or AUC (8.8, 6.7) of diarrhea. In the absence of simvastatin, the GII.4/2006b variant we used to challenge the pigs is highly infectious (Bui et al., 2013; Kocher et al., 2014). The P particle vaccine did not reduce the incidence of virus shedding in either non-simvastatin (83%) or simvastatin (83%) fed pigs; however, P+S+ pigs still had significantly lowered numbers of days with virus shedding (1.7 vs. 4.2) and slightly decreased AUC of shedding (–2.3-fold) compared to control pigs. Hence, even with the presence of simvastatin, the P particle vaccine can still shorten the duration of virus shedding. Simvastatin feeding did not increase fecal virus shedding titers, which is consistent with the previous report in Gn pigs (Jung et al., 2012). We noted a peculiar association between increased severities of diarrhea and decreased viral shedding titers. Increased watery diarrhea among simvastatin-fed pigs might have diluted the viruses or affected the quality of the fecal material collected by rectal swabs, leading to an underestimation of the amount of HuNoV shed.

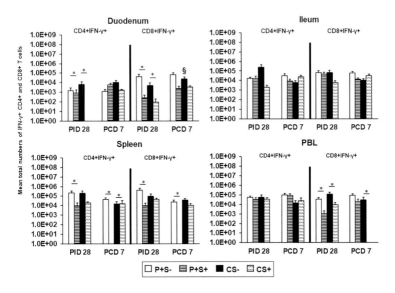

Figure 9.5 Total numbers of IFN-γ+ expressing CD4+ and CD8+ T cells in Gn pigs vaccinated with P particles with or without simvastatin feeding pre- and post-challenge. MNCs were gated as previously described (Kocher et al., 2014) following *in vitro* NoV antigen stimulation and analyzed by flow cytometry. Total numbers of IFN-γ producing CD4+ and CD8+ T cells following subtraction of isotype control and mock-stimulated background numbers are calculated. Data presented are total numbers plus standard errors of the means (n = 6 to 10) pre-challenge and post-challenge in intestinal (duodenum, ileum) and systemic (spleen, PBL) tissues. See Figure 9.2 legend for statistical analysis and an explanation of the symbols indicating statistical significance.

Another goal of this study was to identify the immune-modulatory effects of simvastatin on P particle vaccine-induced T cell immunity. To better illustrate the impacts of simvastatin on T cell subsets, we presented both total number and frequency data of flow cytometry. When comparing the frequencies of the T cell subsets among different treatment groups, the trends are mostly the same as comparing the total numbers. However, there are some statistically significant differences observed only in either the number or the frequency data. For example, frequencies of the CD4+IFN-γ+ and CD8+IFN-γ+ T cells did not show any statistically significant difference among any treatment groups, whereas the total numbers are significantly different in duodenum, spleen, and blood between the simvastatin-fed and non-simvastatin-fed groups. This discrepancy emphasizes the necessity to examine both total number and frequency data of flow cytometry in immunophenotypic analysis when testing an immune-modulating agent.

Simvastatin significantly reduced numbers of all analyzed cell types in the duodenum of vaccinated pigs at PID 28. Simvastatin further reduced the numbers of effector/memory CD4+IFN-γ+ and CD8+IFN-γ+ T cells in duodenum and effector/memory CD8+IFN-γ+ T cells in PBL at PID 28 compared to non-simvastatin-fed pigs. Since simvastatin does not affect the constitutive expression of MHC II (Kwak et al., 2000), P particles were likely able to stimulate the initial innate and primary adaptive immune responses, but not memory immune responses. The latter is evident in the lack of Th and CTL responses in the spleen (Figure 9.3A and 9.3B) and the complete lack of splenic virus-specific IFN-γ+ producing T cells in P+S+ pigs following challenge (Figure 9.5), which indicates a lack of HuNoV-specific memory T cell response.

The most important effects of simvastatin were determined to be on the development of total MNCs and proliferating T cells. Simvastatin reduced the total numbers of isolated MNCs in duodenum from vaccinated pigs pre-challenge and in PBL of both vaccinated and control pigs post-challenge, which likely contribute to the significantly lower numbers of all

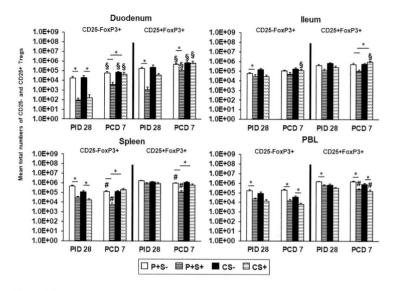

Figure 9.6A Total numbers of CD25– and CD25+ Tregs in Gn pigs vaccinated with P particles with or without simvastatin feeding pre- and post-challenge. Tregs were gated as described in (Kocher et al., 2014) following staining freshly on the day of cell isolation. Total numbers of FoxP3+CD25– and FoxP3+CD25+ Tregs plus standard errors of the means (n = 6 to 10) among total MNCs pre-challenge and post-challenge in intestinal (duodenum, ileum) and systemic (spleen, PBL) tissues are presented. See Figure 9.2 legend for statistical analysis and an explanation of the symbols indicating statistical significance.

T cell types analyzed in duodenum at PID 28. Similarly, simvastatin-fed pigs had lower or significantly lower frequencies and numbers of Tregs in PBL at PCD 7. Simvastatin appears to impair the development of all MNCs, presumably including impaired antigen-presenting cells, T and B cells.

Serum HBGA blocking antibodies are believed to be correlates of protection (Reeck et al., 2010; Zhu et al., 2013), but innate immunity has also been shown to be critical for control of MNV infection (Karst et al., 2003). Norovirus immunity requires intact innate and adaptive immune responses; simvastatin's impairment of total MNC development and immune activation is likely responsible for the abolishment of vaccine-induced partial protection. Further, simvastatin decreased the frequencies of non-specific proliferating T cells in the blood and CD8+ T cells in the duodenum. Simvastatin's effects on HuNoV infection-induced and vaccine-induced B cell and innate immunity need to be investigated in future studies.

There are several limitations to this study. First, the immaturity of the immune systems of neonatal Gn pigs may limit the direct extrapolation of the findings to elderly human populations. Second, we focused on studying T cell responses and did not measure the impact of simvastatin on antibody responses which are known to play important role in the immunity against HuNoV. More preclinical studies in animals and human clinical trials will need to be conducted to thoroughly determine how simvastatin impacts HuNoV pathogenesis and impairs vaccine immunogenicity and protective efficacy. These questions all have important practical implications. The impacts of concurrent treatment with simvastatin during HuNoV infection on the pathogenesis and the impact of simvastatin on humoral immune responses should be studied using the Gn pig model of HuNoV infection and diarrhea in the future.

In conclusion, simvastatin erased the partial protection conferred by the P particle vaccine and inhibited T cell development in the duodenum of vaccinated pigs. A robust immune response including T cells and B cells are required for viral clearance (Chachu et al., 2008; Tomov et al., 2013). Since the elderly and aging are one of the primary target populations for a HuNoV

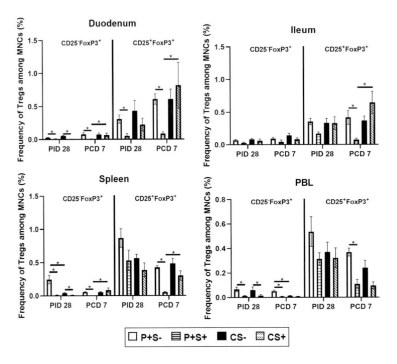

Figure 9.6B   Frequencies of CD25− or CD25+ Tregs in Gn pigs vaccinated with P particles with or without simvastatin feeding pre- and post-challenge. Data presented are mean frequencies +/− SEM (n = 6–10) of CD25− or CD25+ cells expressing FoxP3+ among total MNCs in intestinal (duodenum, ileum) and systemic (spleen, PBL) tissue at PID 28 or PCD 7. An asterisk indicates statistical significance among groups for the same cell type at the same time point ($p \leq 0.05$, Kruskal-Wallis rank-sum test).

vaccine (Aliabadi et al., 2015) and the primary consumers of simvastatin-type drugs, the implications of our findings that simvastatin can increase HuNoV-induced diarrhea and decrease the development of overall T cell responses are important in the development of preventive and therapeutic strategies against HuNoV gastroenteritis, and maybe for other pathogens (Wickert et al., 2014). Simvastatin likely affects immune responses against many pathogens; the benefits and side effects of its use should be carefully analyzed and balanced.

10

Dissecting Importance of B Cells versus T Cells in Rotavirus Vaccine-Induced Immunity Using Gene Knockout Gn Pigs

This study was led by my Ph.D. student Ke Wen (Wen et al., 2016). The project was funded by an R01 award from National Center for Complementary and Alternative Medicine (NCCAM), NIH titled "Mechanisms of Immune Modulation by Probiotics" (R01AT004789. 2009–2014, PI: Lijuan Yuan). This chapter includes most of the contents of the paper Wen et al. "B-Cell-Deficient and CD8 T-Cell-Depleted Gnotobiotic Pigs for the Study of Human Rotavirus Vaccine-Induced Protective Immune Responses." *Viral Immunol.* 2016. 29:112–127, with copyright permission from the publisher Mary Ann Liebert, Inc.

10.1 Introduction

For rotavirus research, various gene knockout adult mice (i.e., Rag-2 mice devoid of both T and B cells, β2m mice that lack cytotoxic T cell responses, JHD mice that lack B cell responses, and IgA-knockout mice that have no detectable IgA in the serum or any secretions) have been extensively used in studying determinants of protective immunity against rotavirus infection (Blutt et al., 2012; Dharakul et al., 1991; Franco and Greenberg, 1999; McNeal et al., 2002; VanCott et al., 2001). These studies have produced important observations regarding the roles of various components of humoral and cellular immunity (IgA antibody, CD4, or CD8 T cell) in the resolution of primary infection or protection against chronic rotavirus infection. However, adult mice do not develop diarrhea after murine rotavirus infection. Also, the pathogenesis of rotavirus infection in mice is very different from that in humans as we have discussed earlier in this book. Moreover, different genetic backgrounds of mice lead to different, even conflicting, results (Franco and Greenberg, 1997, 2000). One study suggested that CD4 T cells are the only lymphocytes needed to protect mice against rotavirus shedding after immunization with rotavirus VP6 protein (McNeal et al., 2002). Others have suggested that 1) neither CD4+, CD8+ T cells, nor antibodies were essential for protection against rotavirus primary infection in mice, 2) B cell responses were necessary for the development of immunity against rotavirus reinfection, and 3) the importance of each lymphocyte population as effectors of protection was found to be dependent on the immunogen (live, inactivated, or virus-like particles) and the route of immunization (oral or intranasal) (Blutt et al., 2008; Ward, 2003).

In contrast to adult mice, the neonatal Gn pig model of HRV infection and diarrhea more closely recapitulate rotavirus gastroenteritis in human infants and young children (Ward et al., 1996a). The Gn pig is the only animal model susceptible to HRV diarrhea for up to eight weeks of age and the model has been extensively used for investigating rotavirus vaccine-induced protective immunity against both infection and diseases as we have discussed in previous chapters. Our previous studies found that numbers of HRV-specific intestinal IgA ASC and frequencies of intestinal IFN-γ producing T cells were significantly correlated with protection rates against rotavirus diarrhea, indicating the role of both B and T cells in the protection against rotavirus disease (Yuan et al., 1996; Yuan et al., 2008). However, these studies could not separate the humoral and cellular protective immunity from each other. Recently, Revivicor, Inc. (www.revivicor.com/), located near the Virginia Tech campus at Blacksburg, Virginia, has generated total B cell–deficient pigs. These genetically modified pigs have the gene encoding for the immunoglobulin (Ig) heavy chain knockout (HCKO) using Revivicor's proprietary techniques. These pigs are impaired in their germinal center development and are completely incapable of producing antibodies, but theoretically maintain intact innate immune cell and adaptive T cell responses. As such, these animals offer a unique model to elucidate the mechanisms of vaccine-induced protective immunity when challenged with VirHRV.

The uniquely modified HCKO pigs provided by Revivicor, Inc. have the gene encoding for the Ig heavy chain knocked out to impair the VDJ rearrangement using Revivicor's proprietary techniques (US patent application number 20080026457 and 20060130157). Briefly, the poly (A)-trap gene targeting technology was applied for disrupting the single functional heavy-chain joining region (J_H) in a porcine primary fetal fibroblast (PPFF) cell line, and the SCNT strategy was used to produce genetically modified piglets as described in previous studies (Dai et al., 2002). The J_H- female pigs were outbred with wild-type male pigs of a similar large white breed to generate F1 offspring. The $J_H\pm$ F1 pigs (male and female) were subsequently cross-bred to obtain the F2 generation (the ratio of J_H-, $J_H\pm$, and J_H+ genotypes approximated the expected 1:2:1 Mendelian ratio). F2 fetuses with confirmed JH− genotype from one sow were used to derive the JH− PPFF cells and produced reconstructed embryos, which were used to produce HCKO pigs through the SCNT technology.

We characterized the development of cellular immune responses to the AttHRV vaccine in the HCKO pigs compared to wild-type pigs at challenge and post-challenge. We also depleted CD8 cells in a subset of HCKO pigs by intravenously injecting anti-pig CD8 monoclonal antibodies prior to the time of challenge, examined the influence of the CD8 cell depletion on cellular immune responses, and assessed the contribution of CD8 T cells in protective immunity against rotavirus diarrhea conferred by the AttHRV vaccine (Wen et al., 2016).

Near-term wild-type and cloned HCKO Gn pigs were derived by hysterectomy and maintained in germ-free isolators. Gn pigs (both males and females) were randomly assigned to the following groups: 1) wild-type Gn pigs with AttHRV (n = 10, AttHRV/WT), 2) HCKO Gn pigs with AttHRV (n = 12; six pigs for AttHRV/HCKO and six pigs for AttHRV/HCKO/CD8−), and 3) control HCKO Gn pigs (n = 5, HCKO). Pigs in the AttHRV groups were orally inoculated with 5×10^7 FFU/dose of AttHRV at both 5 and 15 days of age (PID 0 and 10, respectively). Pigs not inoculated with AttHRV were given an equal volume of diluent as a control. At PID 28, subsets of pigs from AttHRV/WT (n = 5), AttHRV/HCKO (n = 3), AttHRV/HCKO/CD8− (n = 3), and HCKO (n = 3) groups were orally challenged with 10^5 FFU VirHRV as previously described (Wen et al., 2012b). Post-challenge, pigs were examined daily for clinical signs, and fecal swabs were collected daily as was previously described (Yuan et al., 1996).

A subset of HCKO pigs was intravenously injected with a purified anti-pig CD8 monoclonal antibody (clone 76-2-11, Southern Biotech; 1.4 mg per kilogram of body weight in PBS with pH 7.4) at 28 days of age to generate transiently CD8 T cell-depleted pigs (designated as HCKO/CD8−). The concentration of the anti-pig CD8 antibody was 1.0 mg/ml and the injected maximum volume of the antibody was up to 5 ml. For examining the efficacy of CD8 cell depletion, peripheral blood was collected from pigs in the AttHRV/HCKO/CD8− group at 31 days of age (three days after the CD8 antibody injection and two days before the VirHRV challenge) and PBMCs were isolated. MNCs were also isolated from the ileum, spleen, and peripheral blood of pigs euthanized on PID 28 or PCD 7 as was previously described (Wen et al., 2012b; Yuan et al., 1996) to examine the efficacy of CD8 cell depletion and cellular immune responses by flow cytometry (Wen et al., 2016). In addition, MNCs from the ileum, spleen, and peripheral blood of AttHRV/HCKO pigs were subjected to the examination of rotavirus-specific IgM, IgA, and IgG ASC responses using the ELISPOT assay previously described (Yuan et al., 1996).

10.2 Immunophenotyping of B cell–deficient and CD8 T cell–depleted Gn pigs

10.2.1 HCKO pigs lack immature B cells and Ig secreting cells (IgSC) and are totally incapable of producing antibodies

To characterize the immunophenotype of the B cell-deficient HCKO (JH−) pigs, normal wild type (n = 4 and 6 for postpartum day [PPD] 0–7 and PPD 28, respectively) and cloned HCKO Gn pigs (n = 4 and 2 for PPD 0–7 and PPD 28, respectively) were used to examine the presence of immature B cells and IgSC in various lymphoid tissues. Gn pigs were mono-associated with *Lactobacillus acidophilus* NCFM® strain to stimulate the development of the neonatal immune system (Butler et al., 2009a) as was described previously (Wen et al., 2012b). All HCKO pigs had almost no detectable immature B cells and IgSC in all tissues examined at PPD 0–7 or PPD 28 except for a few IgSC in the thymus of one pig at PPD 0–7 (Figure 10.1). In contrast, wild-type pigs had higher or significantly higher frequencies of immature B cells and IgSC in duodenum, ileum, spleen, MLN, and blood compared to HCKO pigs at PPD 0–7 and PPD 28. The lack of B

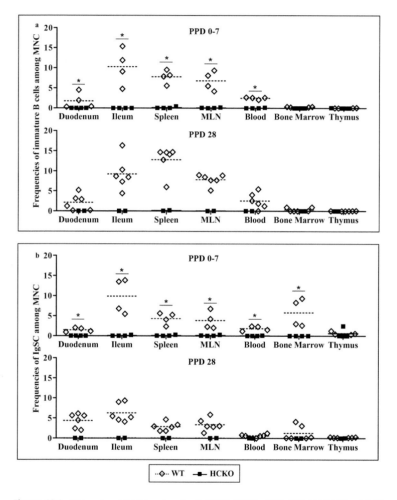

Figure 10.1   Frequencies of total immature B cells (CD79a+SWC3a+CD21+CD2+/–IgM+) (a) and IgSC (CD79a+SWC3a–CD21+/–CD2+/–IgM–) (b) among total gated MNC. The top figures show the frequencies of immature B cells (Figure 10.1A) among MNCs at PPD 0–7 and PID 28, and the bottom ones show IgSC (Figure 10.1B). Bars indicate group means. The sign "*" indicates significant differences among groups (Kruskal–Wallis test, $p < 0.05$). PPD, post-partum day; WT, wild type; HCKO, heavy chain knockout; MNC, mononuclear cells. Figures 10.1–10.7 are reproduced from Wen et al. "B-Cell-Deficient and CD8 T-Cell-depleted Gnotobiotic Pigs for the Study of Human Rotavirus Vaccine-Induced Protective Immune Responses." *Viral Immunol.* 2016. 29:112–127 with permission from Mary Ann Liebert, Inc.

cells is associated with the relatively smaller size of the spleen of HCKO pigs compared to wild-type pigs. Rotavirus-specific IgM, IgA, and IgG antibody-secreting cells were also not detected in the ileum, spleen, and peripheral blood of any AttHRV/HCKO pigs at post-AttHRV inoculation day (PID) 28 and post-VirHRV challenge day (PCD) 7.

10.2.2 Efficiency of CD8 T cell depletion by a single injection of anti-pig CD8 antibody

The anti-pig CD8 monoclonal antibody was intravenously injected in HCKO pigs five days before VirHRV challenge on PID 23 (PPD 28). Peripheral blood was collected three days later (PPD 31) to examine the efficiency of CD8 cell depletion by using flow cytometry. The representative dot plot in Figure 10.2A shows the frequencies (0.92%) of CD8 cells among lymphocytes in the blood. On PID 28/PCD0 (PPD 33) and PID35/PCD 7 (PPD 40), the efficiency of CD8 depletion was examined in the blood, spleen, and ileum. The representative dot plots are presented in Figure 10.2B (blood) and Figure 10.2C (spleen). As shown, anti-pig CD8 antibody injection resulted in a 97.6% reduction (32.72% to 0.78%) of CD8 cells in the blood (Figure 10.2B upper panel) and an 81% reduction (33.87% to 6.43%) of CD8 cells in the spleen (Figure 10.2C upper panel) on the day of VirHRV challenge (PPD 33). Increasing numbers of cells started to re-express the CD8 marker 12 days after injection by PPD 40; however, the frequencies of CD8 cells are still much lower compared to untreated pigs. Among in vitro semi-purified AttHRV antigen-stimulated cells, more cells re-expressed CD8 marker after 17 hrs of cell culture (Figures 10.2B and 10.2C, lower panels). Our result concurred with the previous report that after a single intravenous injection of the anti-pig CD8 antibody (clone 76-2-11 as ascites; 0.7 ml/kg), the frequencies of CD8 T cells in the blood of the pigs were reduced to the lowest levels (0–5% of total cells) at four to six days post-injection (Suzuki et al., 1990). The effects of anti-pig CD8 antibody injection on the frequencies of CD4 and CD8 T cells, Treg cells, NK cells, and $\gamma\delta$ T cells in all the lymphoid tissues examined on PID 28 (PPD 33) and PCD 7 (PPD 40) are summarized (Figures 10.3A, 10.3B, 10.4, and 10.5) and compared with AttHRV/WT and/or AttHRV/HCKO pigs in the following sections.

10.2.3 HCKO pigs had increased CD4 and CD8 T cell population but decreased Treg cell population

To assess the influence of B cell deficiency on other lymphocyte populations, we examined the frequencies and total numbers of NK and T cell subsets by flow cytometry. B cell deficiency had no significant effect on NK cell frequencies but altered the frequencies of CD4 and CD8 T cells, and Treg cells (Figure 10.3A). AttHRV/HCKO pigs had significantly higher frequencies of CD4 T cells in the ileum, spleen, and blood and significantly higher frequencies of CD8 T cells in the ileum and spleen at both PID 28 and PCD 7 compared to AttHRV/WT pigs (Figure 10.3A). In addition, AttHRV/HCKO pigs had significantly higher frequencies of CD4 T cells in intraepithelial lymphocytes (IEL) and significantly higher frequencies of CD8 T cells in IEL and blood at PCD 7 compared to AttHRV/WT pigs (Figure 10.3A). Changes in the absolute numbers of the T cell populations were also observed, although to a lesser degree. AttHRV/HCKO pigs had significantly higher numbers of CD4 and CD8 T cells in the blood at PID 28 and significantly higher numbers of CD4 T cells in IEL at PCD 7 compared to AttHRV/WT pigs (Figure 10.3B). Interestingly, AttHRV/HCKO pigs had significantly lower numbers of CD4 and CD8 T cells in the spleen at both PID 28 and PCD 7 compared to AttHRV/WT pigs (Figure 10.3B). AttHRV/HCKO pigs had significantly lower frequencies of Treg cells in the ileum and IEL at both PID 28 and PCD 7 and significantly lower frequencies of Treg cells in the blood at PID 28 compared to AttHRV/WT pigs (Figure 10.3A). AttHRV/HCKO pigs also had lower or significantly lower numbers of Treg cells in all tissues compared to AttHRV/WT pigs.

10.2.4 CD8 depletion in B cell–deficient pigs significantly decreased CD8 T cells and NK cells

As expected, the CD8 cell depletion procedure of intravenous injection of anti-pig CD8 antibodies decreased the population of CD8 T cells in AttHRV/HCKO/CD8– pigs. Compared to non-CD8 depleted pigs, AttHRV/HCKO/CD8– pigs had significantly lower frequencies of CD8 T cells in the spleen at PID 28 and in all tissues at PCD 7 than AttHRV/HCKO pigs (Figure 10.3A). In addition, AttHRV/HCKO/CD8– pigs had significantly lower numbers of CD8 T cells in the spleen and blood at PID 28 and PCD 7 than AttHRV/HCKO pigs (Figure 10.3B).

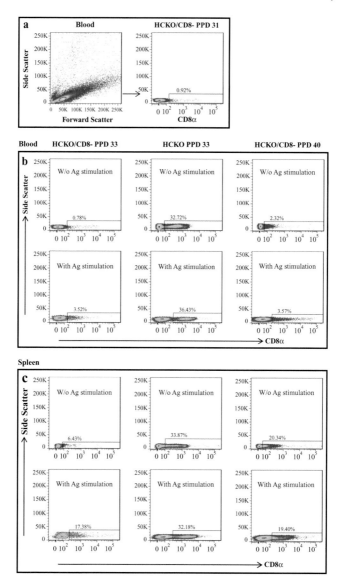

Figure 10.2   The efficiency of transient CD8 cell depletion by a single injection of anti-pig CD8 antibody. Anti-pig CD8 monoclonal antibody was given at 28 days of age (PPD 28) and, to examine the efficacy of CD8 depletion, peripheral blood was collected from pigs in the AttHRV/HCKO/CD8– group three days later (PPD 31). MNCs were also isolated from the ileum, spleen, and peripheral blood of pigs euthanized on PID 28 (PPD 33) or PCD 7. Figure 10.2A shows representative dot plots collected from blood at PPD 31. The figures in the top of 2B and 2C show those collected from blood and spleen, respectively, without semi-purified AttHRV antigen stimulation, and those in the bottom show the representative dot plots with semi-purified AttHRV antigen stimulation for 17 hrs. The first and second columns in the 2B and 2C show the dot plots from HCKO/CD8– and HCKO pigs, respectively, at PPD 33. The last one shows those from HCKO/CD8– pigs at PPD 40. The numbers above the rectangles in dot plots are the frequencies of CD8 cells among lymphocytes.

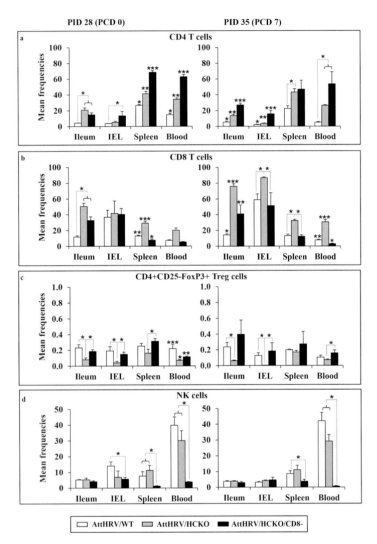

Figure 10.3 Mean frequencies of CD4 T cells (a), CD8 T cells (b), Treg cells (c), and NK cells (d) among lymphocytes (3A) and mean numbers per tissue (3B). MNCs were mock stimulated for 17 hrs (Brefeldin A was added for the last 5 hrs) and then subjected to intracellular staining and flow cytometry for collecting the frequency and number data of CD4 T cells (a), CD8 T cells (b), and NK cells (d) (Figure 10.3) and served as background controls for the data of virus-specific IFN-γ producing cell frequencies (Figure 10.4). MNCs were stained freshly for Treg cells (c) without *in vitro* stimulation. The absolute numbers of CD4 and CD8 T cells, Treg cells, and NK cells per tissue were calculated based on their frequencies and the total number of MNCs isolated from each tissue. The left figures within each panel show the pre-challenge data at PID 28 and those in the right show the post-challenge data at PCD 7. Data are presented as mean frequency/number ± standard error of the mean (n = 3 or 5). The asterisk "*" indicates a significant difference in frequencies/numbers among groups for the same cell type and tissue; a different number of "*" indicates significant differences among groups (Kruskal–Wallis test, *p* < 0.05).

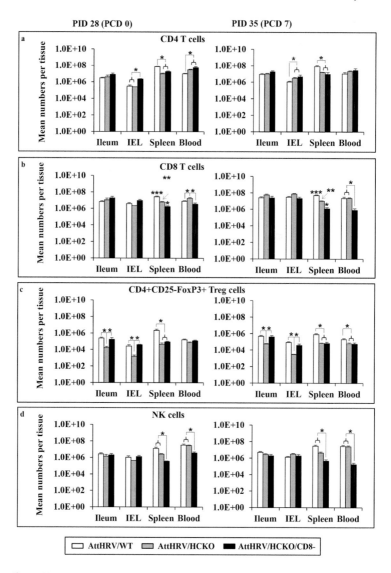

Figure 10.3   (Continued)

Compared to AttHRV/WT pigs, AttHRV/HCKO/CD8− pigs had significantly lower frequencies of CD8 T cells in the spleen at PID 28 and in the blood at PCD 7 than AttHRV/WT pigs (Figure 10.3A). AttHRV/HCKO/CD8− pigs also had significantly lower numbers of CD8 T cells in the spleen at both PID 28 and PCD 7 and in the blood at PCD 7 than AttHRV/WT pigs (Figure 10.3B). However, AttHRV/HCKO/CD8− pigs still had significantly higher frequencies of CD8 T cells in ileum at both PID 28 and PCD 7 than AttHRV/WT pigs due to the large increase in the total CD8 T cell frequencies in the B cell–deficient pigs compared to the wild-type pigs (Figure 10.3A).

CD8 cell depletion also decreased the population of NK cells because they too express the CD8 surface marker in swine (Gerner et al., 2009). CD8 depletion greatly reduced the NK cell population in the spleen and blood at both PID 28 and PCD 7 (Figure 10.3A). AttHRV/HCKO/CD8– pigs had significantly lower frequencies of NK cells in spleen and blood at PID 28 and in blood at PCD 7 than AttHRV/WT pigs and AttHRV/HCKO pigs (Figure 10.3A). AttHRV/HCKO/CD8– pigs also had significantly lower frequencies of NK cells in IEL at PID 28 and in spleen at PCD 7 than AttHRV/WT pigs (Figure 10.3A). Similar to the reduced frequencies, AttHRV/HCKO/CD8– pigs had significantly lower numbers of NK cells in spleen and blood at both PID 28 and PCD 7 than AttHRV/WT pigs and AttHRV/HCKO pigs (Figure 10.3B).

10.2.5 CD8 depletion in B cell–deficient pigs further increased CD4 T cells but restored the level of Treg cell population

AttHRV/HCKO/CD8– pigs had overall the highest frequencies (with the exception in ileum at PID 28) (Figure 10.3A) and the highest numbers (with the exception in spleen) (Figure 10.3B) of CD4 T cells among three groups. CD8 cell depletion in the HCKO pigs significantly increased the frequencies of CD4 T cells in the spleen and blood at PID 28 and in ileum and IEL at PCD 7 (Figure 10.3A). AttHRV/HCKO/CD8– pigs also had significantly increased numbers of CD4 T cells in IEL at PID 28 (Figure 10.3B). AttHRV/HCKO/CD8– pigs had similar frequencies of Treg cells as in AttHRV/WT pigs (with the exception in blood at PID 28) and significantly higher frequencies of Treg cells in all tissues at PID 28 and IEL and blood at PCD 7 compared to AttHRV/HCKO pigs (Figure 10.3A). AttHRV/HCKO/CD8– pigs also had similar numbers of Treg cells in ileum and IEL as in AttHRV/WT pigs but significantly higher frequencies of Treg cells in ileum and IEL compared to AttHRV/HCKO pigs (Figure 10.3B).

10.2.6 B cell–deficient pigs had higher frequencies of IFN-γ producing CD4 and CD8 T cells and NK cells

AttHRV/HCKO pigs had overall higher frequencies of IFN-γ producing CD4 and CD8 T cells, and NK cells in all tissues compared to AttHRV/WT pigs (Figure 10.4). Notably, before challenge, AttHRV/HCKO pigs had significantly higher frequencies of IFN-γ producing CD4 and CD8 T cells in the ileum, spleen, and blood compared to AttHRV/WT pigs at PID 28 (Figures 10.4A and 10.4B). AttHRV/HCKO pigs had significantly higher frequencies of IFN-γ producing NK cells (Figure 10.4C) and CD8 T cells in all tissues and IFN-γ producing CD4 T cells in spleen and blood at PCD 7 (Figures 10.4A and 10.4B).

10.2.7 CD8 depletion in B cell–deficient pigs significantly decreased the IFN-γ producing CD4 and CD8 T cells in most tissues

Compared to AttHRV/HCKO pigs, AttHRV/HCKO/CD8– pigs had significantly lower frequencies of IFN-γ producing CD8 T cells in ileum, spleen, and blood at PID 28 and in IEL, spleen, and blood at PCD 7 (Figure 10.4B). Interestingly, AttHRV/HCKO/CD8– pigs also had significantly lower frequencies of IFN-γ producing CD4 T cells in spleen and blood at PID 28 and in the spleen at PCD 7 compared to AttHRV/HCKO pigs (Figure 10.4A), suggesting that IFN-γ producing CD4 T cells in pig systemic lymphoid tissues are likely double-positive cells (CD4+CD8+).

10.2.8 CD8 depletion in B cell–deficient pigs further enhanced IFN-γ producing NK cells in all tissues and IFN-γ producing CD4 and CD8 T cells in ileum at PCD 7

AttHRV/HCKO/CD8– pigs had the highest frequencies of IFN-γ producing NK cells in all tissues among all three groups at PCD 7 (Figure 10.4C). AttHRV/HCKO/CD8– pigs had significantly higher frequencies of IFN-γ producing NK cells in ileum at PCD 7 compared to AttHRV/HCKO pigs (Figure 10.4C). In addition, AttHRV/HCKO/CD8– pigs had the highest frequencies of IFN-γ producing CD4 and CD8 T cells in ileum among all three groups at PCD 7 (Figure 10.4A

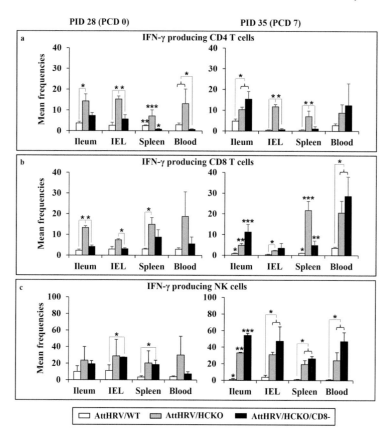

Figure 10.4   IFN-γ producing CD4 and CD8 T cell and NK cell responses. MNCs were stimulated with semi-purified AttHRV antigen *in vitro* for 17 hrs (Brefeldin A was added for the last 5 hrs to block IFN-γ secretion) and then subjected to intracellular staining and flow cytometry for collecting the frequency data of IFN-γ producing CD4 (a) and CD8 (b) T cells and NK cells (c). Data are presented as mean frequency ± standard error of the mean (n = 3 or 5). See Figure 10.3 legend for panel description and statistical analysis.

and 10.4B). AttHRV/HCKO/CD8– pigs had significantly higher frequencies of IFN-γ producing CD4 T cells in the ileum at PCD 7 compared to AttHRV/HCKO pigs (Figure 10.4A).

10.2.9 CD8 depletion in B cell–deficient pigs reduced IFN-γ producing CD8+ γδ T cells but enhanced IFN-γ producing CD8– γδ T cells

Compared to AttHRV/HCKO pigs, CD8 depletion overall had no effects on γδ T cell numbers but slightly increased γδ T cell frequencies (one exception in the spleen at PID 28) (Figures 10.5A and 10.5B) and significantly enhanced IFN-γ producing CD8– γδ T cells in ileum and blood at PCD 7 (Figure 10.5C). As expected, CD8 depletion reduced IFN-γ producing CD8+ γδ T cells in all tissues, with significant reductions in the spleen at both PID 28 and PCD 7 (Figure 10.5D).

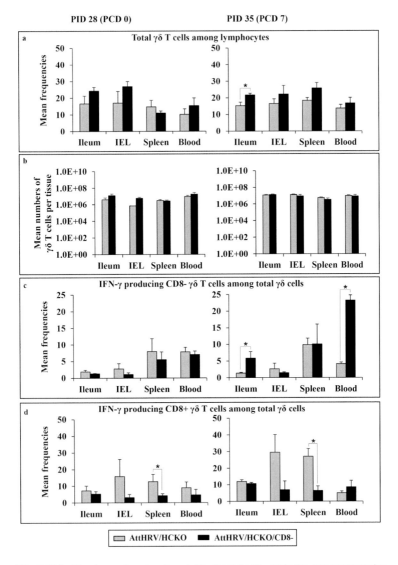

Figure 10.5   Mean frequencies and numbers of γδ T cells and the IFN-γ production. Data are presented as mean frequency/number ± standard error of the mean (n = 3). See Figures 10.3 and 10.4 legends for detection of rotavirus-specific IFN-γ producing cells, panel description, and statistical analysis. Data were not collected in the AttHRV/WT pigs.

10.3 Cellular immune responses to AttHRV vaccine and protection against VirHRV challenge in HCKO pigs

10.3.1 AttHRV vaccination increased CD4 and CD8 T cell and NK cells population and decreased Treg cell population in HCKO pigs

The MNCs were not stimulated with antigen or mitogen *in vitro* to investigate the sole effect of AttHRV vaccination on the cell population *in vivo*. Compared to HCKO pigs without vaccination, AttHRV/HCKO pigs had the overall trend of increased frequencies of CD4 and CD8 T cells and NK cells and their numbers in the ileum, spleen, and blood (Figure 10.6), suggesting that AttHRV vaccination provided sufficient antigen stimulation in neonatal Gn pigs for activation of immune cells. AttHRV/HCKO pigs had lower frequencies and numbers of Treg cells (one exception for the Treg cell mean numbers in spleen) compared to non-vaccinated pigs (Figure 10.6C).

10.3.2 B cells contribute to protective immunity against rotavirus infection and diarrhea

To dissect the protective role of lymphocyte responses elicited by the AttHRV vaccine, we compared clinical signs and rotavirus fecal shedding in immunized wild-type pigs and B cell–deficient HCKO pigs. Forty percent (2 out of 5) AttHRV/WT pigs did not have diarrhea or shed rotavirus after virulent HRV challenge whereas all Mock/WT, Mock/HCKO, and AttHRV/HCKO pigs developed diarrhea and virus shedding (Table 10.1). AttHRV/HCKO pigs had a significantly earlier onset of fecal virus shedding (1.7 versus 5.2 days) and prolonged duration of virus shedding (4.0 versus 0.8 days) compared to AttHRV/WT pigs (Table 10.1). AttHRV/HCKO pigs also had earlier onset of diarrhea (Table 10.1), increased daily shedding titers (Figure 10.7a) compared to AttHRV/WT pigs. But these differences were not statistically significant, with one exception (the significantly higher shedding titers of AttHRV/HCKO pigs at PCD 1 in Figure 10.7a). AttHRV/HCKO pigs also had more infectious virus particles compared to AttHRV/WT pigs from PCD 2 to PCD 4 (Figure 10.7b).

10.3.3 In the absence of B cells, vaccine-induced adaptive immune responses still provided partial protection

When comparing between AttHRV/HCKO and Mock/HCKO pigs, the vaccinated pigs had significantly shorter mean duration (1.3 versus 3.3 days) and lower cumulative scores (7.5 versus 11.2) of diarrhea, as well as significantly lower (1,669-fold) peak titers of virus shed than Mock/HCKO pigs (Table 10.1). In addition, AttHRV/HCKO pigs had lower (from PCD 2 to PCD 7) or significantly lower (from PCD 3 to PCD 5) virus shedding titers than Mock/HCKO pigs (Figure 10.7A). AttHRV/HCKO pigs also had fewer (from PCD 2 to PCD 6) or significantly fewer (PCD 4 and PCD 5) infectious virus particles than Mock/HCKO pigs (Figure 10.7B). The data suggest that in the absence of B cells, vaccine-induced other lymphocyte responses (T cells) are actively involved in providing partial protection. Notably, when comparing between Mock/HCKO and Mock/WT pigs, Mock/HCKO pigs shed significantly more infectious rotavirus particles on PCD 4 than the Mock/WT (Figure 10.7B). However, Mock/WT pigs shed significantly more rotavirus antigen (measured by enzyme-linked immunosorbent assay [ELISA] OD value) from PCD 3 to PCD 5 than Mock/HCKO pigs (Figure 10.7A).

10.3.4 AttHRV vaccine-induced CD8 T cells played a role in shortening the duration of diarrhea and decreasing virus shedding titers

VirHRV challenge of the AttHRV/HCKO/CD8– pigs was performed five days after the injection of anti-pig CD8 antibodies when the CD8 cells decreased to the lowest frequencies among lymphocytes. The CD8 cell depletion extended the mean duration of rotavirus diarrhea (3.7 versus 1.3 days) and increased the fecal cumulative scores (12.0 versus 7.5) in AttHRV/HCKO/CD8– pigs

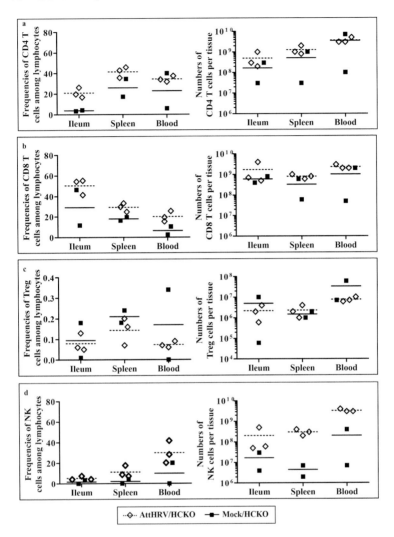

Figure 10.6 Frequencies of CD4 T cells (a), CD8 T cells (b), Treg cells (c), and NK cells (d) among gated lymphocytes and the numbers per tissue in vaccinated HCKO pigs and HCKO control pigs at PID 28. See Figure 10.3 legend for data collection. Because there were only two pigs in the Mock/HCKO group, no statistical comparison between the groups was performed. Bars indicate group means.

compared to AttHRV/HCKO pigs (Table 10.1). Notably, pigs in the AttHRV/HCKO/CD8– group had almost the same mean duration of diarrhea (3.7 versus 3.3 days) and mean cumulative score (12.0 versus 11.2) as those in the Mock/HCKO control group (Table 10.1). Depletion of CD8 lymphocytes also substantially increased the peak virus shedding titers (43 versus 944 fluorescent focus-forming units [FFU]/ml) (Table 10.1), significantly increased viral antigen shedding on PCD 2 and 7 (Figure 10.7A), and significantly increased infectious virus particles shedding from PCD 3 to PCD 7 (Figure 10.7B). The data suggest that the CD8 T cell response induced by the AttHRV vaccine in AttHRV/HCKO pigs was largely responsible for the partial protection against diarrhea and for controlling rotavirus replication after infection.

Table 10.1 Clinical Signs and Rotavirus Fecal Shedding in Gn Pigs after VirHRV Challenge

Treatments	n	Clinical signs				Fecal virus shedding (by CCIF and/or ELISA)			
		% with diarrhea[*],[a]	Mean days to onset[**]	Mean duration days[**],[b]	Mean cumulative score[**],[c]	% shedding virus[*]	Mean days to onset[**]	Mean duration days[**]	Mean peak titer (FFU/ml)[**],[e]
AttHRV/WT	5	60 (3/5)[a]	5.4 (1.3[d])[a]	1.0 (0.4)[c]	8.5 (0.5)[b]	60 (3/5)[a]	5.2 (1.2)[a]	0.8 (0.4)[b]	5 (40)[c]
AttHRV/HCKO	3	100 (3/3)[a]	3.0 (0.0)[a]	1.3 (0.3)[bc]	7.5 (1.3)[b]	100 (3/3)[a]	1.7 (0.3)[b]	4.0 (1.2)[a]	43 (66)[c]
AttHRV/HCKO/CD8-	3	100 (3/3)[a]	3.3 (0.7)[a]	3.7 (0.9)[ab]	12.0 (1.9)[ab]	100 (3/3)[a]	1.7 (0.7)[ab]	5.3 (0.7)[a]	944 (8,666)[bc]
Mock/HCKO	3	100 (3/3)[a]	3.0 (0.0)[a]	3.3 (0.3)[a]	11.2 (0.2)[a]	100 (3/3)[a]	2.3 (0.3)[ab]	5.0 (0.0)[a]	71,765 (29,956)[a]
Mock/WT	4	100 (4/4)[a]	2.8 (0.8)[a]	3.8 (0.9)[ab]	11.9 (0.8)[a]	100 (4/4)[a]	2.0 (0.0)[b]	5.3 (0.5)[a]	7,601 (2,070)[ab]

Note:

[a] Pigs with daily fecal scores of ≥2 were considered diarrheic. Fecal consistency was scored as follows: 0, normal; 1, pasty; 2, semiliquid; and 3, liquid.

[b] For durations of diarrhea and virus shedding, if no diarrhea or virus shedding until the euthanasia day (PCD7), the duration (days) was recorded as 0 and the onset (days) was as 8 for statistical analysis.

[c] Mean cumulative score calculation included all the pigs in each group.

[d] Standard error of the mean.

[e] FFU, fluorescent focus-forming units. Geometric mean peak titers were calculated among pigs that shed the virus.

[*] Fisher's exact test or [**] Kruskal–Wallis rank-sum test was used for comparisons. Different letters indicate significant differences among treatment groups ($p < 0.05$), while shared letters indicate no significant difference.

This table is reproduced from Wen et al. "B-Cell-deficient and CD8 T-Cell-depleted gnotobiotic pigs for the study of human rotavirus vaccine-induced protective immune responses." *Viral Immunol.* 2016. 29:112–127 with permission from Mary Ann Liebert, Inc.

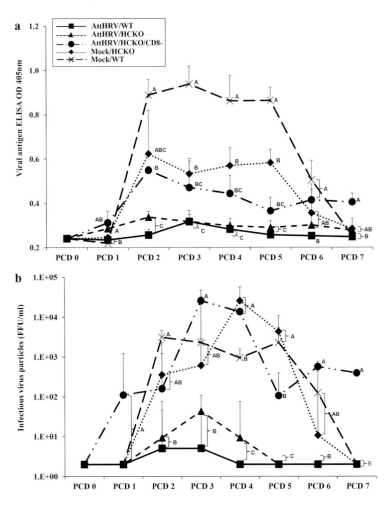

Figure 10.7   ELISA mean optical density (OD) values for rotavirus antigen (a) and infectious virus particles measured by CCIF (b) in fecal samples of wild-type Gn pigs, HCKO Gn pigs, and CD8 depleted (CD8−) HCKO Gn pigs. Pigs were vaccinated with AttHRV or mock vaccinated and challenged with VirHRV. Rectal swabs were collected for seven days after pigs were challenged at PID 28. Fecal samples from mock-infected pigs were used as negative controls. Fecal rotavirus antigen was measured by ELISA, and results are expressed as OD units. The OD values were adjusted based on the average OD values of the negative controls from different ELISA plates. Data are presented as mean OD values/FFU/ml ± standard error of the mean (n = 3–5). Different letters on each time point indicate significant differences among groups (Kruskal–Wallis test, $p < 0.05$), while shared letters indicate no significant difference.

10.3.5 Vaccine-induced other lymphocytes may have contributed to resolving rotavirus shedding

Compared to Mock/HCKO pigs, pigs in the AttHRV/HCKO/CD8− group had significantly lower peak virus shedding titers (944 versus 71,765 FFU/ml) (Table 10.1), shed a lower amount of viral antigen from PCD 2 to PCD 5 (Figure 10.7A), and shed significantly lower infectious virus particles at PCD 5 (Figure 10.7B). Since B cells were absent in pigs of both groups and CD8 cells were mostly depleted in AttHRV/HCKO/CD8– pigs, the results suggest that AttHRV vaccine-induced other lymphocytes (mainly CD4 T cells), may have functioned in controlling virus shedding post-challenge. Interestingly, the AttHRV/HCKO/CD8– pigs shed significantly more infectious virus particles at PCD 7 compared to Mock/HCKO pigs (Figure 10.7B).

10.4 Summary

The total B cell–deficient HCKO pigs offer a unique animal model to elucidate the mechanisms of rotavirus protective immunity. In this study, we first characterized this model by immune phenotyping and then unitized the model to address the question "which lymphocyte population is the most critical in protection against rotavirus infection and disease?" The lack of B cell immune responses increased the incidence of rotavirus diarrhea, demonstrating the critical role of B cell responses in protection against rotavirus diarrhea. The lack of B cells also increased the incidence and the duration of rotavirus shedding, which is consistent with findings from the previous studies of different gene knockout mice, showing that B cell–dependent humoral immunity was the main protective arm against rotavirus infection (Franco and Greenberg, 1999; VanCott et al., 2001). However, B cell deficiency only increased the quantities of rotavirus shedding to a relatively small degree compared to CD8 depletion suggesting that B cells contribute to, but are not essential for, protection against rotavirus shedding. A previous study using T cell knockout mice found the exclusive role of B cell responses in completely resolving rotavirus infection (Franco and Greenberg, 1997).

Depletion of CD8 cells in AttHRV-vaccinated HCKO pigs increased the duration and severity of diarrhea (cumulative fecal score) confirming the critical role of CD8 T cells in protection against rotavirus diarrhea. Moreover, CD8 cell depletion significantly increased virus shedding quantities and resulted in rotavirus shedding persisted beyond PCD 7, suggesting that CD8 T cells are required for completely resolving rotavirus infection. This is consistent with the previous observation in severely immunodeficient mice, in which the adoptive transfer of immune CD8 T cells completely cleared rotavirus infection (Offit and Dudzik, 1990). Our study also found that immune CD4 T cells contributed to controlling the quantities of rotavirus shedding because AttHRV/HCKO/CD8– pigs had 76-fold lower mean peak virus shedding titers compared to non-vaccinated Mock/HCKO pigs. Depletion of CD8 cells led to persistent rotavirus infection and consequently significant increases in IFN-γ producing NK cell response in the ileum and the IFN-γ producing CD8– (pro-inflammatory) γδ T cell response in the ileum and blood post-challenge, reflecting enhanced intestinal innate immune responses to the increased viral antigen load in the AttHRV/HCKO/CD8– pigs (Wen et al., 2012a). Hence both arms of the adaptive immune responses are critical in protection against rotavirus diarrhea. Effective rotavirus vaccines should include antigenic epitopes and adjuvants for stimulating and enhancing all adaptive immune response effector cells (B cells, CD4, and CD8 T cells).

The redundancy of vaccine-induced immune responses is well recognized (Casadevall and Pirofski, 2003). Our current study also found that B cell deficiency drastically increased CD4 and CD8 T cell population, and increased IFN-γ producing CD4 and CD8 T cell responses. The increased IFN-γ producing CD4 and CD8 T cell responses at PID 28 partially compensated for the lack of B cell responses in protection against rotavirus challenge. Studies of B cell–deficient mice (C57BL/6 μMT) infected with Brucella abortus also showed that the μMT mice had significantly increased percentages of CD4+ IFN-γ and CD8+ IFN-γ cells and a reduction in IL-10 producing cells compared with that of infected WT mice (Goenka et al., 2011).

Injection of an anti-CD8 antibody not only temporarily reduced the frequencies of all CD8+ cells (including CD8 T cells, NK cells, and CD8+ γδ T cells), it also potentially affects other lymphocyte populations (cortical thymocytes and subsets of dendritic cells). Since the direct role of the latter cell populations in protective immunity against rotavirus has not been identified, their potential effects as confounders on the observed results are unknown. CD8 depletion increased the CD4 T cell population in most lymphoid tissues of the HCKO pigs but down-regulated IFN-γ producing CD4 T cell responses. CD4 T cells essentially shape the magnitude and quality of CD8 T cell responses (Wiesel and Oxenius, 2012); however, CD8 T cells are also believed to reflectively regulate CD4 T cell responses, although its investigation and understanding have much lagged behind those for the effect of CD4 T cells on CD8 T cells. The cytokines IFN-γ and IL-12 produced by CD8+ T cells mainly stimulate proliferation and differentiation of IFN-γ producing Th1 CD4 T cells (Yamane and Paul, 2013). Therefore, depletion of CD8+ cells (CD8 T cells, NK cells, etc.) will inevitably lower the production of IFN-γ and IL-12 and thus down-regulate IFN-γ producing CD4 T cells. It is interesting to note that B cell–deficient pigs had decreased CD4+CD25– Treg cell population in all tissues, probably due to the strong IFN-γ cytokine milieu in the HCKO pigs that down-regulated Treg cell population. The reduced IFN-γ cytokine milieu in the CD8-depleted and B cell–deficient HCKO pigs at PID 28 led to the increased Treg cell responses and the recovering of the Treg cell population.

In the current study, we had a low number of HCKO pigs in some treatment groups because using somatic cell nuclear transfer (SCNT) and cloning technology to generate genetically modified pigs, such as the HCKO pigs, can be challenging at times. These processes are labor-intensive and inefficient. In addition, pigs generated through SCNT often show developmental defects, which is the reason for the high mortality rate at birth observed in HCKO pigs in this study (data not shown) and other studies with genetically modified pigs using SCNT. However, in recent years the explosive development in porcine gene-editing technologies based on meganucleases (Zinc Finger Nucleases, Tal effector nucleases, CRISPR/Cas9) has provided new tools that may help to get around the need for SCNT for developing genetically modified pigs with much higher efficiency at lower costs (Lee et al., 2014; Sato et al., 2014; Wei et al., 2013). Genetically modified pigs are becoming more readily available for the study of mechanisms of protective immunity against human viral diseases. In addition to using the B cell-deficient HCKO Gn pigs, future studies using other genetically modified pigs, such as IgA-deficient, IgG-deficient, CD8-deficient (CD8α–/–), αβ T cell–deficient (αβ TCR –/–), γδ T cell–deficient (γδ TLR –/–), αβ and γδ T cell–deficient (αβ/γδ TLR –/–), and SCID pig (Rag2 –/– and IL2RG –/–) will greatly further advance the understanding of mechanisms of rotavirus protective immunity and the field of viral immunity research in general. After this study using HCKO Gn pigs, we have also used RAG2-/IL2RG– Gn pigs that were generated with CRISPR/Cas9 technology in a study of HuNoV infection and immunity (Lei et al., 2016b).

In the past, some of the gene knockout mouse models have been used in studies of protective immunity against rotavirus infections (Blutt et al., 2012; Dharakul et al., 1991; Franco and Greenberg, 1999; McNeal et al., 2002; VanCott et al., 2001); however, the determinates of rotavirus protection immunity against rotavirus diarrhea in human infants still have not been fully resolved. Compared to mice and other animals, pigs are the closest relatives to humans (except for non-human primates). During thousands of years of domestication, they have developed many similarities to humans. For example, they have developed digestion and absorption systems more similar to humans than to other members of the *Artiodactyla* clade. They share many diseases with humans and are susceptible to many of the same infectious viral and bacterial pathogens, including rotavirus gastroenteritis. The findings from pig models of human diseases are more relevant for the development of therapeutics, vaccines, and antivirals for humans (Butler et al., 2009a; Kim, 2007; Meurens et al., 2012; Swindle et al., 2012; Verma et al., 2011; Vodicka et al., 2005; Walters and Prather, 2013). Findings from studies using gene knockout pig models will greatly improve our understanding of specific immune effectors in rotavirus vaccine-induced protective immunity.

11
Human Gut Microbiota– Transplanted Gn Pig Models for HRV Infection

Humanized microbiota models, such as germ-free animals transplanted with human feces, are valuable for studying microbiota composition change due to external factors by minimizing confounding variables (Fei and Zhao, 2013; Turnbaugh et al., 2009). Gn pigs provide an excellent model for isolating microbiota as an environmental factor in disease models (Zhang et al., 2013a) because pigs and humans share high genome homology (98%), similar intestinal anatomy, physiology, immune systems, nutritional requirements, and food transit times (Meurens et al., 2012). Transplantation of gut microbiota from human to Gn pigs was first shown to be feasible in Pang et al. (2007). Since then, human microbiota-associated Gn pigs have been increasingly used to elucidate microbe-host interactions in human health and diseases (Wang and Donovan, 2015). We established a human gut microbiota (HGM)–transplanted Gn pig model to study HRV, microbiome and host interactions, the impact of the microbiome on rotavirus infection and immunity, and the beneficial effects of probiotics on improving HRV vaccine efficacy (Twitchell et al., 2016; Wang et al., 2016a; Wen et al., 2014a; Zhang et al., 2014). This chapter includes two studies using this model to investigate the interactions of HGM and HRV (Zhang et al., 2014) and the impacts of healthy versus unhealthy microbiome on rotavirus immunity (Twitchell et al., 2016). The first author of the first study, Dr. Husen Zhang, was an assistant research professor and data scientist at Virginia Tech at that time. The animal model and virology experiment were led by my post-doc research associate Ke Wen and the microbiome data analysis was led by Husen (Harry) Zhang. The next section of the chapter will review studies of the immune-modulatory effects of probiotics in HGM Gn pigs on innate and adaptive immune responses induced by HRV vaccines (Wang et al., 2016a; Wen et al., 2014a). This series of studies were funded by the R01 award from NCCAM, NIH (R01AT004789, 2009–2014, PI: Lijuan Yuan) titled "Mechanisms of Immune Modulation by Probiotics." A similar human microbiota-associated pig model is later used by other researchers in studies of the impact of malnutrition on rotavirus vaccine-induced immunity and vaccine efficacy (Miyazaki et al., 2018; Srivastava et al., 2020; Vlasova et al., 2017).

11.1 HRV infection changes the microbial community structures and probiotics prevent the change in HGM Gn pigs

HRV infection is still one of the leading causes of gastroenteritis in infants and children in developing countries even after the introduction of the universal rotavirus vaccination program in many countries starting from 2006 (Tate et al., 2016). The burden is exacerbated for infants and young children in low-income countries, because many of them have a weak immune response to oral rotavirus vaccines due to repeated infections with various enteric pathogens, leading to lower protective efficacies of the vaccines compared to that in the developed world (O'Ryan, 2017). The gut microbiota's response to rotavirus infection and vaccination has not been systematically investigated. Microbiota disruption (intestinal dysbiosis) may be a risk factor for long-term adverse effects of HRV infection and low vaccine efficacy.

Rotavirus infection in humans was found to be associated with an increase in *Bacteroides fragilis* and decreased *B. vulgatus* and *B. stercoris* in a clone-library study (Zhang et al., 2009). However, a comprehensive analysis is lacking for the effect of therapeutic interventions such

DOI: 10.1201/b22816-11

as vaccination and probiotics on rotavirus-infected gut microbiota. Orally administered probiotic bacterium *Lactobacillus rhamnosus* GG (LGG) has been tested in numerous clinical trials to prevent or shorten rotavirus-induced diarrhea (Isolauri et al., 1991; Szjawska et al., 2011). Supplementation with LGG for four weeks after acute rotavirus infection reduced intestinal permeability in children with rotavirus diarrhea, reduced the number of subsequent diarrheal episodes, and increased IgG antibody response (Sindhu et al., 2014). We hypothesize that the beneficial effects of LGG against rotavirus-related diarrhea may be a result of modulating the intestinal microbiota towards a healthier profile.

We aimed to determine how rotavirus infection affects the transplanted HGM in Gn pigs and whether probiotic LGG can prevent the disruption of the microbiota. We transplanted human infant fecal microbiota to newborn Gn pigs. The pigs were treated with or without a daily dose of LGG for two weeks, vaccinated with an oral attenuated HRV vaccine, and subsequently challenged or not challenged with virulent HRV. We investigated: 1) how efficiently different bacterial species in the HGM colonize neonatal germ-free pigs, 2) the associative changes in the microbiota in response to HRV challenge, and 3) whether LGG, HRV, and interactions between LGG and HRV have effects on the gut microbial community structure.

11.1.1 Establish the HGM Gn pig model of HRV infection and analysis of microbiome

11.1.1.1 Transplantation of HGM into Gn pigs – Multiple stool samples from a cesarean-section-delivered and exclusively breastfed healthy infant at 17–23 days of age were collected and made into an inoculum pool to generate HGM pigs. Briefly, the daily collected fresh stool was diluted 20-fold in sterile pre-reduced PBS (pH 7.2) and glycerol (15% by volume) and stored at −80°C under an atmosphere of nitrogen. Week-long multiple stool samples were pooled and homogenized for use in the entire course of the experiment. Before inoculation, the human stool was screened for pathogens as described elsewhere (Wen et al., 2014a). The screening showed no hemolytic activity. Next-generation sequencing at Viral Diagnostics and Discovery Center at UCSF showed no known viruses in the sample. Pigs were orally inoculated with the human stool inoculum (1 ml of 5% stool suspension in PBS) once daily starting at 12 hours after birth for three days to establish the HGM in Gn pigs. The timing of HGM inoculation is to mimic the natural microbial colonization of the newborn's gut (within hours after birth). The pigs were fed ultra-high-temperature sterilized milk (Hershey) throughout the experiment.

11.1.1.2 Inoculation of Gn pigs with attenuated HRV vaccine, virulent HRV, and probiotic LGG – The cell-culture-adapted attenuated HRV Wa strain was used as the vaccine at a dose of 5 × 10⁷ FFU. The virulent HRV Wa strain was used for challenge of Gn pigs at a dose of ~10⁵ FFU. All the HGM pigs received two doses of the oral attenuated HRV vaccine at five and 15 days of age. The purpose of the vaccination is to study the effects of LGG in enhancing the immunogenicity of rotavirus vaccines in the HGM pigs, which was the objective of another concurrent study and will be discussed in the next chapter (Wen et al., 2014a). The HGM-transplanted and vaccinated pigs were divided into four groups: a) no LGG feeding, no virulent HRV challenge (−LGG-HRV, n = 4), b) no LGG feeding, with HRV challenge (−LGG+HRV, n = 4), c) with LGG feeding, no HRV challenge (+LGG-HRV, n = 4), and d) with both LGG feeding and HRV challenge (+LGG+HRV, n = 3, one pig was euthanized prior to the scheduled time due to health problems). Probiotic LGG (ATCC# 53103) was propagated in *Lactobacilli* MRS broth and the bacterial counts (CFU/ml) were titrated on MRS plates. Prior to feeding, the bacteria stored at 80°C in 15% glycerol were thawed and washed two times with 0.1% peptone water by centrifuging at 2,000 rpm/min for 10 min at 4° C. LGG was diluted and fed to pigs in 3 ml of 0.1% peptone water. Daily LGG feeding started at three days of age for 14 days (3–16 days of age) with a ten-fold incremental LGG dose-increase every day (from 10³ to 10⁹ CFU/dose) as previously described (Liu et al., 2014). Non-LGG-fed pigs were given 3 ml of 0.1% peptone water. Pigs in the +HRV groups were challenged with the virulent HRV at post-attenuated HRV inoculation day (PID) 28. Pigs were euthanized at PID 28 before challenge (−LGG-HRV and +LGG-HRV groups) or at post-challenged day (PCD) 7 (−LGG+HRV and +LGG+HRV groups). The pig body weight at euthanasia did not differ among different treatment groups. Colonic contents and serum samples were collected at euthanasia as previously described (Wen et al., 2014a). After the virulent HRV challenge, rotavirus diarrhea and fecal virus shedding were monitored from PCD 1 to 7.

11.1.1.3 Microbial community analysis – Colonic contents from the pigs' large intestine were collected at euthanasia and stored at −80°C. DNA from human stools and pig intestinal contents was extracted with the QIAamp stool mini kit following the manufacturer's instructions. The 16S rRNA gene amplicons were generated by PCR with 515F and barcoded 806R primers (Caporaso et al., 2012). Purified amplicons were sequenced with Illumina MiSeq™.

Sequencing reads were processed with Quantitative Insights into Microbial Ecology (QIIME) (Caporaso et al., 2010). High-quality reads with Phred quality score ≥ 20 (corresponding to a sequencing error rate ≤ 0.01) were clustered into operational taxonomic units (OTUs) with the program UCLUST (Edgar, 2010). Chimeric sequences were identified with CHIMERASLAYER (Haas et al., 2011) and removed from further analysis. Bacterial taxonomy was assigned by using a naïve Bayes classifier (Wang et al., 2007b) against reference databases and bacterial taxonomy maps at Greengenes (McDonald et al., 2012). A phylogenetic tree was constructed (Price et al., 2010) from PyNAST-aligned sequences representing each OTU. Principle coordinate analysis on stool samples was based on UniFrac distances (Lozupone and Knight, 2005). Distance-based redundancy analysis for the effect of HRV on community structures was performed with the *Vegan* package (*Vegan: Community Ecology Package*, 2013). Shannon and Simpson diversity indices and a rank abundance curve were both generated with QIIME.

The nucleotide sequences have been deposited to MG-RAST (Meyer et al., 2008) with the accession number 4547774.3. Comparison of data for LGG and HRV treatment was done with unpaired t-tests, Mann-Whitney test without assuming normal distributions, one-way or two-way ANOVA.

11.1.2 HRV infection caused a phylum-level shift from Firmicutes to Proteobacteria, and LGG prevented the changes in microbial communities caused by HRV in HGM pigs

A total of 5,616,353 non-chimeric high-quality sequences from the feces of the human donor and the recipient pigs were analyzed with QIIME. We analyzed the sequences at the operational taxonomic unit (OTU) level (Schloss and Handelsman, 2005). The recipient pigs carried microbiota that is similar to the human donor's microbiota (Figure 11.1A), despite that all pigs had received the attenuated HRV vaccine. Two bacterial phyla, Firmicutes and Proteobacteria, representing over 98% of total bacterial sequences in each subject, dominated the microbiota of both the human donor (delivered by C-section) and the recipient pigs. The most abundant genera within Firmicutes were *Streptococcus*, *Enterococcus*, *Veillonella*, and *Staphylococcus* (Figure 11.1A). The rank abundance curve showed that the gut microbiota of the human donor and pigs had a long tail of rare OTUs (data not shown). For example, around 900 OTUs (species rank from 100 to 1,000) each accounted only 0.01% (10^{-4}) to well below 0.001% (10^{-5}) of total bacteria.

We observed a shift in the microbiota composition at phylum, genus, and OTU levels. At the phylum level, the relative abundance of Proteobacteria and Firmicutes in transplanted pigs was affected by HRV challenge (Figure 11.1B). HRV-challenged pigs had 16% fewer Firmicutes and accordingly 16% more Proteobacteria than non-challenged pigs. The significance of the shift was confirmed by the Mann-Whitney test ($p < 0.05$ for both phyla). The phyla Proteobacteria harbor many aerobes and facultative anaerobes and could serve important roles in removing oxygen diffused from the gut epithelium (Pedron et al., 2012). We also found that HRV challenge was a factor in changes in overall community structures at the OTU level (Figure 11.3), based on results from distance-based redundancy analysis (db-RDA). At the genus level, two abundant genera, *Enterococcus* and *Streptococcus*, were affected by LGG feeding and HRV challenge. In the absence of LGG feeding, HRV-challenged pigs had significantly elevated *Enterococcus* over non-challenged pigs ($p < 0.01$, Figure 11.1C). This effect was absent for pigs fed with LGG, indicating that LGG prevented HRV's effect on *Enterococcus*. In HRV-challenged pigs, LGG-treated pigs again had significantly reduced *Enterococcus* ($p < 0.01$, Figure 11.1C). HRV decreased *Streptococcus* in pigs without LGG feeding ($p < 0.05$, Figure 11.1D), but the effect disappeared in LGG-fed pigs, suggesting again that LGG prevented microbiota perturbation by rotavirus infection. The combination of LGG and HRV also affected less abundant genera including *Veillonella* and *Aeromonas* (data not shown). Similar to our findings, an increase in Proteobacteria was reported in norovirus-infected humans (Nelson et al., 2012).

We were interested in whether HRV and/or LGG changed overall community richness and evenness. We found that HRV challenge had no significant effect on Shannon or Simpson diversity

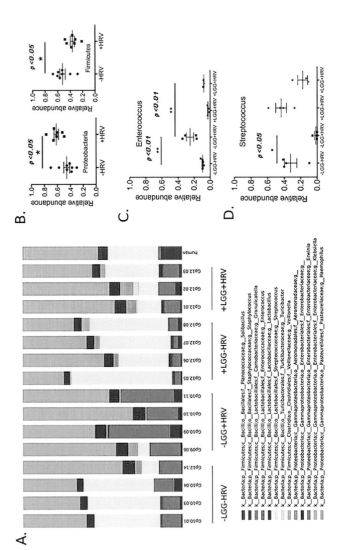

Figure 11.1 Bacterial taxonomic summary for the human donor and the recipient Gn pigs. Panel A: taxonomy breakdown at the genus level (for genera collectively accounting for more than 0.5% of total community) for all subjects. Probiotic treatment and virulent human rotavirus (HRV) challenge are designated as ±LGG or ±HRV, respectively. For example, −LGG−HRV means no LGG treatment and no virulent HRV challenge. Panel B: HRV infection changes the relative abundance of Proteobacteria and Firmicutes regardless of LGG. Error bars represent the standard error of the mean. The *p* values are based on the Mann–Whitney test. The statistical significance was the same when we used the unpaired t-test to log-transformed data. Panel C: combination of LGG and HRV changes relative abundance of *Enterococcus*. Panel D: combination of LGG and HRV changes relative abundance of *Streptococcus*. The *p* values in panels C and D were based on two-way ANOVA.

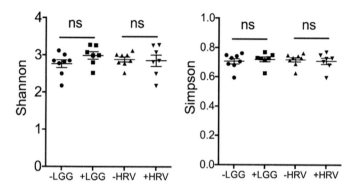

Figure 11.2 Shannon (H) and Simpson (E) diversity indices of pig microbiota with respect to LGG and HRV treatment. Values for the human donor microbiota are H = 3.017 and E = 0.757. ns: not significant based on unpaired *t*-tests.

indices. LGG-treated microbiota appeared to be slightly more diverse than untreated microbiota, but the effect was not significant (Figure 11.2).

To analyze LGG's effect on microbial community structures, we performed principal coordinate analysis (PCoA) on weighted UniFrac distances. The results showed that LGG-treated pig microbiota was distinct from those receiving no LGG (Figure 11.3A), supported by a permutational multivariate analysis (PERMANOVA) with a *p*-value of 0.005 at 999 permutations (Anderson, 2008). The human microbiota appeared to cluster closer with −LGG pigs. Figure 11.3B showed that the extent to which HRV changed microbiota depended on LGG. The HRV-caused microbiota change, measured by UniFrac distances between +HRV and −HRV pigs, was smaller for LGG-treated pigs than for no-LGG treated pigs ($p < 0.001$, Figure 11.3B), suggesting an interaction between LGG and HRV on the microbiota structure. Overall, LGG treatment could resist the change of microbial community structures caused by HRV challenge.

11.1.3 Summary

In this study, we demonstrated that human gut microbiota could be transplanted to and colonize Gn pigs. The resulting "humanized pigs" share the majority of human donor microbiota, albeit the human donor microbiota appeared to be more even than the colonized pig gut microbiota. Virulent rotavirus challenge changed *Enterococcus* and *Streptococcus* abundance in the humanized pig microbiota. Adding probiotic LGG prevented these changes.

The human infant microbiota used in our study was a composite sample from a week-long daily collection of feces. Our intention was to reduce the temporal variations and dynamics known for human microbiota of this age (Koenig et al., 2011; Palmer et al., 2007). We acknowledge that the effect of inter-personal microbiota variations on the colonization of Gn pigs would need to be evaluated through the use of fecal samples from a higher number of infants. Moreover, the inter-pig variations in HGM colonization could be minimized through the use of highly inbred pigs. Overcoming these limitations will allow us to quantify how stable the microbiota is in recipient pigs.

We chose a C-section-delivered HGM donor in this study because of the increasing popularity of this delivery mode. The dominance of Firmicutes and Proteobacteria while lacking Actinobacteria and Bacteroidetes is typical for C-section-delivered babies and different from vaginally delivered babies (Biasucci et al., 2008; Dominguez-Bello et al., 2010). In particular, Bifidobacteria, a prominent group in the phylum Actinobacteria, were absent in microbiota from C-section-delivered infants, but was found abundant in vaginally delivered infants (Biasucci et al., 2008). In another high-throughput sequencing study, the mode of delivery was found to be the main determinant of a newborn's microbiota (Dominguez-Bello et al., 2010). In that study, C-section-delivered babies lacked mothers' vaginal species within Actinobacteria; Bacteroidetes were found to be

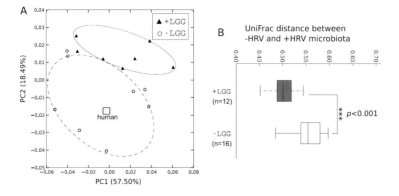

Figure 11.3 Phylogenetic dissimilarities among transplanted pigs and the effect of LGG treatment. Panel A: a principal coordinate (PCoA) analysis of weighted UniFrac distances among all pigs. Only the first two axes (PC1 and PC2) that explain the largest variations among samples are plotted. Open circles are pigs receiving no LGG, and filled triangles are pigs treated with LGG. The open square indicates the human microbiota. Significant grouping by LGG was tested by PERMANOVA described in the text. Panel B: the UniFrac distances between HRV-challenged pig microbiota versus non-challenged pig microbiota. These distances were divided into two groups according to whether LGG was added. The test of significance was performed using the Mann-Whitney test.

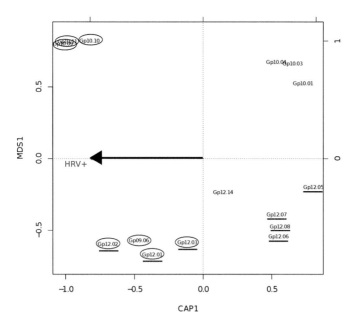

Figure 11.4 Distance-based redundancy analysis (db-RDA) of pig microbiota in response to HRV challenge. CAP1 and MDS1 are constrained and unconstrained axes, respectively. HRV-challenged pigs are indicated by ovals, and LGG supplemented pigs are indicated by underlines. The arrow indicates source of variation explained by HRV.

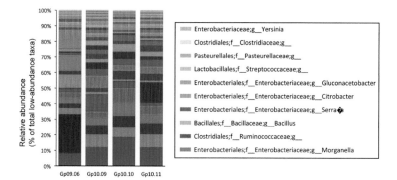

Relative abundance
(% of total low-abundance taxa)

Gp09.06   Gp10.09   Gp10.10   Gp10.11

Enterobacteriaceae;g__Yersinia

Clostridiales;f__Clostridiaceae;g__

Pasteurellales;f__Pasteurellaceae;g__

Lactobacillales;f__Streptococcaceae;g__

Enterobacteriales;f__Enterobacteriaceae;g__Gluconacetobacter

Enterobacteriales;f__Enterobacteriaceae;g__Citrobacter

Enterobacteriales;f__Enterobacteriaceae;g__Serratia

Bacillales;f__Bacillaceae;g__Bacillus

Clostridiales;f__Ruminococcaceae;g__

Enterobacteriales;f__Enterobacteriaceae;g__Morganella

Figure 11.5   Low abundant bacterial taxa (collectively accounted for <0.5% of total community). Gp09.06 and Gp 10.09 were protected from HRV infection by attenuated HRV vaccines. Gp10.10 and Gp10.11 were unprotected.

mainly in the vaginally delivered babies. The same study also reported that *Staphylococcus* species appeared in C-section-delivered babies, which agreed with our results (Figure 11.1A, *Staphylococcus* in dark blue color).

Previous studies have demonstrated that the oral HRV vaccine does not alter the gut microbiota in older children (Garcia-Lopez et al., 2012). This provided rationale for us not to include HMG pigs without an HRV vaccine. All the Gn piglets in our study were vaccinated with attenuated HRV, yet the microbiota in the vaccinated pigs was still significantly altered by virulent HRV (Figure 11.1B and Figure 11.3). We speculate that the extent of microbiota alterations caused by virulent HRV would have been larger if the Gn pigs were not vaccinated. This hypothesis warrants further investigation, as most children in developing countries are not vaccinated against HRV. We found LGG could prevent certain changes in microbiota induced by HRV, suggesting potential interacting effects of LGG and HRV on microbiota, although the exact nature of such interactions is unclear.

Szajewska et al. (2011) demonstrate that LGG improves HRV-induced diarrhea. That study used subjects one month to 18 years old but did not specify the delivery type. It is possible that LGG may not have the same protective effects on C-section-delivered infants which contain a different gut microbiota compared with vaginally delivered infants. Although all the pigs studied received the oral attenuated HRV vaccine, the protection against virulent HRV-induced diarrhea was only partial (Wen et al., 2014a). There were no significant differences in protection rate against diarrhea or virus shedding, the severity of diarrhea, or the titer of virus shedding (data not shown); therefore, we could not evaluate quantitatively whether the changes in the microbiome correlate with protection against diarrhea or virus shedding. Fifty percent (2/4) of the pigs in the −LGG+HRV group were protected from infection upon virulent HRV challenge. There are no apparent differences in the abundant taxa between protected (Gp09.06 and Gp10.09) versus unprotected (Gp10.10 and Gp10.11) pigs (Figure 11.1A, group "−LGG+HRV"). However, the low-abundance bacterial taxa (Figures 11.5 and 11.6), which accounted for less than 0.5% of the microbiota, showed that one of the two protected pigs (Gp09.06) harbored unique bacteria such as *Ruminococcaceae*. The other protected pig (Gp10.09) shared similar low-abundance taxa with unprotected pigs. Further studies with more animals would be needed to identify rare taxa potentially associated with viral protection.

Several relevant issues that were not resolved in the present study will be addressed in future studies. These include a) the effect of vaccination on the gut microbiota. b) Microbiota in human infants varies among individuals and changes during the first few months of life, especially at weaning. Our study used HGM from only one C-section-delivered newborn; an HGM mixture from multiple older children matching the age of rotavirus vaccination (e.g., two to six months) will be a better model for vaccine evaluation. Additional data from both C-section-derived and conventional birth-derived infants would be highly desirable to further evaluate the HGM pig model. c) The LGG feeding in this study did not significantly improve the protection conferred

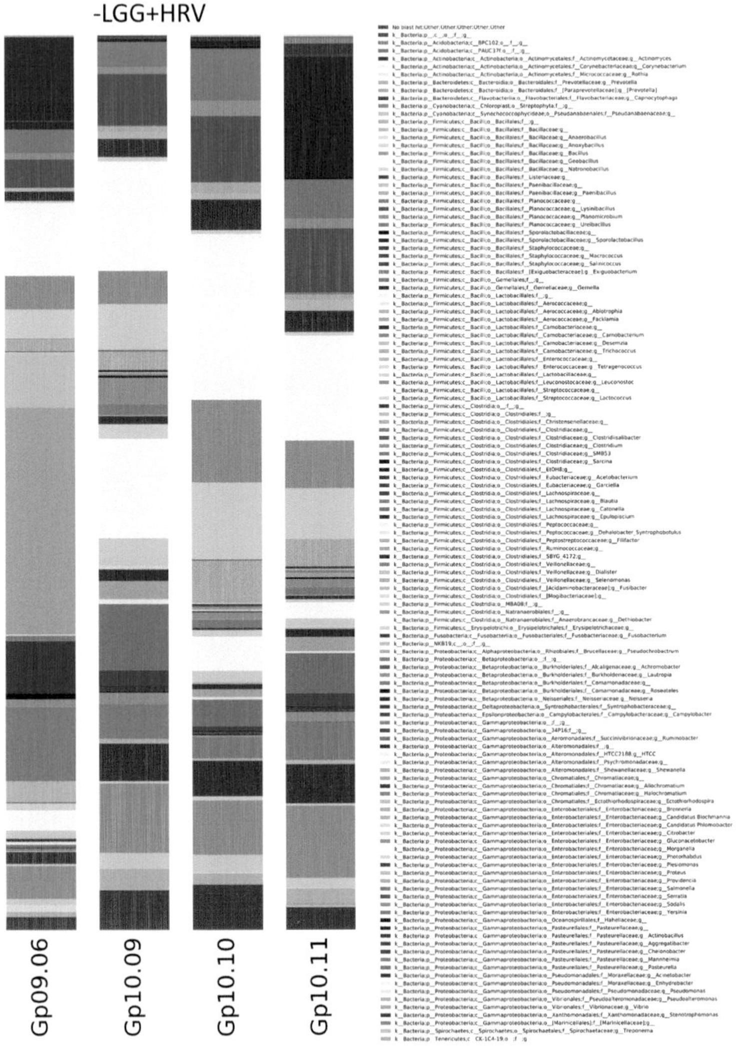

Figure 11.6   Rare taxa (OTUs <0.5% of total bacteria) in the -LGG+HRV group.

by the rotavirus vaccine. Changes in the microbiome due to LGG feeding were not associated with a change in the protective efficacy of the vaccine. Adjustment of the dose and dosing regimen of LGG may be needed for LGG to exert its adjuvant effect on the rotavirus vaccine. The microbiome structure and composition that may favor stronger immunogenicity and protective efficacy of rotavirus vaccines require further studies to identify.

In conclusion, the HGM-transplanted Gn pig presents a useful model for testing interventions using probiotics and vaccines to prevent or treat infantile diarrhea and improve enteric health and immunity. A future study using a mixture of feces from multiple older children that match the age of rotavirus vaccination will improve this model for vaccine evaluation.

11.2 Modeling human enteric dysbiosis and rotavirus immunity in HGM Gn pigs

This study was led by my Ph.D. student Erica Twitchell (Twitchell et al., 2016). The study was funded by an award we received from Bill and Melinda Gates Foundation Grand Challenges Exploration Grant Phase I titled "Gnotobiotic Pig Model for Dysbiosis and Enteric Immunity" (OPP1108188, PI: Lijuan Yuan, May 2014–October 2015) in collaboration with Sylvia Becker-Dreps (University of North Carolina School of Medicine at Chapel Hill, North Carolina) and Samuel Vilchez (National Autonomous University of Nicaragua, León, Nicaragua).

11.2.1 Introduction

Rotavirus gastroenteritis-associated deaths are mostly from low-middle-income countries (LMICs) (2013; Tate et al., 2012). Oral vaccines for rotavirus, poliovirus, cholera, and shigellosis are less efficacious in children from LMICs than in children from higher-income countries (Gilmartin and Petri, 2015; Valdez et al., 2014). Specifically, regarding oral rotavirus vaccination, the two commercially available vaccines (RotaTeq® and Rotarix®) only have 39–70% efficacy, with an average of 50–60% efficacy in LMICs; whereas rotavirus vaccines are 80–90% effective in high-income countries (Armah et al., 2010; Gilmartin and Petri, 2015; Glass et al., 2014; Zaman et al., 2010). Differences in rotavirus vaccine efficacy may be due to a combination of factors including environmental enteric dysfunction (EED), variations in the gut microbiome, an altered gut microbiota composition (dysbiosis), high maternal antibody titers transferred through the placenta or breast milk, malnutrition, or influence of concurrent enteropathogens (Becker-Dreps et al., 2015b; Gilmartin and Petri, 2015; Glass et al., 2014; Valdez et al., 2014; Zaman et al., 2010).

EED, also known as environmental enteropathy is a structural and functional disorder of the small intestine that is most frequently seen in children from low-income settings (Ali et al., 2016). EED leads to gut barrier disruption, nutrient malabsorption, impaired gut immune function, and ultimately, oral vaccine failure, growth stunting, and delayed cognitive development (Ali et al., 2016; Crane et al., 2015; Naylor et al., 2015). However, children with EED do not always have diarrhea (Crane et al., 2015). Histologic features include villous blunting, crypt hyperplasia, and lymphocytic infiltration of the epithelium and lamina propria in the small intestine (Ali et al., 2016). The cause of this condition is unknown, but theories include constant exposure to enteropathogens in food, water, or the environment, imbalance of gut microbiota, and an altered immune response triggered by intestinal microbes (Ali et al., 2016; Crane et al., 2015; Valdez et al., 2014).

The gastrointestinal microbiota is important for the development of enteric immunity, prevention of enteropathogen colonization, and utilization of dietary energy (Grzeskowiak et al., 2012; Valdez et al., 2014). The composition of the gut microbiota in children is influenced by the method of delivery (vaginal or C-section), environmental hygiene, and nutritional status (Valdez et al., 2014). Studies have shown that the composition of gut microbiota is significantly different between African and northern European infants and between malnourished and well-nourished children (Crane et al., 2015; Grzeskowiak et al., 2012; Monira et al., 2011).

The goal of this study was to create a Gn pig model of human enteric dysbiosis to evaluate the influence of the gut microbiota on immune responses to an AttHRV vaccine. Since pigs and humans share high genomic and protein sequence homologies, an omnivorous diet, colonic fermentation, and similar immune systems, pigs serve as valuable models in biomedical research (Wang and Donovan, 2015). To this effect, the neonatal Gn pig model is a proven model of HRV disease and immunity (Ward et al., 1996a). Protection against rotavirus diarrhea in the Gn pig model is positively correlated with frequencies of intestinal IFN-γ producing T cells, rotavirus-specific serum IgA, intestinal IgA and intestinal IgG antibody titers, and IgA ASC in the intestine (To et al., 1998; Yuan et al., 1996; Yuan et al., 2008).

The study of microbiota on virus infectivity is facilitated by the germ-free condition in Gn pigs (Lei et al., 2016c; Wen et al., 2014a; Zhang et al., 2014). In addition, our previous studies of HGM transplanted Gn pigs, demonstrated that HGM from a healthy C-section delivered infant was able to colonize newborn Gn pig intestine, and similar microbiota was observed between HGM from the infant and colonized Gn pigs (Wen et al., 2014a; Zhang et al., 2014). In this study, we attempted to develop a Gn pig model of human enteric dysbiosis by colonizing Gn pigs with unhealthy HGM (UHGM). UHGM designates the HGM samples from children with evidence of gut inflammation and permeability (based on enteropathy scores [ES] and an impaired immune response to the rotavirus vaccine RV5 [RotaTeq®]), suggesting an unhealthy gut. HGM from children with low ES and robust immune response to RV5, suggesting a normal gut, was designated as healthy HGM (HHGM). We hypothesized that HHGM in Gn pigs would induce a stronger immune response after vaccination than the dysbiosis pigs and that certain components of the HGM may be correlated with impaired immune responses.

11.2.2 HGM transplantation, vaccination, and challenge of Gn pigs

11.2.2.1 Stool samples for HGM transplantation – Stool samples were obtained and analyzed from infants as described previously (Becker-Dreps et al., 2017). Briefly, infants from León, Nicaragua were recruited one day before their first pentavalent rotavirus (RV5) immunization, had blood drawn and a stool sample collected from a soiled diaper. A second blood sample was collected one month after the first dose of the vaccine. Stool specimens were diluted 20-fold and homogenized in sterile pre-reduced anaerobic saline-0.1 M potassium phosphate buffer (pH 7.2) containing 15% glycerol (v/v). The samples were snap-frozen in liquid nitrogen in the field, immediately transferred to –80°C, shipped on dry ice to our lab, and stored at –80°C until assayed. All stool samples underwent 16S rRNA amplicon sequencing for characterization of the gut microbiome, as previously described (Becker-Dreps et al., 2015a). ELISAs for four biomarkers of enteropathy (α-1 antitrypsin, neopterin, myeloperoxidase, and calprotectin) were run on the samples (George et al., 2015; Kosek et al., 2013). Results from each marker were divided into quantiles with 1 assigned to the first quantile, 2 for the second quantile, 3 for the third quantile, and 4 for the fourth quantile. The numbers were then added to create a combined enteropathy score (ES) for these four biomarkers ranging from 4–16. Rotavirus-specific serum IgA titers were measured on all blood samples. Results of the infant studies were detailed in a previous publication (Becker-Dreps, submitted for publication). Compared to infants who were seroconverted, infants who did not seroconvert had higher concentrations of both myeloperoxidase and calprotectin in stool samples, and higher mean ES. The UHGM sample (ID PM25) was collected from a nine-week-old female child with poor seroconversion to the RV5 vaccine, high ES, and low alpha diversity (AD). Conversely, the HHGM sample (ID SV14) was collected from an eight-week-old male child with a strong IgA response to the RV5 vaccine, low ES, and high AD. Both children were delivered vaginally and breastfed. They came from different homes; both homes had municipal indoor piped water and indoor toilets.

11.2.2.2 Selection and preparation of infant samples for HGM Gn pig transplantation – The UHGM stool sample was selected from an infant who did not seroconvert following receipt of the first dose of RV5 (seroconversion defined as >four-fold increase in rotavirus-specific serum IgA titers one month after RV5 immunization), low phylogenetic diversity of the gut microbiome, and high ES. The HHGM stool sample was selected from a child with high RV5 immunogenicity (experienced a >four-fold increase in rotavirus-specific serum IgA titers after RV5 immunization), high phylogenetic diversity of the gut microbiome, and low ES.

The selected HGM were screened and confirmed negative for rotavirus, astrovirus, norovirus, sapovirus, adenovirus, and *Klebsiella* spp. via PCR prior to oral transplantation into the Gn pigs. *Klebsiella* spp. were specifically targeted because previous studies have lost HGM transplanted pigs to *Klebsiella* infection from the inocula (Wei et al., 2008; Wen et al., 2014a). Primers for the *Klebsiella* spp. PCR targeted the *gyr*A gene and was based on previous publications (Brisse and Verhoef, 2001; Chander et al., 2011). The universal 16S rRNA primers, 27F and 1492R, were included in the PCR protocol to confirm the presence of amplifiable DNA. DNA was extracted with ZR Fecal DNA MiniPrep (Zymo, Irvine, CA) from 150 µl of sample per manufacturer's instructions. *Klebsiella pneumoniae*, obtained from the Virginia Tech Animal Laboratory Services bacteriology laboratory, was used as a positive control sample.

PCR for norovirus genogroups I and II, adenovirus, sapovirus, and astrovirus were carried out at St. Jude Children's Research Hospital. DNA and RNA were extracted from 175 µl of a sample using the MagMax Total Nucleic Acid Isolation kit (ThermoFisher Scientific). PCR buffer for the RNA samples was Taqman fast virus one-step (ThermoFisher Scientific). Taq core kit (Qiagen, Valencia, CA) was used for DNA samples. Norovirus I and II, sapovirus, adenovirus, and astrovirus real-time PCR primers, as well as probes and conditions, were based on previous publications (Gu et al., 2015; Jothikumar et al., 2005; Oka et al., 2006; Stals et al., 2009).

RT-PCR for rotavirus gene segment 6 was used to screen samples for rotavirus. RNA was extracted from samples using TRIZOL LS (ThermoFisher Scientific) per the manufacturer's instructions. The mixture for cDNA synthesis included 2 μmol of primer 1581 and 11 μl of RNA. RNA was denatured at 95° C for 5 min then cooled to 4° C. The mixture for reverse transcription contained 4 μl of 5X buffer, 1μl 0.1 M DTT, 1 μl Superscript III reverse transcriptase (ThermoFisher Scientific), and 1 μl of RNase free water. The cDNA sample was combined with the mixture for reverse transcription and incubated at 50° C for 60 min then 70° C for 15 min and held at 4° C. One μl of RNaseH (New England Biolabs, Ipswich, MA) was added and the sample was incubated at 37° C for 20 min. The PCR mix contained 10 μl of buffer, 1 μl of MyTaq DNA polymerase (Bioline, Taunton, MA), 27 μl of ddH2O, 1μl of each 20 μM primer (158 [forward] and 15 [reverse]), and 10 μl of DNA. The sequence for primer 158 was GGC TTT AAA ACG AAG TCT TCG AC and that for primer 15 was GGT CAC ATC CTC TCA CTA. The sample was incubated at 95° C for 3 min followed by 35 cycles of 95° C 20 sec, 47.5° C 30 sec, and 72° C 90 sec. The final extension was at 72° C for 7 minutes. AttHRV was used as a positive control.

11.2.2.3 Inoculation of Gn pigs with HGM, vaccination with AttHRV, and challenge with VirHRV – HGM was prepared for oral inoculation into the Gn pigs as described previously (Wen et al., 2014a). The sample was washed with a ten-fold volume of sterile PBS to remove glycerol and then centrifuged at 2,000 rpm for 10 min at 4° C. The pellet was resuspended to the original volume with sterile PBS. Pigs received 400–700 μl of HGM sample at five, six, and seven days of age.

Near-term pigs (Yorkshire crossbred) were derived by hysterectomy and maintained in sterile isolator units as described previously (Meyer et al., 1964; Yuan et al., 2017). The estimated sample size to obtain greater than 0.9 power is n ≥6 based on a power analysis using an ANOVA or ANCOVA model by the Virginia Tech Department of Statistics. Sterility was confirmed by culturing isolator swabs and porcine rectal swabs on blood agar plates and in thioglycolate media three days after derivation. Commercial ultra-high temperature-treated sterile cow milk was used throughout the study. The pigs were housed with a 12-hour light–dark cycle at 80–93°F (based on age) and provided with a toy for environmental enrichment. In total, 24 pigs were randomly assigned to the four groups (HHGM-PID28/PCD0 $n = 5$, HHGM PCD7 $n = 7$, UHGM PID28/PCD0 $n = 6$, and UHGM PCD7 $n = 6$). Intraperitoneal (IP) injections of gamma-irradiated, commercially available porcine serum (Rocky Mountain Biologicals Inc., Missoula, MT) were given to the pigs in an attempt to provide immunoprotection against potentially pathogenic bacteria in the HGM transplants. A total volume of 60 ml was given at three time points, six hours apart, starting on the day of or day after derivation. In the commercial serum, rotavirus-specific virus neutralization titer was measured at 64, rotavirus-specific IgG titer at 65,536, and rotavirus-specific IgA titer at 64. At PCD 0, the geometric mean titers for rotavirus-specific antibodies among all pigs were 16.7 for virus-neutralizing antibodies, 10,935 for IgG, and 75.9 for IgA. These titers mimic normal maternally acquired rotavirus antibody titers in conventionally derived suckling pigs or human infants (Nguyen et al., 2006b). The cell culture-adapted HRV Wa strain derived from the 35th passage in MA104 cells was used as the AttHRV vaccine for oral inoculation at a dose of 5×10^7 FFU on PID 0 (seven days of age), PID10, and PID20 (Yuan et al., 1996). A subset of pigs ($n = 5$–6) from each group was euthanized before challenge PID 28/PCD 0 for evaluation of immune responses induced by AttHRV vaccination. The remaining pigs ($n = 6$–7) were orally challenged with 10^5 FFU of VirHRV Wa strain and monitored from PCD 0–PCD 7 to evaluate virus shedding and diarrhea before euthanasia on PCD 7.

11.2.3 Results

11.2.3.1 Antibody response in HHGM and UHGM pigs – HHGM and UHGM samples were orally inoculated into a subset of newborn Gn pigs, respectively. After AttHRV inoculation and VirHRV challenge, antibody titers were measured in SIC, LIC, and serum. Rotavirus-specific immunoglobulin titers in intestinal contents from HHGM pigs consistently trended higher than titers from the UHGM pigs, before and after VirHRV challenge, including IgG and IgA in SIC and IgA in LIC (Figure 11.7A). However, rotavirus-specific IgA, IgG, and virus-neutralizing antibody responses in serum did not differ between the two groups at any time point (Figure 11.7B). These data suggest that HHGM could have an influence toward a stronger mucosal immune response to oral rotavirus vaccine than UHGM.

11.2.3.2 Virus-specific effector T cell response – Frequencies of IFN-γ+CD8+T cells among total CD8+ T cells in the ileum, spleen, and blood of HHGM pigs were significantly

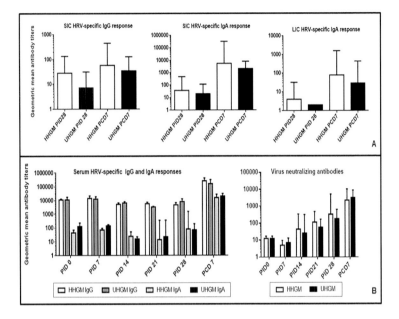

Figure 11.7   Rotavirus-specific antibody responses. Rotavirus-specific IgG and IgA antibody responses in small and large intestinal contents (a) and IgG, IgA, and virus-neutralizing antibody response in serum (b). *Error bars* are represented as the standard error of the mean. The Kruskal–Wallis rank-sum test was used for comparisons. There are no significant differences between the groups.

higher than those in UHGM pigs at the time of VirHRV challenge (Figure 11.8). Frequencies of IFN-γ+CD4+ T cells among total CD4+ T cells in the ileum and blood were also significantly higher in HHGM pigs than in UHGM pigs at the time of VirHRV challenge. After challenge, IFN-γ+CD8+ and IFN-γ+CD4+ T cell responses did not differ significantly between the two groups in any tissue. The data demonstrate that the AttHRV vaccine induced significantly stronger anti-viral effector T cell immune responses in pigs colonized with HHGM than those with UHGM. The significantly higher virus-specific effector T cell responses at the time of challenge were associated with increased protection against rotavirus shedding and clinical signs (Figure 11.8 and Table 11.1).

11.2.3.3 Clinical signs and virus shedding – After challenge with VirHRV, HHGM pigs had significantly reduced incidence and shorter duration of viral shedding, and lower mean peak virus titer than UHGM pigs (Table 11.1). HHGM pigs had a slightly lower incidence, delayed onset, shorter duration of diarrhea, and lower cumulative diarrhea score compared to the UHGM pigs. These results suggest that HHGM is associated with less severe clinical signs and viral shedding than UHGM in vaccinated pigs subsequently challenged with VirHRV, indicating that HHGM facilitates the development of a stronger protective immunity.

11.2.3.4 Microbiome analysis – Alpha diversity, measured by Shannon index, phylogenic diversity, observed species, and Chao 1 were compared between HHGM and UHGM pig groups (Table 11.2).

Measurements of alpha diversity in HHGM pigs were significantly lower than those in UHGM pigs at PID 28 and PCD 7. In addition, alpha diversity measurements decreased in HHGM pigs from PID28 to PCD7. There were no significant differences before or after challenge for the UHGM pigs. These results suggest that VirHRV challenge caused a greater disruption to the microbiota in HHGM pigs than in UHGM pigs. Beta diversity analysis was visualized with a PCoA plot based on unweighted UniFrac. Regardless of the time point, the microbiota from

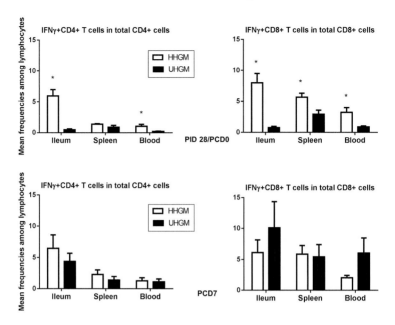

Figure 11.8   Frequencies of IFN-γ producing CD8+ and CD4+ T cells. Frequencies of IFN-γ producing CD8+ and CD4+ T cells among total CD3+CD8+ and CD3+CD4+ cells on PID28/PCD0 (*upper panel*) and PCD7 (*lower panel*) in ileum, spleen and blood of HHGM verses UHGM-colonized pigs. *Error bars* indicate standard errors of the mean. *Asterisks* indicate significant differences when compared to UHGM pigs (Kruskal–Wallis rank-sum test, $p < 0.05$; $n = 5$–7).

pigs with HHGM clustered in one group while samples from UHGM pigs formed another group (Figure 11.9).

In HGM transplanted pigs, phyla represented in UHGM pigs were similar to those in the human infant samples, with Firmicutes being the most abundant. Firmicutes were also the most abundant phylum in the healthy human infant stool sample. Conversely, Proteobacteria or Bacteroidetes was the most abundant phyla in HHGM pigs with Firmicutes being second or third in mean relative abundance. There were significantly more Firmicutes in UHGM pigs than in HHGM pigs at PID28. After VirHRV challenge on PCD7, the phyla Firmicutes, Proteobacteria, and Tenericutes had significantly higher mean relative abundance in the UHGM pigs while the mean relative abundance of Bacteroidetes was significantly higher in HHGM pigs. When evaluating microbiome shifts in HHGM pigs before and after VirHRV, Firmicutes, Proteobacteria, and Verrucomicrobia were shown to be significantly decreased while Bacteroidetes significantly increased in mean relative abundance. There were no significant changes in gut phyla composition in UHGM between pre- and post-VirHRV challenge samples (Twitchell et al., 2016).

Differences at the genera level were also evaluated between the two groups. On PID28, the mean relative abundance was significantly higher for the following OTUs in the HHGM group compared to the UHGM group; *Enterococcus*, *Collinsella*, *Stenotrophomonas*, *Pseudomonas*, and unclassified members of Lactobacillales and Enterococcaceae. The mean relative abundance of the following OTUs was significantly higher in UHGM pigs compared to HHGM pigs on PID28; *Clostridium*, *Streptococcus*, *Ruminococcus*, *Anaerococcus*, *Propionibacterium*, *Blautia*, and unclassified members of Clostridiales, Bacillales, and Lachnospiraceae (Twitchell et al., 2016).

At PCD7, mean relative abundance was significantly higher for *Bacteroides*, *Collinsella*, and unclassified members of Clostridiaceae and Erysipelothichaceae in HHGM pigs compared

Table 11.1 Clinical Signs and Rotavirus Fecal Shedding in Gn Pigs after VirHRV Challenge

Group	N	Clinical signs				Fecal virus shedding (by CCIF and/or ELISA)			
		% with diarrhea[a],*	Mean days to onset[b],**	Mean duration days[e],**	Mean cumulative score[c],**	% shedding virus*	Mean days to onset[b],**	Mean duration days[e],**	Mean peak titer (FFU/ml)
HHGM	7	57.1	4.3 (0.23)[d]	1.3 (0.57)	7.7 (0.77)	42.85*	1.7 (0.2)	1.6 (0.7)	657.4
UHGM	6	83.3	2.3 (1.1)	2.0 (0.89)	11.2 (1.2)	100*	2.4 (0.3)	3.0 (0.63)	1,683.7

Note:

a Pigs with daily fecal scores of ≥2 were considered diarrheic. Fecal consistency was scored as follows: 0, normal; 1, pasty; 2, semiliquid; and 3, liquid.
b In the groups where some but not all pigs had diarrhea or shedding, the onset of diarrhea or shedding for non-diarrheic/shedding pigs was designated as 8 for calculating the mean days to onset.
c Mean cumulative score calculation included all the pigs in each group.
d Standard error of the mean.
e For days of diarrhea and virus shedding, if no diarrhea or virus shedding until the euthanasia day (PCD7), the duration days were recorded as 0.
* Fisher's exact test was used for comparisons. Asterisk indicates significant differences among groups (n = 6–7; $p < 0.05$).
** Kruskal–Wallis rank-sum test was used for comparisons. No statistically significant differences were observed between the groups.

Table 11.2 Mean Alpha Diversity Parameters in the Gut Microbiome of HGM Colonized Gn Pigs

HHGM	Comparison between time points			Comparison between groups			
	PID28	PCD7	p**	PID 28	HHGM	UHGM	p
Shannon index	2.366	1.236	0.009	Shannon index	2.366	2.743	0.006
Phylogenetic diversity	6.806	6.108	0.016	Phylogenetic diversity	6.806	6.580	0.855
Observed species	92.460	65.180	0.009	Observed species	92.460	90.550	0.855
Chao 1	168.399	125.946	0.009	Chao 1	168.399	189.092	0.201

UHGM	Comparison between time points			Comparison between groups			
	PID28	PCD7	p	PCD7	HHGM	UHGM	p
Shannon Index	2.743	2.780	0.749	Shannon index	1.236	2.780	0.006
Phylogenetic diversity	6.580	6.988	0.262	Phylogenetic diversity	6.108	6.988	0.045
Observed species	90.550	102.583	0.173	Observed species	65.180	102.583	0.006
Chao 1	189.092	206.175	0.631	Chao 1	125.946	206.175	0.006

Note:
** Kruskal-Wallis rank-sum test was used for comparisons. $p < 0.05$ is considered significant.

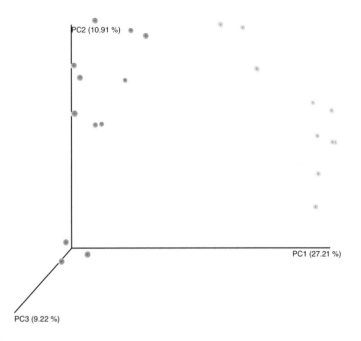

Figure 11.9 PCoA plot of the microbial communities in the large intestinal contents of Gn pigs. Communities were plotted based on unweighted UniFrac. Light gray dots represent HHGM and dark gray dots represent UHGM.

to that in UHGM pigs. UHGM pigs had a significantly higher mean relative abundance of *Ruminococcus*, *Streptococcus*, *Clostridium*, *Bifidobacterium*, *Staphylococcus*, *Turicibacter*, *Propriobacterium*, *Haemophilus*, *Moraxella*, *Blautia*, *Prevotella*, *Granulicatella*, and unclassified members of Enterobacteriaceae, Bacillales, Lachnospiraceae, and Clostridiales than the HHGM pigs (Twitchell et al., 2016).

Spearman's correlation coefficients were determined for frequencies of rotavirus-specific IFN-γ producing T cells in ileum, spleen, and blood, and OTUs at the genus level on PID28 (PCD0) and PCD7 (Table 11.3). There were significant positive correlations between *Collinsella* and CD8+ T cells in blood and ileum, as well as CD4+ T cells in the blood at PCD0. At PCD0, significant negative correlations existed between *Clostridium* and *Anaerococcus*, and ileal CD8+ and CD4+ T cells. At this time point, CD8+ T cells in blood were negatively correlated with *Propionibacterium*, *Blautia*, and an unclassified member of Bacillales, while CD4+ T cells in blood were negatively correlated with two unclassified members of Clostridiales.

At PCD7, there were significant positive correlations between unclassified members of Clostridiales and Mycoplasmataceae and ileal CD8+ T cells. Splenic CD8+ and CD4+ T cells and ileal CD8+ T cells were negatively correlated with *Anaerococcus*. A negative correlation also existed between splenic CD4+ T cells and *Staphylococcus*.

These data suggest the influence of specific bacterial OTUs on the vaccine-induced T cell responses, as well as the selective impact of IFN-γ producing T cell responses on the gut microbiome. Further investigations are needed to determine which OTUs are most important for influencing the immune response and the mechanism by which they do so.

11.2.3.5 Enteropathy biomarkers, histopathology, pig weights – Concentrations of α-1-antitrypsin, myeloperoxidase, and regenerating islet-derived protein 1 beta (REG1B) in SIC

Table 11.3 Spearman's Rank Correlation Coefficients between Specified OTUs and Rotavirus-Specific IFN-γ+CD8+ or IFN-γ+CD4+ T Cells among All Gn Pigs

Positive correlations	OTU	Tissue	T cell type	ρ	p-value	adj. p-value
PCD0	*Collinsella*	Ileum	CD8+	0.91	<0.01	0.001
	Collinsella	Blood	CD8+	0.83	<0.01	0.016
	Collinsella	Blood	CD4+	0.89	<0.01	0.002
PCD7	Clostridiales (unclassified)	Ileum	CD8+	0.89	<0.01	0.002
	Mycoplasmataceae (unclassified)	Ileum	CD8+	0.80	0.01	0.021
Negative correlations	**OTU**	**Tissue**	**T cell type**	**ρ**	**p-value**	**adj. p-value**
PCD0	*Clostridium*	Ileum	CD8+	−0.90	<0.01	0.002
	Anaerococcus	Ileum	CD8+	−0.88	<0.01	0.002
	Propionibacterium	Blood	CD8+	−0.88	<0.01	0.002
	Bacillales (unclassified)	Blood	CD8+	−0.87	<0.01	0.003
	Blautia	Blood	CD8+	−0.81	<0.01	0.009
	Clostridium	Ileum	CD4+	−0.83	<0.01	0.013
	Anaerococcus	Ileum	CD4+	−0.86	<0.01	0.006
	Clostridiales (unclassified)	Blood	CD4+	−0.89	<0.01	<0.01
	Clostridiales (unclassified)	Blood	CD4+	−0.82	<0.01	0.014
PCD7	*Anaerococcus*	Ileum	CD8+	−0.84	<0.01	0.010
	Anaerococcus	Spleen	CD8+	−0.83	<0.01	0.016
	Anaerococcus	Spleen	CD4+	−0.87	<0.01	0.005
	Staphylococcus	Spleen	CD4+	−0.81	<0.01	0.020

Table 11.4   Characterization of HGM Samples Used for Oral Inoculation of Gn Pigs

	SV14 (healthy gut)	PM25 (unhealthy gut)
Enteropathy score	4	11
Myeloperoxidase	0.37 ug/ml	2.08 ug/ml
α-1-antitrypsin	22.9 ug/ml	141.4 ug/ml
Neopterin	74.7 nmol/l	412.4 nmol/l
Calprotectin	148.5 ug/g	220.6 ug/g
Pre-vaccination IgA titer	1:20	1:50
Post-vaccination IgA titer	1:3,200	1:100
Fold increase in IgA titer	160	2
Phylogenetic diversity	7.8	5.7
Shannon index	4	3.3
Observed species	107.4	67.3

and LIC were measured using porcine-specific ELISAs. Alpha-1-antitrypsin is a serum protein that is not present in the stool unless there is increased gut permeability (Ali et al., 2016). Myeloperoxidase is released from activated neutrophils and is an indicator of inflammation (Ali et al., 2016). REG1B is a pro-proliferative antiapoptotic protein secreted by damaged enterocytes and is involved in tissue repair, cell growth, and regeneration (Donowitz et al., 2016; Peterson et al., 2013). There were no significant differences between the two pig groups at PID28 or PCD7. When compared to those of the two human infants (Table 11.4), the α-1-antitrypsin and myeloperoxidase concentrations were even lower than the HHGM, suggesting that no enteropathy developed in the pigs.

Histopathologic evaluation of villus length, crypt depth, villus to crypt ratio, villus width, and mitotic index in the duodenum, jejunum, and ileum did not reveal any significant differences between the two groups at the two time points. Thus, the enteropathy status of the UHGM infant donor, as indicated by the higher ES of the UHGM samples, was not recapitulated in the UHGM-colonized Gn pigs.

Weights of pigs did not differ between the groups at any of the time points. This suggests the microbiota did not differentially affect nutrient assimilation or growth. The lack of significant differences between the groups in regard to enteropathy biomarkers, histopathology, or weight indicates enteropathy did not develop in this Gn model within the timeframe of this experiment.

11.2.4 Summary

In this study, we demonstrated that Gn pigs colonized by UHGM can serve as a model system for enteric dysbiosis and impaired immunity. We used this neonatal pig model system to evaluate the effects of gut microbiota on immune responses to oral rotavirus vaccination. We also evaluated the impact of the rotavirus challenge on the gut microbiota. In this model system, the only variable was the different HGM; thus, any observed difference in the results could be attributed to differences in the gut microbiota. We demonstrated that pigs colonized with HHGM had a significantly stronger effector T cell immune response to vaccination compared to UHGM pigs. Additionally, we found significant correlations between some bacterial OTUs and frequencies of effector T cells, and we also observed differences in the gut microbiome composition and diversity both before and after VirHRV challenge.

When the two groups were compared, UHGM had weaker virus-specific T cell immune responses to AttHRV and trended lower rotavirus-specific antibody titers in intestinal contents. These data indicate that HHGM facilitated a stronger adaptive immune response to oral rotavirus vaccine than did UHGM. Evidently, UHGM pigs had higher viral shedding and more severe clinical signs compared to HHGM pigs after challenge with VirHRV. Our findings are consistent with a study using a microbiota- and diet-dependent mouse EED model, in which EED induces regulatory T cells that inhibit small intestinal CD4+ T cell responses and impair oral vaccine efficacy (Bhattacharjee et al., 2021).

The rotavirus challenge caused significant decreases in alpha diversity indices in the HHGM pigs, while no significant changes were detected in UHGM pigs. The decreased alpha diversity in HHGM pigs after VirHRV challenge was expected. A study evaluating microbiome and diarrhea found that diarrhea was associated with decreased phylogenetic diversity (Becker-Dreps et al., 2015a). Although no samples were available for time points after PCD7, it is expected, given our data on enteropathy biomarkers, that diversity of HHGM pigs would return to normal levels. The lack of significant differences in alpha diversity parameters in UHGM pigs from PID28 to PCD7 may be due to an already abnormal microbiome and/or previous exposure to rotavirus based on pre-vaccination serum IgA titer in the infant donor.

The mean relative abundance of phyla between UHGM pig LIC samples and the infant stool sample (PM25) was similar, which was as expected. However, HHGM pig samples differed from their infant donor (SV14). Community structure may have been different between the human donor and pigs because of the influence of different environments, differences in diet, lack of natural microbial succession in the pigs, or differences in host genetics (Schmidt et al., 2011). The phylum- and genus-level differences between HHGM and UHGM pigs at both time points warrant further study to assess the influences of these OTUs on the host immune response. In HHGM pigs, after VirHRV challenge, there was a decrease in the relative abundance of Firmicutes, similar to previous observations (Zhang et al., 2014). Although we did not sample pigs without AttHRV vaccination, human studies have shown that rotavirus vaccination does not have any major effects on the gut microbiota of children (Ang et al., 2014; Garcia-Lopez et al., 2012). Similar to a human study, HHGM pigs had an increased mean relative abundance of *Bacteroides* after rotavirus infection (Zhang et al., 2009). *Bacteroides* and *Lactobacillus* species have been shown to modify cell-surface glycans in human intestinal cultured cells, effectively blocking rotavirus infection (Varyukhina et al., 2012). This may partially explain why HHGM pigs had decreased viral shedding when compared to UHGM pigs. In agreement with a previous study, we also observed a decrease in levels of *Streptococcus* in HHGM pigs after the rotavirus challenge (Zhang et al., 2014). There is limited data on mechanisms by which microbiota directly influences enteric virus infectivity. Microbiota may modify the cell surface or bind to pathogens. *In vitro* experiments have demonstrated soluble factors from *Bacteroides thetaiotaomicron* and *Lactobacillus casei* can increase cell-surface galactose and block rotavirus infection (Varyukhina et al., 2012). Poliovirus binds bacterial surface polysaccharides, which enhances virion stability and cell attachment, and may enhance transmission (Robinson et al., 2014). Gnotobiotic pigs colonized with *E. coli Nissle* 1917 (EcN) had lower viral shedding titers, which the authors speculated was because EcN bound to HRV particles (Kandasamy et al., 2016).

There were significant positive and negative correlations between OTUs and IFN-γ producing T cell responses in the ileum, blood, and spleen. The biological relevance of these findings needs to be explored further. Previous studies have shown that the HGM promotes the development of the neonatal immune system as evidenced by the significantly enhanced IFN-γ producing T cell response and decreased regulatory T cells in AttHRV-vaccinated pigs when comparing HGM colonized Gn pigs to non-HGM colonized Gn pigs (Wen et al., 2014a). Human studies have shown correlations between microbiome components and response to vaccination. In Bangladeshi infants, Actinobacteria, Coriobacteriaceae, and *Bifidobacterium* abundance were positively correlated with T cell responses to the oral polio vaccine, while Pseuodmonadales, Clostridiales, and Enterobacteriales had a negative correlation (Huda et al., 2014). Concurring with the human study, in this present study *Collinsella*, a member of the Coriobacteriaceae, was strongly and positively correlated with intestinal and circulating rotavirus-specific IFN-γ producing CD8+ T cell responses, which are known to correlate with protection against rotavirus diarrhea (Yuan et al., 2008).

We successfully generated a model of dysbiosis, but not enteropathy. There were no differences between the pig groups in regard to gut permeability, assessed by α-1-antitrypsin, gut inflammation, assessed by myeloperoxidase, or epithelial regeneration assessed by REG1B. Histopathologic parameters did not differ between the groups. Our Gn pig model did not exhibit small intestinal villous blunting, crypt hyperplasia, or lymphocytic infiltration, which are the histologic lesions of EED. The pigs also did not have microscopic lesions of rotavirus infection, the lack of which may be partially due to timing. At PCD7, the small intestine has recovered from rotavirus infection and there are no histopathologic lesions in Gn pigs (Ward et al., 1996a). Although the nutritional status of the pigs was not assessed in this study, the lack of significant differences in body weights between the groups suggests that nutrient assimilation was not altered. We can speculate that other factors such as malnutrition or concurrent enteropathogen infections are needed to induce the EED phenotype. Additionally, six weeks may not have provided sufficient time to develop EED; however, it is logistically difficult to keep Gn pigs for extended periods.

A recent paper describing a Gn mouse model of EED demonstrated that both malnutrition and gut microbiome were key to developing the model (Brown et al., 2015). Our pigs received a nutritionally adequate diet. Thus, we suspect incorporation of malnutrition, and potentially enteropathogens into our dysbiosis model will be required to establish a Gn pig model of EED. A mouse study demonstrated that protein-energy malnutrition alone does not impair vaccine efficacy or increase the severity of infection (Maier et al., 2013). Another recent paper showed that EED, systemic inflammation, and poor maternal health were associated with the under-performance of oral rotavirus vaccine (Rotarix®) and oral poliovirus vaccine but not parenteral vaccines in Bangladeshi children (Naylor et al., 2015). Systemic inflammation and poor maternal health were also predictive of malnutrition (Naylor et al., 2015). In the future, we can use the Gn pig model to address the effects of malnutrition and systemic inflammation on gut immunity.

In future studies, if we succeed in creating histologic changes and altered biomarker concentrations in our attempt to create an enteropathy pig model, additional assays can be utilized to further characterize the model. Tight junction and adherens junction proteins such as occludin, claudins, E-cadherin, and catenins can provide further insight into intestinal permeability. Systemic inflammation can be assessed with cytokines and acute-phase proteins such as IL-1β, IL-4, IL-5, TNFα, C reactive protein (CRP), ferritin, and soluble CD14 (sCD14) (Naylor et al., 2015). Endotoxin core antibody IgG (Endocab) in serum can be used to evaluate bacterial translocation as well as systemic inflammation (Ali et al., 2016). The lactulose to mannitol ratio assay is a standard test used in children to assess intestinal permeability and impaired tight junctions (Ali et al., 2016). We did not utilize this test as the requirement for two-to-five-hour urine collection and the need to use a closed urinary collection system in the pigs is not technically feasible with Gn pig isolators or behavior of pigs (Ali et al., 2016).

There are limitations to this study. Due to the size and weight constraints of the pig isolators, long-term studies are not feasible. This limitation did not allow us to determine if structural or functional changes occur after a longer time. Regarding the HGM, we used one HHGM and one UHGM sample, both from breastfed, vaginally delivered, Nicaraguan children living in homes with piped water and indoor toilets. As stated previously, numerous factors contribute to gut microbiome composition and environmental enteropathy. We do not know if the same results would be obtained with different HHGM and UHGM samples. Ideally, future studies will evaluate multiple HHGM and UHGM from children exposed to different variables such as diet, access to clean water, method of delivery, and country of a habitation with equal numbers of male and female samples. Despite the limitations, the Gn pig model of dysbiosis provides a valuable tool for future studies to investigate the immunomodulating mechanisms of the gut microbiota on the immune system and disease pathogenesis.

Further studies are needed to characterize the roles of specific bacterial species on enteric immunity and rotavirus infectivity. Additionally, by adding malnutrition and enteropathogens to this model, we may be able to recapitulate EED. A pig model of EED and rotavirus vaccination will be valuable since mouse models of subclinical enteric viral infections often do not predict vaccine efficacy against disease in humans. Rotavirus pathogenesis and immunity are similar in pigs and humans but different in mice (Yuan and Saif, 2002).

In conclusion, we established an HGM transplanted Gn pig model of dysbiosis and evaluated the influence of the gut microbiota on vaccine immunogenicity and microbiota response to VirHRV challenge. We demonstrated that impaired enteric immunity in human infants can be recapitulated in the Gn pig model with UHGM transplantation. Our findings indicate that the gut microbiota has a major impact on vaccine immunogenicity. This animal model will be valuable in the evaluation of various strategies (i.e., probiotics, prebiotics, nutritional supplementation) to modulate the intestinal microbiome to enhance the immune responses to rotavirus vaccines as well as other vaccines, and enteropathogens such as norovirus, enterotoxigenic *E. coli*, and *Shigella*. Results from this study provide a stepping stone for further research involving the influence of specific bacterial species on enteric immunity and interaction with rotavirus. Additionally, by adding additional factors to this model such as malnutrition, it may be possible to emulate EED. Enteric dysbiosis, malnutrition, and EED interact in a complex and ill-defined way to negatively impact the health and immunity of young children in the developing world, leading to widespread morbidity and likely increased mortality.

12

Probiotics Modulate Adaptive Immune Responses to Oral HRV Vaccines in HGM Transplanted Gn Pigs

12.1 Introduction

In the previous chapter, we discussed the effect of probiotic LGG on preventing the microbiome structural changes caused by VirHRV challenge in pigs that were all vaccinated with the AttHRV. In this chapter, we will describe the impacts of LGG at two different doses on the development of the host immune system and adaptive immune responses induced by the AttHRV vaccination and VirHRV challenge. This study was led by Ke Wen (Wen et al., 2014a), who worked as a post-doctoral research associate in my lab during that time. The studies were funded by an R01 award from NCCAM, NIH (R01AT004789, 2009–2014, PI: Lijuan Yuan) titled "Mechanisms of Immune Modulation by Probiotics."

Interactions between probiotics and host immune responses can be influenced by gut microbiota, which is lacking in Gn pigs. A pioneering study of HGM pigs showed that the gut microbiota composition in HGM transplanted pigs resembles that of the human donor (Pang et al., 2007). *Bifidobacteria* from human stool could not colonize the gut of HGM transplanted ex-germfree mice (Hirayama, 1999) but succeed in the colonization of HGM transplanted pigs (Pang et al., 2007). Furthermore, the colonization and evolutional development of microbiota in the gut of HGM transplanted pigs was similar to that observed in humans (Pang et al., 2007; Zhang et al., 2014). Of most importance to human health, HGM transplanted pigs had a healthy gut immune system and similar immunity to pathogens as those in humans (Che et al., 2009), which supports the applicability of HGM transplanted pigs in studying immune responses to human pathogens and vaccines. Therefore, to verify the functions of probiotic adjuvants demonstrated in the previous studies, the current study utilized the HGM transplanted neonatal Gn pig model of HRV infection and diarrhea to closely mimic the real context of rotavirus vaccination in infants and to test LGG's dose effects on the immune response profiles induced by the AttHRV vaccine (Wen et al., 2014a).

12.2 Preparation of HGM inoculum and generation of HGM transplanted Gn pigs

12.2.1 Human gut microbiota inoculum

Human gut microbiota was obtained from stool samples of a healthy male infant born in the United States who was delivered by C-section, solely breastfed, and did not exhibit any signs of digestive disorders or receive any medication before stool sample collection. A freshly passed stool was diluted 20-fold and homogenized in sterile and pre-reduced 0.1 M potassium phosphate buffer (PBS, pH 7.2) containing 15% glycerol (v/v) to produce a 5% human fecal suspension, which was then aliquoted into 15 ml sterile tubes and injected with nitrogen for 1 min and stored at −80 C. After completing sample collection daily between 17 to 23 days of age, the fecal samples were then thawed, mixed, and homogenized to make a large stock for the entire experiment. The stock was divided into 15 ml aliquots, injected with nitrogen, and stored at −80 C. To remove glycerol before feeding to piglets, the fecal samples were thawed and washed with at least ten-fold volumes of PBS, centrifuged at 2,000 rpm/min for 10 min at 4° C, and then diluted to the original volume with PBS.

Death of neonatal pigs caused by opportunistic bacteria infection after transplanting HGM from an apparently healthy 11 years old human donor has been reported previously (Wei et al., 2008). Therefore, before the study, the stool samples collected from the solely breastfed, healthy infant at 17–23 days of age were subjected to multi-step safety testing. First, the stool suspension was plated on blood agar plates to observe any sign of hemolytic activity which can be an indication of the presence of pathogenic bacteria. No hemolytic activity was observed. The stool sample was then screened by Viral Chips and the genomes were sequenced on an Illumina MiSeq™ at the Viral Diagnostics and Discovery Center, University of California, San Francisco. The sequencing results confirmed that no known viruses are present in this sample. Lastly, a bioassay was conducted. High doses of the stool inoculum (2.5, 5, and 10 times the normal dosage of the stool inoculum) were tested in newborn Gn pigs and no adverse effect, except for mild diarrhea, was observed.

12.2.2 Treatment groups and inoculation of Gn pigs

Gn pigs were derived and maintained as we previously described (Wen et al., 2012b). The pigs in the HGM groups were orally fed with 1 ml/dose of the 5% fecal suspension starting at 12 hrs after birth daily for three days to establish HGM transplanted Gn pigs. HGM transplanted Gn pigs were randomly assigned to five treatment groups: 1) mock control with no treatment, 2) 14 doses of LGG (LGG14X) alone, 3) AttHRV alone, 4) AttHRV plus nine doses of LGG (AttHRV+LGG9X), and 5) AttHRV plus 14 doses of LGG (AttHRV+LGG14X). Detailed probiotic dosing schedules for the intermediate dose (LGG9X) and high (LGG14X) dose were described previously (Liu et al., 2014). Pigs from both non-HGM and HGM transplanted groups were vaccinated with the AttHRV vaccine following the same protocol as we described elsewhere (Wen et al., 2012b). At PID 28, subsets of HGM pigs from the three AttHRV groups and the non-HGM AttHRV group, all HGM pigs from the mock control and LGG14X groups were orally challenged with 10^5 FFU VirHRV. After HGM inoculation from 0 to 12 days of age and from PCD 0 to 7, pigs were examined daily for clinical signs (diarrhea), and fecal swabs were collected daily as we previously described (Yuan et al., 1996). MNCs were isolated from the ileum, spleen, and peripheral blood of pigs euthanized on PID 28 or PCD 7 as we previously described (Wen et al., 2012b; Yuan et al., 1996).

Feces and LIC were collected and processed as described previously (Azevedo et al., 2005; Liu et al., 2010). Bacterial DNA was extracted using the Wizard Genomic DNA Purification Kit (Promega, Madison, WI) according to the manufacturer's instructions. The extracted DNA was then used for enumerating LGG numbers by real-time PCR. Pure LGG with known numbers quantified as described in the previous study (Zhang et al., 2008d) was used as the standard to calculate relative numbers of LGG in the samples. Real-time quantitative PCR was carried out with RT² SYBR® Green qPCR Mastermix (QIAGEN Inc., Valencia, CA). PCR cycling conditions were 5 min at 94° C, and then 40 cycles of 15 s at 94° C, 15 s at 60° C, and 30 s at 72° C. The reaction mixtures each contained 2 μl of the sample DNA and 18 μl of the master mix, which included the appropriate sense and antisense primers. The sense primer for LGG is 5'-CGCCCTT AACAGCAGTCTTCAAAT-3' and the antisense primer is 5'-ACGCGCCCTCCGTATGCTTAAACC-3' (Ahlroos and Tynkkynen, 2009). Detection of fecal rotavirus shedding was performed as described in previous chapters.

12.3 Results

12.3.1 Similar clinical signs in AttHRV-vaccinated pigs with or without HGM transplantation after VirHRV challenge

Comparisons of clinical signs and fecal virus shedding of AttHRV-vaccinated pigs with or without HGM transplantation after VirHRV challenge are summarized in Figure 12.1A. Compared to the Gn pigs, except for the significantly higher cumulative fecal scores and virus shedding titers in the HGM transplanted Gn pigs, both pig groups had a similar incidence of diarrhea (67%

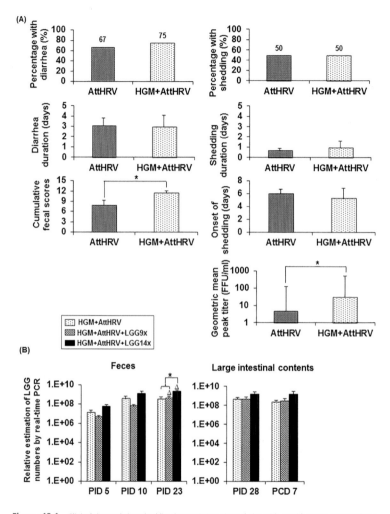

Figure 12.1   Clinical sign and virus shedding in AttHRV-vaccinated pigs with or without HGM transplantation (A) and LGG shedding in fecal samples and large intestinal contents of HGM transplanted Gn pigs fed with or without LGG (B). After VirHRV challenge, pigs were monitored for seven days for incidence of diarrhea, fecal score, and virus shedding. Data are presented as mean ± standard error of the mean (n = 12 for AttHRV group; n = 4 for HGM+AttHRV group). The sign "*" in (A) indicates a significant difference between groups (Kruskal–Wallis test, $p < 0.05$). LGG amounts at different time points were determined by strain-specific real-time PCR and are presented as mean counts/ml ± standard error of the mean (n = 7–10 for fecal samples and n = 3–6 for large intestinal content samples). The sign "*" in (B) indicates significant differences between groups at the same time points and the symbol "Δ" indicates significant increases in LGG numbers compared to PID 5 for the same group (Kruskal–Wallis test, $p < 0.05$).

and 75% for Gn and HGM pig groups, respectively) and fecal virus shedding (50% for both pig groups), similar mean duration of diarrhea and fecal virus shedding, as well as a similar onset of fecal virus shedding. Both pig groups shed low titers of virus (<200 FFU/ml); however, the HGM pigs shed significantly higher virus titers (six-fold) compared to the Gn pigs (Figure 12.1A).

12.3.2 High dose LGG feeding significantly enhanced the fecal and intestinal LGG counts in HGM transplanted Gn pigs

Two LGG dosing regimens were included to investigate the effectiveness of different LGG feeding regimens in modulating the LGG counts in the feces and intestines of the HGM transplanted pigs (Figure 12.1B). AttHRV+LGG14X pigs had significantly higher LGG titers in feces at PID 23 than AttHRV+LGG9X and AttHRV pigs; this high dose LGG feeding also resulted in higher (not significantly) LGG counts in feces at PID 5 and PID 10 and in LIC at PID 28 and PCD 7 compared to the other two groups. AttHRV+LGG9X and AttHRV-only pigs had similar LGG counts from PID 5 to PCD 7. Since AttHRV-only pigs were not fed LGG, the data indicate that the LGG detected in those pigs was a component of the donor HGM with successful colonization of the Gn pig intestine. Notably, the magnitude of increases of LGG counts in the AttHRV+LGG14X group (22-fold) was greater than the AttHRV+LGG9X group (14 fold) from PID 5 to PID 10, reflecting the difference in the LGG dosing regimens.

12.3.3 HGM colonization significantly promoted the development of Th1 type immune responses and down-regulated Treg cell responses

To study interactions among HGM, neonatal immune system, and rotavirus vaccine, and to evaluate the effect of HGM on the development of the neonatal immune system in the context of rotavirus vaccination, we compared the total numbers and frequencies of effector (Figure 12.2) and Treg cell (Figure 12.3) subsets in the intestinal (ileum and intraepithelial lymphocytes [IEL]) and systemic (spleen and blood) lymphoid tissues of the AttHRV-vaccinated pigs with and without HGM. HGM did not significantly alter the total numbers of CD3+CD4+ and CD3+CD8+ T cells in any tissues tested at PID 28, suggesting that AttHRV provided sufficient antigen stimulation for the expansion of the T cell compartment of the neonatal immune system in Gn pigs without HGM. However, AttHRV-vaccinated pigs with HGM had significantly higher frequencies of IFN-γ+CD4+ T cells among CD3+ cells in the blood and IFN-γ+CD8+ T cells among CD3+ cells in all tissues at PID 28 compared to AttHRV-vaccinated pigs without HGM. HGM+AttHRV pigs also had significantly higher numbers of IFN-γ+CD3+CD8+ T cells in the spleen at PID 28 than AttHRV-only pigs. Post-challenge, HGM significantly increased total numbers of CD3+CD8+ cells in the spleen and significantly enhanced IFN-γ producing CD8+ T cell responses measured in both frequencies and numbers in the spleen and blood of AttHRV-vaccinated pigs (Figure 12.2).

Consistent with the enhanced Th1 type immune responses, HGM down-regulated or significantly down-regulated the frequencies of CD4+CD25-FoxP3+ Treg cells in all tissues of AttHRV-vaccinated pigs. HGM+AttHRV pigs had significantly lower numbers of Treg cells in the spleen and blood at PID 28 and in the ileum, IEL, and blood at PCD 7 compared to AttHRV-only pigs. Furthermore, HGM decreased or significantly decreased the frequencies of IL-10 or TGF-β producing Treg cells in all tissues (with one exception in IL-10 producing Treg cells in the blood at PCD7) of AttHRV-vaccinated pigs (Figure 12.3).

12.3.4 High dose LGG significantly enhanced rotavirus-specific IFN-γ producing T cell responses but did not affect Treg cells in AttHRV-vaccinated pigs with HGM

To investigate the dose-effect of LGG as an adjuvant in enhancing the immunogenicity of the rotavirus vaccine, we measured the virus-specific effector T cell responses in AttHRV-vaccinated HGM pigs fed with or without the 9 and 14 doses of LGG at challenge (PID 28) and post-challenge (PCD 7) (Figure 12.4).

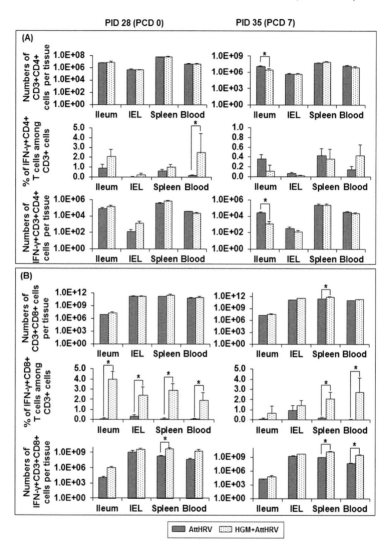

Figure 12.2   T cell responses in AttHRV-vaccinated pigs with or without HGM transplantation. MNCs were stimulated with semi-purified AttHRV antigen *in vitro* for 17 hrs. Brefeldin A was added for the last 5 hrs to block the secretion of cytokines produced by the T cells. HRV-specific IFN-γ producing CD4+ and CD8+ T cells was detected by intracellular staining and flow cytometry as we previously described (Yuan et al., 2008). The frequencies of IFN-γ+CD4+/CD8+ T cells were expressed as percentages among total CD3+ T cells (A and B, middle panel). All mean frequencies are reported after subtraction of the background frequencies. The absolute numbers of CD3+CD4+/CD8+ cells and IFN-γ+CD3+CD4+/CD8+ cells per tissue (A and B, top and bottom panels) were calculated based on the frequencies of CD3+CD4+/CD8+ cells and IFN-γ+CD3+CD4+/CD8+ cells, respectively, and the total number of MNCs isolated from each tissue. Data are presented as the mean number or frequency ± standard error of the mean (n = 4–12). The sign "*" indicates the significant difference between groups (Kruskal–Wallis test, $p < 0.05$).

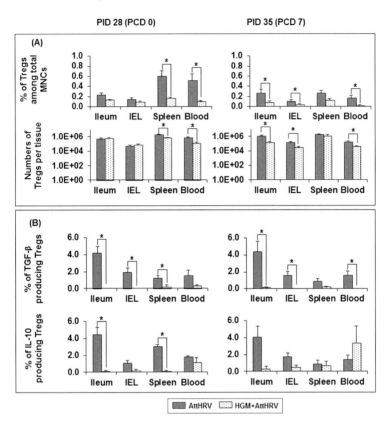

Figure 12.3 Treg responses in AttHRV-vaccinated pigs with or without HGM transplantation. MNCs were stained freshly without *in vitro* stimulation. The frequencies of Tregs were expressed as the percentages among gated MNCs (A, top panel). The absolute numbers of Tregs per tissue were calculated based on the frequencies of Tregs and the total number of MNCs isolated from each tissue (A, bottom panel). The frequencies of IL-10+ or TGF-β+ Tregs were expressed as the percentages of IL-10+ or TGF-β+ cells among the Tregs (B). Data are presented as the mean number or frequency ± standard error of the mean (n = 4–9). See Figure 12.2 legend for statistical analysis.

The AttHRV+LGG14X pigs had significantly higher frequencies of rotavirus-specific IFN-γ producing CD4+ T cells in IEL and spleen and IFN-γ producing CD8+ T cells in the ileum, IEL, and blood compared to both the AttHRV and AttHRV+LGG9X pig groups at PID 28. The AttHRV+LGG14X pigs also had significantly increased frequencies of rotavirus-specific IFN-γ producing CD4+ T cells in ileum and blood compared to the AttHRV+LGG9X pigs at PID 28. Post-challenge, AttHRV+LGG14X pigs had significantly higher frequencies of IFN-γ producing CD4+ T cells in the ileum than the AttHRV and AttHRV+LGG9X pigs. The AttHRV+LGG14X pigs also had significantly increased frequencies of rotavirus-specific IFN-γ producing CD8+ T cells in ileum and IEL compared to the AttHRV-only pigs at PCD 7. The AttHRV+LGG9X pigs had slightly (not significantly) increased IFN-γ producing CD4+ and CD8+ T cell responses in all tissues at PCD 7.

No significant differences were observed in the frequencies of CD4+CD25-FoxP3+ Treg cells and the frequencies of their cytokine production in any tissues among the three AttHRV-vaccinated HGM pig groups with or without LGG feeding at PID 28 or PCD 7 (data not shown).

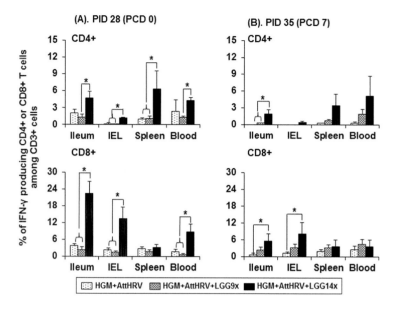

Figure 12.4   Rotavirus-specific IFN-γ producing T cell responses in HGM transplanted Gn pigs fed with different doses of LGG. Data are presented as mean frequency ± standard error of the mean (n = 4–6). See Figure 12.2 legend for detection of rotavirus-specific IFN-γ producing T cell and statistical analysis.

12.3.5 Similar rotavirus-specific antibody responses associated with similar protection rate against rotavirus infection among all three groups of AttHRV-vaccinated HGM pigs with or without LGG feeding

In the HGM pigs, AttHRV vaccination with and without LGG induced significantly higher rotavirus-specific serum IgA responses at PID 22, PID 28, and PCD 7 and serum IgG responses from PID 12 to PCD 7 compared to the unvaccinated pig groups (Figure 12.5). Of interest, pigs from all three AttHRV groups with and without LGG feeding had similar rotavirus-specific serum IgA and IgG titers at each examined time point from PID 0 to PID 28. Consistent with this data, LGG feeding also did not significantly alter rotavirus-specific IgA responses in small intestinal contents at PID 28, although AttHRV+LGG14X pigs had a trend for higher IgA titers than the AttHRV+LGG9X and AttHRV-only pigs at PID 28. Post-challenge, however, the AttHRV+LGG14X and AttHRV+LGG14X pigs had significantly higher rotavirus-specific IgA titers in serum; AttHRV+LGG9X pigs also had significantly higher rotavirus-specific IgA titers in small intestinal contents compared to the AttHRV-only pigs at PCD 7. Serum IgG titers did not differ significantly among the three AttHRV groups at PCD 7.

When analyzing the dynamics of antibody responses, pigs from all three AttHRV-vaccinated groups had the highest rotavirus-specific serum IgA and IgG titers at PCD 7, which were significantly higher than those examined at all previous time points. All AttHRV pigs from the three groups also had significantly higher rotavirus-specific serum IgA and IgG responses at PID 22 and PID 28 compared to PID 0 or PID 12 and had significantly higher rotavirus-specific serum IgG titers at PID 12 compared to those at PID 0 (Figure 12.5). Thus, LGG feedings did not significantly influence the magnitude and dynamics of rotavirus-specific antibody responses induced by the AttHRV vaccine in HGM pigs at PID 28, but LGG enhanced the anamnestic antibody responses upon VirHRV challenge. The similar antibody responses were associated with similar corresponding protection against rotavirus fecal shedding observed in the three AttHRV pig groups upon VirHRV challenge (Table 12.1). There were no statistically significant differences

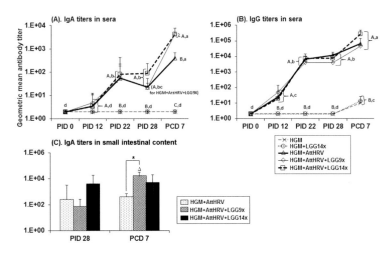

Figure 12.5 Rotavirus-specific serum IgA and IgG antibody responses (A and B) and rotavirus-specific IgA antibody responses in small intestine contents (C) of Gn pigs transplanted with HGM and fed different doses of LGG. Rotavirus-specific antibody titers were measured by an indirect isotype-specific antibody ELISA and presented as geometric mean titers for each treatment group + standard error of the mean (n = 3–10 for serum samples and n = 3–6 for small intestine content samples). Different capital letters (A, B, or C) indicate significant differences in antibody titers compared among different groups at the same time points; different lower-case letters (a, b, c, d) indicate a significant difference between different time points in the same group (Kruskal–Wallis test, p < 0.05), whereas shared letters indicate no significant difference. The sign "*" indicates significant differences in IgA titers in small intestine contents between groups at the same time points and the symbol "Δ" indicates significant increases in IgA titers at PCD 7 compared to PID 28 for the same group (Kruskal–Wallis test, p < 0.05).

in the percent of diarrhea and fecal virus shedding, mean days to onset, and mean duration of diarrhea and virus shedding among the three AttHRV pig groups, likely due to the small number of pigs in each group. However, there was a trend for a shorter mean duration of diarrhea and lower cumulative fecal scores in the LGG fed pigs, which is consistent with the moderate effect of LGG in reducing rotavirus diarrhea observed in Gn pigs (Liu et al., 2013) and human infants (Szajewska et al., 2011).

12.3.6 Safety of the HGM in newborn Gn pigs

Although the multi-step safety testing was done to confirm the safety of the collected stool samples, two to three pigs in each litter became ill (poor appetite, lethargic, fever, vomiting, and/or diarrhea) during the experiments and had to be euthanized prior to the scheduled time point, despite treatment with 50% dextrose (8 ml/dose) by mouth every 8 hrs. The mean mortality rate in the HGM pigs was 20.9%. The necropsy reports from the Anatomic Pathology Laboratory, Virginia Tech Animal Laboratory Services, gave the final diagnoses as mild enteritis and mild hepatitis. The bacterial culture identified *Klebsiella oxytoca*, *Enterococcus faecalis*, and *Staphyloccus epidermdis* in the pigs. *Klebsiella oxytoca* was considered the likely primary pathogen. These early euthanized pigs were excluded from this study. In future studies, *Klebsiella spp* in stool samples for the preparation of HGM inoculum is detected by real-time PCR and only negative stool samples are used (Twitchell et al., 2016).

12.4 Summary

Gnotobiotic pigs offer distinct advantages for investigating enteric virus infections and vaccines and for dissecting the immunomodulatory functions of probiotics. Gn status prevents confounding factors from preexisting gut microflora and maternal antibodies that are present in conventionally reared animals or humans. However, Gn animals also have some disadvantages.

Table 12.1 Clinical Signs and Rotavirus Fecal Shedding in AttHRV-Vaccinated HGM Pigs after VirHRV Challenge

Treatment group	n	Clinical signs				Fecal virus shedding (by CCIF and/or ELISA)			
		% with diarrhea[*]	Mean days to onset[**]	Mean duration (days)[**]	Mean cumulative scores[**]	% virus shedding[*]	Mean days to onset[**]	Mean duration (days)[**]	Geometric mean peak titer (FFU/ml)[**],[a]
AttHRV+LGG9x	4	100[a]	2.3 (1.3)[a]	1.0 (0.0)[a]	9.6 (0.2)[b]	50[a]	5.5 (1.4)[a]	0.5 (0.3)[a]	7.5[a]
AttHRV+LGG14x	5	60[a]	4.0 (1.6)[a]	2.2 (1.1)[a]	10.3 (1.3)[ab]	60[a]	5.0 (1.3)[a]	1.2 (0.6)[a]	4.4[a]
AttHRV	4	75[a]	2.8 (1.8)[a]	3.0 (1.1)[a]	11.5 (0.6)[a]	50[a]	5.3 (1.6)[a]	1.0 (0.6)[a]	29.9[a]

Note:

[a] Geometric mean peak titers were calculated among pigs that shed the virus.

[*] Fisher's exact test or [**] Kruskal–Wallis rank-sum test was used for comparisons. Different letters indicate significant differences among treatment groups ($p < 0.05$), while shared letters indicate no significant difference.

The major drawback is that the germfree status causes underdeveloped intestinal lymphatic constituents and leads to the decreasing number of gut-associated lymphoid tissues (Pleasants et al., 1986; Stepankova et al., 1998; Tlaskalova-Hogenova et al., 1983). Intestinal coloniza-tion of germfree animals with commensal microbes significantly promoted the development of mucosal and systemic immune systems (Atarashi et al., 2011; Gaboriau-Routhiau et al., 2009; Kunisawa and Kiyono, 2011). Our current study demonstrated that HGM transplantation sig-nificantly promoted the activation of Th1 effector cells in both intestinal and systemic lym-phoid tissues as evidenced by the increased numbers and significantly increased frequencies of IFN-γ producing CD8+ T cells in the ileum, IEL, spleen, and blood of the AttHRV-vaccinated pigs at PID28. However, HGM did not significantly alter the total numbers of CD3+CD4+ and CD3+CD8+ T cells in any tissues of the AttHRV-vaccinated pigs at PID28, suggesting that expo-sure to the live AttHRV vaccine alone provided sufficient antigen stimulation for the expansion of the T cell compartment of the neonatal immune system in Gn pigs without HGM.

Our previous studies (Liu et al., 2014; Wen et al., 2012b) using mono-associated Gn pigs dem-onstrated that probiotic L. acidophilus regulated T cell and B cell immune responses in a dose-dependent manner. We also found that the probiotic LGG dose-dependently regulated T cell and B cell immune responses in Gn pigs (Wen et al., 2015). Due to competition between LGG and other bacterial species for colonizing the gut in HGM transplanted pigs, LGG doses higher than that used for Gn pigs were needed to effectively regulate immune responses in HGM pigs. We found that 14 LGG feedings (up to 10^9 CFU/dose), but not nine feedings (up to 10^6 CFU/dose), enhanced virus-specific IFN-γ producing T cell responses in AttHRV-vaccinated pigs transplanted with HGM; AttHRV+LGG14X pigs also had a trend for higher IgA titers than the AttHRV+LGG9X and AttHRV-only pigs at PID28, which suggests that an even higher dose of LGG may be needed to achieve the desired adjuvant effect in the HGM pigs (Liu et al., 2014). In a recent study (Zhu et al., 2014), Zhu et al. fed piglets with the total LGG (ATCC 7469) dose of 7×10^{10} CFU, which was called "low dose" but higher than the highest dose (total 2.2×10^9 CFU) of LGG used in our studies. They found that 7×10^{10} CFU of LGG down-regulated ileal IL-17A, enhanced ileal TGF-β1 and IL-10 mRNA, but had no effect on IFN-γ, IL-12, and IL-4 mRNA expression in the small intestine (Zhu et al., 2014), an immunoregulatory profile suggesting that the LGG dose was too high for an immunostimulatory profile desired by our purpose of using LGG as an adjuvant (Liu et al., 2014; Wen et al., 2012b). Collectively, these studies suggest that all probiotic strains differentially regulate immune responses in each specific narrow dose range. Further studies are needed to identify the appropriate LGG dose and dosing regimen as a vaccine adjuvant to effectively enhance virus-specific antibody responses in the HGM pigs. The interplay between HGM, probiotics, and host immune system is complex, yet it reflects the real-world situation of human infants vaccinated with rotavirus vaccines and treated with probiotics. The results from the HGM pigs provide more relevant references for human clinical practice. However, using the HGM pig model, like human clinical trials, is more difficult than the Gn pig model to reach clear conclusions, and future experiments will be required to include a larger number of subjects in each treatment group to establish the correlations between T cell and B cell responses and protection.

Because Gn pigs derived by hysterectomy (un-suckled) are devoid of maternal antibodies due to the impervious nature of the sow placenta to immunoglobulin, it is expected that Gn pigs are highly prone to opportunistic pathogen infections from transplanted HGM. One early study reported that in two litters of HGM pigs, 17 of 24 died due to the opportunistic pathogen Klebsiella pneumoniae present in the HGM from an apparently healthy human donor of 11 years of age (Wei et al., 2008). With these safety concerns in mind, the HGM inoculum we used in this study was screened carefully; however, 20.9% of the pigs still became ill within several days after the oral administration of the HGM from the infant donor. Bacterial cultures from the euthanized pigs identified Klebsiella oxytoca as the likely cause of the illness. To build safe inoc-ulum pools of HGM from donor stools, a more comprehensive screening procedure is needed. In addition to the screening steps we performed in this study, high-depth and high-throughput sequencing analysis and species-specific PCR for detection of potentially pathogenic bacterial species for pigs such as Klebsiella spp. should be performed.

AttHRV vaccination with or without HGM transplantation conferred similar levels of protection in Gn pigs against diarrhea, virus shedding, and overall clinical signs upon VirHRV challenge, even though HGM pigs had significantly higher Th1 type effector T cell responses and lower Treg cell responses compared to Gn pigs. Furthermore, high dose LGG feeding increased the fecal and intestinal LGG counts, suggesting that sufficient doses of LGG fed to pigs with preex-isting HGM and prior colonization by LGG were able to further enhance the fecal and intestinal counts of LGG. High dose LGG feeding also significantly enhanced the intestinal and circulating virus-specific effector T cell responses but did not significantly enhance intestinal or serum virus-specific antibody responses in HGM pigs at the time of challenge (PID 28) and did not significantly enhance the protective efficacy of the AttHRV vaccine. The magnitude of virus-specific serum and intestinal IgA antibody responses induced by rotavirus vaccines is an important indicator of the protective efficacy of the vaccines (Liu et al., 2014; To et al., 1998). The results from this study highlighted again the importance of antibody responses. It is worth noting that only the AttHRV+LGG9X pigs had significant increases in the intestinal IgA antibody response from PID 28 to PCD 7. The AttHRV+LGG14X pigs had significantly higher intestinal virus-specific IFN-γ producing CD4+ T cell responses than the AttHRV+LGG9X pigs at PID 28 and PCD 7, indicating that AttHRV+LGG14X pigs had a skewed Th1 type response in the intestine that suppressed the virus-specific intestinal IgA antibody responses. Nonetheless, there is a disagreement between the increased immunogenicity (as indicated by the increased level of intestinal IFN-γ producing T cells) of the AttHRV vaccine in the HGM over the Gn pigs and the similar levels of protection upon VirHRV challenge. A plausible explanation is that certain strains of bacteria in the HGM facilitated the replication of VirHRV in the gut leading to increased challenge virus load. We noticed that the mean peak virus shed in feces post-challenge in the HGM pigs was six-fold higher than the Gn pigs. Our study of a porcine epithelial cell line (IPEC-J2 cells) showed that treatment of the cells with *L. acidophilus* prior to rotavirus infection significantly increased the amount of rotavirus antigens and the virus titers (Liu et al., 2010). In two other previous studies, Gn pigs inoculated with the VirHRV and fed with *L. acidophilus* plus *L. reuteri* or inoculated with the AttHRV and fed with *L. acidophilus* shed higher titers of the viruses in feces than the pigs not receiving lactobacilli feeding (Zhang et al., 2008a; Zhang et al., 2008b). Several other studies have also shown that gut microbiota enhanced the replication or virus entry of enteric viruses and their pathogenesis (Kane et al., 2011; Kuss et al., 2011; Uchiyama et al., 2014) and that elimination of microbiota delayed rotavirus infection and significantly reduced rotavirus infectivity in mice (Uchiyama et al., 2014).

In previously reported studies using HGM pigs, HGM from a healthy three-month-old baby (Zhang et al., 2013a) or older human donors at 10, 14, 27–28, or 50–70 years of age (Che et al., 2009; Pang et al., 2007; Shen et al., 2010; Wei et al., 2008; Zhang et al., 2013a) were used to transplant gut microbiota to newborn pigs. Considering the dramatic shifts in composition and diversity of the fecal microbiota that occur with age from infancy to adulthood (Avershina et al., 2013), our HGM transplanted pig model more closely mimics the colonization and evolution of HGM in infants from a very early age and is a more applicable model for studying HRV, which causes gastroenteritis in infants.

In conclusion, both Gn pig and HGM Gn pig models are applicable models for the evaluation of rotavirus vaccines and therapeutics, but each model has its advantages and drawbacks. The Gn pigs have the advantage of being devoid of confounding factors from the gut microbiota, whereas HGM Gn pigs have the advantage of more closely mimicking real-world situations. This current study demonstrated that the HGM Gn pig model is an appropriate model for studying HRV infection and vaccines and for evaluation of the immunomodulatory effects of probiotics. The findings also have implications in using Gn animal models with HGM for studies of other pathogens and diseases.

13

Probiotics Modulate Cell Signaling Pathway and Innate Cytokine Responses to Oral HRV Vaccine in HGM-Transplanted Gn Pigs

The previous chapter described the impacts of LGG on the development of T cell subset responses and antibody responses in the intestinal and systemic lymphoid tissues induced by the AttHRV vaccination and VirHRV challenge. The study described in this chapter on modulation of innate immune responses by LGG was performed in parallel to the study discussed in the previous chapter but focused on the small intestine, the MNCs from the ileum. This study was led by Haifeng Wang (Wang et al., 2016a), who was a visiting post-doctoral research associate in my lab. The study was partially funded by the R01 award from NCCAM, NIH (R01AT004789, 2009–2014, PI: Lijuan Yuan) titled "Mechanisms of Immune Modulation by Probiotics."

13.1 Introduction

Intestinal microbiota consists of approximately 10^{14} bacteria that can be classified into more than 1,000 species (Toki et al., 2009). Intestinal microbiota impacts mucosal immune responses in human infants (Macpherson and Harris, 2004), yet our understanding of how enteric immunity is modulated by gut microbes is limited because of difficulties in performing such studies in humans, especially in infants due to ethical reasons. Germfree pigs transplanted with HGM provide a model system that is ideal for the study of the manifold effects of human microbiota on health and disease (Che et al., 2009). The human gastrointestinal tract (GI) can be colonized at birth by facultative anaerobes including *Enterobacter*, *Lactobacillus*, and *Streptococcus* in genus level, forming a reducing environment during the first week of life enabling colonization by strict anaerobes such as *Bacteroides*, *Clostridium*, and *Bifidobacterium* in genus level (Favier et al., 2002). This microbial colonization contributes to the recruitment of immune cells to the gastrointestinal tract and is a major contributor to the development of the mucosal and systemic immune systems in neonates (Lefrancois and Goodman, 1989). Colonization in early infancy is crucial to the final composition of the permanent microbiota in adults and also in inducing immunological maturation in the intestine and shaping future immune responses of the host (Hansen et al., 2012).

Many previous studies have demonstrated that probiotic LGG strain has beneficial effects on intestinal function, including stimulating development and mucosal immunity, maintaining and improving intestinal barrier function, and prolonging remission in ulcerative colitis and pouchitis (Tao et al., 2006). Studies have also demonstrated the adjuvant effect of LGG in enhancing the immunogenicity of rotavirus, influenza virus, poliovirus, and *Salmonella typhi* Ty21a vaccines (Licciardi and Tang, 2011). Probiotics modulate immunity in the GI tract by interacting with a range of receptors on intestinal epithelial cells (IEC), M-cells, and dendritic cells (West et al., 2009). Probiotics also enhance immunity beyond the GI tract through interactions with the common mucosal immune system.

Microorganisms can be sensed via pattern recognition receptors (PRRs), such as Toll-like receptors (TLRs), to initiate an innate immune response in the GI tract, thus affecting the development

DOI: 10.1201/b22816-13

of the subsequent adaptive immune response. Due to the heavy bacterial antigen load in the lumen, the expression of PRRs is tightly regulated in IEC (Zeuthen et al., 2010). The TLR pathways activate several different signaling elements, including nuclear factor κB (NF-κB) and extracellular signal-regulated kinase (ERK)/c-Jun-NH2-kinase (JNK)/p38, which regulate many immunologically relevant proteins (Akira et al., 2006). NF-κB activation is essential for eliciting protective antigen-specific immune responses after vaccination (Jan et al., 2012; Ligtenberg et al., 2013). Modulation of the signaling pathway will have a significant impact on vaccine immunogenicity and efficacy.

In this study, we used HGM-transplanted Gn pigs to investigate how two different dosing regimens of LGG (9 and 14 doses) impacted the intestinal bacterial communities and modulated the immune signaling pathway and innate cytokine responses to an oral AttHRV vaccine. The animal study design, HGM transplantation, LGG treatment, AttHRV vaccination, and VirHRV challenge are described in the previous chapter. This knowledge should facilitate the selection of proper dosage of probiotics in their applications as vaccine adjuvants and as treatments of intestinal infectious or inflammatory diseases.

13.2 Bacterial communities in feces of HGM-transplanted pigs

PCR-DGGE analysis was performed to study the diversity and richness of the gut microbial community in the HGM pigs. Primer U968-GC and L140lr (Nubel et al., 1996) were used to amplify V6-V8 regions of 16S rDNA. PCR-DGGE was analyzed as we previously described (Wang et al., 2007a). The Shannon index, H' of general diversity was calculated according to a previously described method (Konstantinov et al., 2003) as a parameter for the structural diversity of the microbial community.

The DGGE profile of the HGM-transplanted Gn pigs at PPD 33 is shown in Figure 13.1. There is no significant difference in species richness (DGGE band number, Figure 13.1A) and Shannon index of diversity (Figure 13.1B) among different treatment groups. However, there is a trend for higher richness and diversity in the AttHRV+LGG9X pigs than the other two groups. The similarity index of the individual pigs ranged from 0.79 to 0.89. Based on the similarity, the bacterial communities of AttHRV+LGG14X pigs (Cluster A) are separated from the bacterial communities

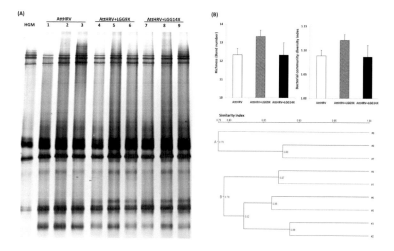

Figure 13.1   DGGE of PCR products of V6–V8 regions of 16S rDNA from bacteria in large intestinal contents of HGM-transplanted Gn pigs (A). Bacterial community richness, diversity, and similarity index of DGGE profiles in the large intestinal contents of HGM-transplanted Gn pigs (B).

of the AttHRV and AttHRV+LGG9X pigs (Cluster B) (Figure 13.1B). Thus, 14 doses of LGG influenced the structure of the transplanted microbiota whereas nine doses of LGG increased its richness and diversity, although not statistically significant.

13.3 Probiotic LGG at nine doses significantly enhanced the innate cytokine and TLR responses at the transcriptional level in the AttHRV-vaccinated HGM pigs

Ileum from all pigs was collected on the day of euthanasia and processed for isolation of MNC as previously described (Yuan et al., 1996). MNCs for quantitative RT-PCR (cytokine and Toll-like receptors expression) and western blot (signal pathway molecular) were subjected to the assays immediately after the isolation of the cells on the same day.

The relative mRNA levels of the selected cytokine and TLRs in ileal MNC were measured by real-time PCR. Briefly, total RNA was extracted from the MNC (2×10^7) using Trizol LS Reagent. cDNA was obtained using a Tetro cDNA Synthesis Kit. The real-time quantitative RT-PCR was done using the SensiMix SYBR & Fluorescein Kit in a final volume 20 ul, which contained 10 ul 2×SYBY mix, 1 ul RT mix, 0.5 ul 10 pmol/ml of each primer for detection of IL-6, IL-8, IL-10, TNF-α, TLR2, TLR4, TLR9, and β-actin which was used as a housekeeping gene. All PCR reactions were done in duplicate on an iQ5 thermocycler (Bio-Rad). The relative levels of different transcripts were calculated using the ΔΔCt method, and results were normalized based on the expression of β-actin within the same experimental setting.

The IL-6 mRNA levels were significantly higher in AttHRV+LGG9X pigs than that in AttHRV pigs (Figure 13.2A). No significant differences were found in IL-8 mRNA levels among different pig groups. The mRNA levels of TNF-α and IL-10 in AttHRV+LGG9X pigs were also significantly higher than those in AttHRV and AttHRV+LGG14X pigs.

The relative mRNA levels of TLR4 were slightly but statistically higher in AttHRV+LGG9X pigs than that in AttHRV pigs (Figure 13.2B). The mRNA levels of TLR9 in AttHRV+LGG9X pigs were significantly higher than those in AttHRV and AttHRV+LGG14X pigs. TLR2 mRNA levels showed the same trend but did not differ significantly among the pig groups. Thus, nine doses, but not 14 doses of LGG significantly enhanced the innate cytokine and TLR responses at the transcriptional level in the AttHRV-vaccinated HGM pigs.

13.4 The signal pathway molecule expression in the ileal MNCs

Signal pathway molecular protein expression in the ileal MNCs was detected by western blot (Wang et al., 2016a). Anti-p38 (no. 8690), p-p38 (no. 4511), ERK (no. 4695), p-ERK (no. 4370), p-NF-κB p65 (no. 3033), IκBα (no. 4814), and β-actin (no. 4970) antibodies were obtained from Cell Signaling Technology (Danvers, MA, USA).

The relative levels of p-p38 were significantly higher in AttHRV+LGG9X pigs than that in AttHRV pigs (Figure 13.3A). There is no significant difference in p38 among different treatment groups. The AttHRV+LGG9X and AttHRV+LGG14X pigs had significantly higher ratios of p-p38/p38 than the AttHRV pigs; however, there was no significant difference between AttHRV+LGG9X and AttHRV+LGG14X pigs

The levels of ERK were significantly higher in AttHRV+LGG9X pigs than in AttHRV pigs (Figure 13.3B). No significant difference was found in p-ERK among different treatment groups. The ratios of p-ERK/ERK were significantly lower in AttHRV+LGG9X and AttHRV+LGG14X pigs than AttHRV pigs; however, there was no significant difference between AttHRV+LGG9X and AttHRV+LGG14X pigs

The levels of p-NF-κB were significantly higher in AttHRV+LGG9X pigs than in AttHRV+LGG14X pigs (Figure 13.3C–E). The level of p-NF-κB in AttHRV pigs did not differ significantly from that

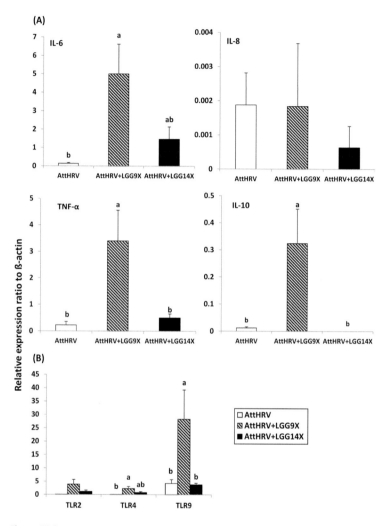

Figure 13.2 Effect of low (AttHRV+LGG9X) or high (AttHRV+LGG14X) doses of LGG on cytokine (A) and TLR (B) levels in ileal MNCs of HGM-transplanted Gn pigs. Different letters on top of bars indicate significant differences compared among groups (Kruskal–Wallis test, $p < 0.05$; n = 3–4), while shared letters or no letters indicate no significant difference.

in AttHRV+LGG9X and AttHRV+LGG14X pigs. AttHRV+LGG9X and AttHRV+LGG14X pigs had significantly higher IκBα levels than AttHRV pigs; IκBα were almost undetectable in AttHRV pigs. No difference was found between the AttHRV+LGG9X and AttHRV+LGG14X pigs.

13.5 Immunohistochemistry for detection of signaling pathway molecules in ileal tissue

Ileum tissue from Gn pigs was fixed in buffered formalin, embedded in paraffin, and cut into serial sections (4 μm) (Fu et al., 2014). Deparaffinized and rehydrated sections were boiled in

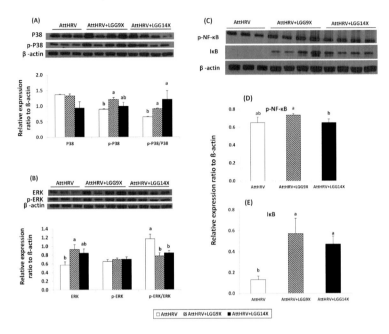

Figure 13.3   Effect of low (AttHRV+LGG9X) or high (AttHRV+LGG14X) doses of LGG on levels of innate immune molecules p38 and p-p38 (A), ERK and p-ERK (B), and p-NF-κB and IκBα (C-E) in ileal MNCs of the HGM-transplanted Gn pigs. See Figure 13.2 legend for statistical analysis.

10 mM sodium citrate buffer (pH = 6.0) for 10 min. After washing twice with Tris-buffered saline with Tween-20 (TBST), sections were blocked with 10% normal goat serum in TBST for 1 h at room temperature. Sections were incubated with primary IFN-γ (1:300 v/v, Cell Sciences, Canton, MA, USA), CD80 (1:200 v/v, Ancell, Bayport, MN, USA), p38, p-p38, ERK, or p-ERK antibodies (1:300 v/v, Cell Signaling Technology, Danvers, MA, USA) overnight at 4° C. After washing three times with TBST, the HRP-conjugated secondary antibody was added and the sections were incubated for 1 h at room temperature. The Diaminobenzidine-HRP detection system was added and sections were incubated at room temperature. Sections were then counterstained with hematoxylin, dehydrated, and cover-slipped. Assessment of positivity of IHC staining (van Diest et al., 1997) was conducted under an ECLIPSE Ti microscope (Nikon Corp., Tokyo, Japan).

Although there is no significant difference in p38 observed among different treatments in immunohistochemistry (Figure 13.4), the AttHRV+LGG9X pigs had significantly higher p-p38 expression levels than the AttHRV and AttHRV+LGG14X pigs. The ERK expression levels of the AttHRV+LGG9X pigs were higher than the AttHRV pigs and significantly higher than the AttHRV+LGG14X pigs. The AttHRV+LGG9X pigs had significantly higher p-ERK expression levels than the AttHRV pigs. There is no significant difference in CD80 or IFN-γ expression levels among different treatment groups in immunohistochemistry (data not shown).

13.6 Summary

Intestinal microbiota plays a critical role in the development of host immune responses (Toki et al., 2009). The composition of gut microbes differentially affects the host intestinal mucosal immunity (Che et al., 2009). High levels of variation are apparent in the microbiota of neonates over time and between individuals (Schwiertz et al., 2003). Studies using Gn animals have the

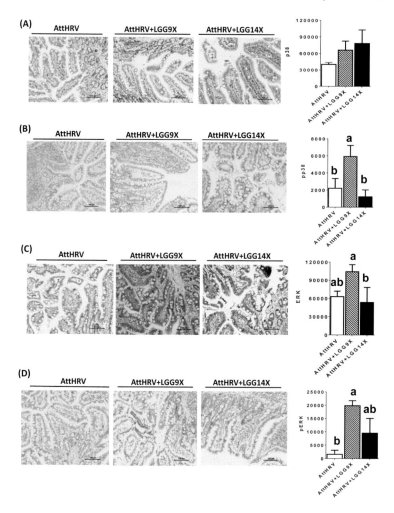

Figure 13.4 LGG treatment modulates specific MAPK family members in ileal tissues of the HGM-transplanted Gn pigs. The levels of p38 (A), p-p38 (B), ERK (C), and p-ERK (D) were evaluated using semi-quantitative histopathology image analysis (ImageJ) following immunohistochemistry staining of paraffin-embedded ileum sections. AttHRV, n = 3; AttHRV + LGG9X, n = 4; and AttHRV + LGG14X, n = 4. See Figure 13.2 legend for statistical analysis.

advantages of highly controlled repeatable experiment design, which reduces inter-individual variation (Laycock et al., 2012). In addition, Gn pigs with a humanized microbiota better mimic the human infants than the germfree pigs without gut microbiota. As described in Chapters 11 and 12 (Wen et al., 2014a; Zhang et al., 2014), HGM from a single healthy newborn infant was successfully transplanted into newborn Gn pigs as shown by microbial 16S rRNA sequencing analysis (Chapter 11). In this study, we confirmed that the bacterial communities in Gn pigs had high similarity to the human donor in DGGE band patterns. Previous research indicated that human flora-associated pigs yielded TGGE (temperature gradient gel electrophoresis) patterns similar to each other as well as to the human donor, but remarkably different from conventionally raised pigs (Pang et al., 2007).

Probiotics are recognized to benefit the host through the improvement of the balance of intestinal microbiota and augmentation of the host defense system (Kim et al., 2006; Plaza-Diaz et al., 2014). Probiotics can modulate the intestinal immune system by either directly affecting immune cell activities or through the positive manipulation of the gut microbiota (Voltan et al., 2007). A clinical study tested the impact of probiotics on the microbiome structure in six-month-old infants fed 1×10^9 CFU/day of LGG and found that communities containing high LGG levels clustered and were associated with a distinct bacterial community composition (Cox et al., 2010). In the current study, 14 doses of LGG feeding influenced the structure of the transplanted microbiota whereas nine doses of LGG slightly increased its richness and diversity. Interestingly, the slightly increased richness and diversity of microbiota in the AttHRV+LGG9X pigs, but not the significantly increased LGG fecal recovery in the AttHRV+LGG14X pigs were associated with the enhanced cytokine and TLR mRNA levels and signaling pathway activation in the ileal MNCs of the HGM Gn pigs. The difference in the two LGG feeding regimens for their timing in relation to AttHRV vaccine inoculation and euthanasia of the pigs may also contribute to the difference besides the LGG dosage. Further studies are needed to evaluate the effect of timing/frequency of probiotic intakes on modulating gut microbiome and immune responses.

Many *in vitro* studies showed that probiotic bacteria stimulate innate immune cells (i.e., dendritic cells, macrophages) to promote expression of various pro- and anti-inflammatory cytokines and TLRs (Plaza-Diaz et al., 2014; West et al., 2009), but evidence for this stimulatory effect *in vivo*, especially in the small intestine, is limited. In this study, we found significantly higher mRNA levels of IL-6, TNF-α, and IL-10 in the ileal MNCs of the pigs fed nine doses, but not 14 doses of LGG compared with the non-LGG-fed pigs. Our results concur with another dose-response study reporting that *Lactobacillus rhamnosus* ATCC 7469 feeding for one week at 1×10^9 CFU/dose, but not 1×10^{14} CFU/dose upregulated mRNA levels of jejunal IL-2, ileal TGF-β1, and ileal IL-10 after F4+ETEC challenge in piglets (Zhu et al., 2014).

Toll-like receptors initiate NF-kB and MAPK cascades, which are the defense-related transcriptional factors. Their activation leads to the production of cytokines (Medzhitov and Janeway, 2000). Excessive immune responses in the intestinal epithelium can be regulated via multiple mechanisms, including modulations of various TLRs expression and localization, or mediation of downstream immune-related cell signaling activation like NF-κB pathway (Wells, 2011). These mechanisms are exerted synergistically to maintain immune responses homeostasis in the GI tract (Wells, 2011).

The role of TLR2 in the induction of innate responses by probiotic lactobacilli including *L. rhamnosus* in immune cells has been extensively demonstrated. TLR2 recognizes gram-positive bacterial lipoteichoic acid, peptidoglycan, and lipoproteins. A previous study showed that LGG enhanced TLR2 mRNA level, and TLR2 was required for NF-κB activation in macrophages (Miettinen et al., 2008). Another recent study confirmed the involvement of TLR2 signaling but not TLR9 in the upregulation of IL-1β, IL-6, IFN-γ, and IL-10 mRNA levels induced by *L. rhamnosus* in porcine intestinal antigen-presenting cells challenged with virus dsRNA analog poly (I:C) (Villena et al., 2014b). In the present study, no significant difference was found in the level of TLR2 mRNA level among treatment groups. TLR9 recognizes bacterial CpG DNA and synthetic unmethylated CpG oligonucleotide mimics (CpG-ODN) (Hemmi et al., 2000). It is known that the genomes of many *Lactobacilli* strains are rich in CpG islands (Rachmilewitz et al., 2004); therefore, *Lactobacilli* may exert a stimulating effect via activation of TLR9 on immune cells. Expression of TLR2, TLR9, and NOD1 mRNA is upregulated in the intestines of pigs pre-treated with a low, but not a high, dose of *L. rhamnosus* (Li et al., 2012). In the present study, TLR9 mRNA level was significantly increased in AttHRV+LGG9X pigs compared to the other two groups. The significantly higher TLR9 mRNA level was associated with the significantly higher IL-6, IL-10, and TNF-α mRNA expression in the AttHRV+LGG9X pigs compared to the other groups, indicating that CpG induced TLR9 signaling is likely one of the pathways that LGG stimulated the secretion of the cytokines. Our results are consistent with the observation that *L. rhamnosus* induced cytokine responses (IL-6, IFN-γ, TNF-α, and IL-10) in a TLR9-dependent manner in human blood MNCs, but the role of TLR2 could not be demonstrated (Plantinga et al., 2011).

TLR4 recognizes lipopolysaccharide from gram-negative bacteria. In this study, the level of TLR4 mRNA in AttHRV+LGG9X pigs was slightly but statistically higher than that in the non-LGG-fed AttHRV pigs. The effect of LGG on TLR4 mRNA levels can only be indirect since LGG does not contain TLR4 ligands. In general, the TLR4 level is down-regulated by anti-inflammatory probiotics (Villena and Kitazawa, 2014). In the present study, the increase in TLR4 mRNA levels is consistent with the increased pro-inflammatory cytokine responses in the AttHRV+LGG9X pigs. The possible reasons for the upregulated TLR4 can be the increased richness and diversity of microbiota in the AttHRV+LGG9X pigs and that certain bacterial species in the microbiota promoted TLR4 expression.

In response to inflammatory signals, the MAPK cascade is activated through phosphorylation of p38, ERK, and JNK, which is associated with the activation and translocation of NF-κB from the cytoplasm to the nucleus (Chon et al., 2010). NF-κB is known to play a central role in inflammatory responses and is involved in the transcriptional regulation of many cytokine genes, including TNF-α (Baeuerle and Baltimore, 1996). Previous research suggested that both NF-κB and p38 MAPK signaling pathways were important for the production of cytokines and chemokines induced by L. acidophilus NCFM (Jiang et al., 2012). Inhibition of MAPKs family pathways, such as ERK, p38, and JNK, alleviates the production of pro-inflammatory cytokines (Sironi et al., 2006). Previous research also indicated that p38 MAPK and ERK-1/2 cross-regulate each other such that inhibition of one enhances activation of the other and the effector functions induced in response to different stimuli (Mathur et al., 2004). We, therefore, examined whether LGG could induce cytokine responses by activation of p38 MAPK and ERK1/2. In western blot analysis, the AttHRV+LGG9X and AttHRV+LGG14X pigs had a significantly higher ratio of p-p38/p38, whereas lower p-ERK/ERK than the AttHRV pigs. This result verified their reciprocal association. The IHC analysis generally verified the result that AttHRV+LGG9X had the highest p-p38 and ERK1/2 in the western blot. There was one exception for p-ERK that the AttHRV+LGG9X pigs had significantly higher p-ERK than the AttHRV pigs in IHC but not in western blot analysis.

ERK pathway plays key regulatory functions in a diverse spectrum of biological processes such as cell proliferation, differentiation, survival, and motility (Kohno and Pouyssegur, 2006) and has an important immunoregulatory role in maintaining homeostasis in the intestine (Blanchette et al., 2001; Villena et al., 2014a). In this study, nine doses of LGG increased levels of ERK and p-ERK as observed in western blotting and IHC, respectively, indicating that the activation of the ERK pathway by LGG could have a protective role against viral infection–induced mucosal injury.

AttHRV+LGG9X pigs had significantly higher levels of p-NF-κB than the AttHRV+LGG14X pigs, but neither pig group significantly differed from the AttHRV pigs. The result indicates that nine doses of LGG further activated the NF-κB, but 14 doses prevented the further NF-κB activation. NF-κB is located in the cytoplasm as an inactive complex bound to IκBα, which is phosphorylated and subsequently degraded, and the degradation of IκBα results in the dissociation of activated NF-κB from IκBα (Baldwin, 1996). Pre-treatment of HT29 and T84 polarized cell monolayers using purified DNA from LGG delayed NF-κB activation, stabilized levels of IκBα, and attenuated IL-8 secretion in response to stimulation by Salmonella DNA or TNF-a (Ghadimi et al., 2010). In the present study, both LGG dosing regimens significantly increased the IκBα level, which indicates that LGG may have inhibited inflammation by increasing the IκBα level to balance the activation of NF-κB (Chon et al., 2010). Therefore, the reduced transcription of TNF-α, IL-6, and IL-10 in the AttHRV+LGG14X pigs may not be due to the lack of impact of LGG on the signaling pathways, rather it reflected the active effect of the higher dose LGG in attenuation of NF-κB activation.

Adequate pro-inflammatory cytokine responses contribute to the clearance of pathogens, but excessive inflammatory immune responses lead to tissue injuries. Therefore, an appropriate balance between pro-inflammatory and anti-inflammatory mediators is crucial for an effective and safe response against infection (Kitazawa and Villena, 2014). IL-10 is a potent immunoregulatory cytokine that might be beneficial in the course of infection by attenuating the excessive host inflammatory response induced by upregulated pro-inflammatory cytokines and thus controlling immunopathology. Several studies have demonstrated that induction of the regulatory IL-10 by probiotic lactobacilli such as L. rhamnosus plays an important role in controlling the inflammatory process upon a viral infection to minimize tissue injury (Zelaya et al., 2014). In the present study, we showed that nine doses of LGG induced higher mRNA levels of both pro-inflammatory cytokines (TNF-α, IL-6) and anti-inflammatory cytokine IL-10 when compared with AttHRV and AttHRV+LGG14X. Therefore, the improved production of IL-10 induced by an appropriate dose of LGG would allow efficient regulation of the inflammatory response and avoid tissue damage during intestinal viral infections. Modulation of the NF-κB and p38 MAPK signaling pathways may also have a significant impact on AttHRV vaccine immunogenicity and efficacy via upregulating cytokine productions. Indeed, the increased IL-6 and TNF-α mRNA levels were associated with significantly enhanced intestinal IgA responses in the AttHRV+LGG9X pigs post-challenge as we reported previously (Wen et al., 2014a). Although the vaccine-induced protection overall did not differ significantly among AttHRV only, AttHRV+LGG9X and AttHRV+LGG14X groups, the AttHRV+LGG9X pigs had the shortest mean duration of diarrhea and virus shedding and significantly lower cumulative fecal diarrhea scores (Wen et al., 2014a).

There is an apparent discrepancy between the significantly increased HRV-specific IFN-γ producing T cell responses in the 14 doses, but not nine doses, LGG-fed HGM pigs (Wen et al., 2014a) and the significantly stronger cytokine, TLR4, TLR9, and p38 MAPK signaling pathway responses in the nine doses, but not 14 doses, LGG-fed HGM pigs. It is important to note the difference in the cell populations studied for the T cell responses (CD3+CD4+ and CD3+CD8+ T cells) versus the signaling pathway responses (total MNC and whole ileum tissues). It might have made interpreting the results and discrepancy easier if our studies were performed using sorted T cells and dendritic cells. An *in vitro* study using human dendritic cells showed that LGG significantly down-regulated p38 expression and negatively regulated NF-κB through a downregulatory effect on miR-146a expression (Giahi et al., 2012). Although the role of the NF-κB pathway in T cell development and function has been well studied (Gerondakis et al., 2014), the relationship between p38 MAPK signaling pathway and T cell development is not so clear. Further studies are needed to explain the observed discrepancy.

In conclusion, the two different LGG doses exerted divergent effects on gut microbiota structure and intestinal immune responses. These results are important since they revealed that the relationship between modulation of gut microbiota and regulation of host immunity using probiotics is complex. More *in vivo* studies are needed to better understand the mechanisms of action of this probiotic strain in its applications as treatment of intestinal infectious or inflammatory disease and as a vaccine adjuvant. An improved understanding of the molecular mechanisms of immunomodulation will facilitate the development of next-generation probiotics and will enhance our understanding of host–microbial interactions.

14
Probiotics as HRV Vaccine Adjuvants in Gn Pigs

14.1 Introduction

Probiotics are defined as viable microorganisms, that when ingested, have a beneficial effect on the health of the host (Senok et al., 2005). Earlier studies on probiotics mainly focused on their benefits to gastrointestinal tract health, especially on reducing diarrheal diseases. Starting a long time ago, investigators described the use, mostly for humans, of many probiotics for prophylactic or therapeutic purposes to prevent or treat diarrhea of different causes (Ouwehand et al., 2002). The use of probiotics in farm animals dates back 75 years; work in the 1960s in pigs has shown that *Lactobacillus* could stimulate significant growth (Reid and Friendship, 2002). Lactic acid bacteria (LAB) (*Lactobacillus acidophilus*, *L. reuteri*, *L. casei*), *Bifidobacterium pseudolongum*, *B. lactis*, and *Bacillus licheniformis* reduce the incidence, severity, or duration of presumable rotavirus (not identified as causes) and/or bacterial diarrhea in newborn calves, pigs (Abe et al., 1995), and weaning pigs (Bomba et al., 1999; Kyriakis et al., 1999; Shu et al., 2001). *L. rhamnosus GG* (*LGG*), *L. acidophilus*, and *L. reuteri* are the three species that have been evaluated for the treatment of diarrheal disease in humans (Majamaa et al., 1995; Shornikova et al., 1997). However, none of these early observational studies clearly identified or quantified the actual diarrhea-inducing agents (often a combination of enteric bacteria and rotaviruses), and the mechanisms of probiotic action were not defined.

In 2005, I received two research grants as PI titled "Immunological Mechanism of Probiotic Lactic Acid Bacteria in Prevention and Treatment of Rotavirus Diarrhea in Neonatal Swine" (OHOA1208 550019, OARDC) and "Effects of Probiotic Lactobacilli on Rotavirus Immunity" (R21 AT002524, NCCAM, NIH) aiming to identify the immune mechanism of probiotic LAB in reducing rotavirus diarrhea. The probiotic research team at that time included Marli Azevedo who just received her Ph.D. degree and stayed as a post-doctoral research associate in Saif Lab, Wei Zhang, and Ke Wen, both of them were master students in the Department of Veterinary Preventive Medicine, The Ohio State Univerisity (Linda Saif and I were co-advisors). In the studies, we used a combination of *L. acidophilus* and *L. reuteri* feeding regimens in Gn pigs. The combination of the two LAB strains did not protect the naive Gn pigs from HRV infection or diarrhea, although immunostimulatory and regulatory effects on T cell, B cell, dendritic cell (DC) and macrophage, γδ T cell, Toll-like receptor, and cytokine responses in the HRV-infected pigs were evident (Azevedo et al., 2012; Wen et al., 2009; Wen et al., 2011; Yuan et al., 2006; Zhang et al., 2008a; Zhang et al., 2008d).

In general, the beneficial effect of probiotics as a cure for diarrhea is low. When studied in children with acute infectious diarrhea, the effect of probiotics was mostly described as shortening the duration of diarrhea by approximately one day (Lievin-Le Moal et al., 2007; Van Niel et al., 2002). Hence, I postulated that probiotics used as immunopotentiators for vaccines may have a more substantial impact than when they are used as biotherapeutic agents to treat infectious diseases. This postulation was based on accumulating evidence that some probiotic LAB strains are strong potentiators of immune responses to viral and bacterial antigens (Mohamadzadeh et al., 2008). By that time, adjuvanticity of LAB in enhancing cellular and humoral immune responses had been reported in influenza virus, poliovirus, and rotavirus vaccines, and rotavirus and *Salmonella typhi* Ty21a infections (de Vrese et al., 2005; Isolauri et al., 1995; Kaila et al., 1992; Link-Amster et al., 1994; Mohamadzadeh et al., 2008; Olivares et al., 2007; Winkler et al., 2005; Zhang et al., 2008c). The following studies described in this chapter confirmed this hypothesis. Over the past 13 years, studies of using probiotics as vaccine adjuvants have grown rapidly, with nearly 100 publications appearing in PubMed from 2008–2021 when using the combined search terms "probiotic adjuvant" and "vaccine." In the most recent review article on this topic, "Probiotics as Adjuvants in Vaccine Strategy: Is There More Room for Improvement?"

DOI: 10.1201/b22816-14

(Peroni and Morelli, 2021), after discussing the mechanisms and reviewing all laboratory and clinical evidence, including our studies, the authors concluded that the use of probiotics as adjuvants in vaccination represents a strategic key and should be considered in future studies, especially in the elderly and children, where vaccine effectiveness and duration of immunity often needs further improvement.

14.2 Probiotic *Lactobacillus acidophilus* enhances the immunogenicity and protective efficacy of the AttHRV vaccine in Gn pigs

When taken around the time of vaccination, probiotics can act as vaccine adjuvants by directly stimulating the host immune system. The novel concept of probiotic adjuvant was established partly based on our study of using *Lactobacillus acidophilus* strain NCFM™ (LA) to enhance the immunogenicity of the AttHRV oral vaccine in Gn pigs (Zhang et al., 2008b). This study was mainly supported by the R21 grant titled "Effects of Probiotic *Lactobacilli* on Rotavirus Immunity" from NCCAM, NIH (R21 AT002524, 2/1/2005–1/31/2008. PI: Lijuan Yuan). Wei Zhang led this study, who was my first graduate student and received her master's degree for the studies of the immunomodulatory effects of probiotics (Zhang et al., 2008a; Zhang et al., 2008b; Zhang et al., 2008d) at The Ohio State University.

Probiotics have been shown to enhance the immunogenicity of viral vaccines in a few earlier studies, e.g., influenza virus and poliovirus (Kukkonen et al., 2006; Olivares et al., 2007). The most commonly used probiotic is *Lactobacillus* sp., which has been used in both humans and animals to prevent as well as treat many gastrointestinal disorders (Ouwehand et al., 2002). One of the mechanisms by which probiotics exert their effects in reducing the severity of infectious diarrhea could be the enhancement of host innate and adaptive immune responses. The goal of the present study was to determine if LA administered orally with a two-dose live AttHRV vaccine could enhance intestinal and systemic ASC and antibody responses to HRV and protection against VirHRV challenge in the neonatal Gn pig model. In addition, the clinical signs after the first dose of AttHRV inoculation in neonatal pigs with or without LA were evaluated to determine if probiotics can prevent side effects (diarrhea) of oral rotavirus vaccines in neonates. During vaccine trials, RotaTeq™ vaccinated children had a small but significant increase compared to the children in the placebo group in developing mild diarrhea shortly after the first dose of the vaccine, a side effect of the live oral vaccine (Ciarlet, 2007). If probiotics orally intake during a period around vaccination are proven effective as adjuvants for live oral AttHRV vaccines, it would likely have the most profound effects in developing countries, where rotavirus diarrhea and mortality are greatest. Probiotics also have the potential to reduce or prevent vaccine-associated diarrhea, if the vaccine taken is not negatively affected. Probiotics thus may serve as inexpensive and safe oral adjuvants to enhance the efficacy of HRV vaccines that will reduce rotavirus diarrhea severity and deaths.

14.2.1 Probiotic LA dosing, AttHRV vaccination, and VirHRV challenge

Gnotobiotic pigs were assigned to four groups as follows: AttHRV-inoculated LA-fed (LA+AttHRV+) (n = 9), AttHRV-inoculated non-LA-fed (LA−AttHRV+) (n = 15), non-AttHRV-inoculated LA-fed (LA+AttHRV−) (n = 4), and non-AttHRV-inoculated non LA-fed (LA−AttHRV−) (n = 19). Pigs were orally dosed with 10^3, 10^4, 10^5, 10^6, and 10^6 CFU of LA in 2 ml of 0.1% of peptone water at 3, 5, 7, 9, and 11 days of age, respectively. Non-LA-fed pigs were given an equal volume of 0.1% peptone water. At five days of age, pigs were orally inoculated with 5 × 10^7 FFU AttHRV and reinoculated with the same dose 10 days later (PID 10). Non-inoculated pigs were given an equal volume of diluent. Four to seven pigs from each group were euthanized at PID 28 for isolation of MNC to measure immune responses (virus-specific ASC and total immunoglobulin-secreting cells [IgSC] by ELISPOT and IFN-γ producing T cell responses by flow cytometry) in intestinal and systemic lymphoid tissues. A subset of pigs (n = 5–11) from LA+AttHRV+, LA−AttHRV+, and LA−AttHRV− groups were orally challenged with 1 × 105 FFU of VirHRV and euthanized at PID 35 (PCD 7) to assess protection rates against HRV shedding and diarrhea. Serum samples were collected at PID 0, 10, 21, and 28 for detection of serum antibodies and PCD 2 for detection of viremia.

14.2.2 LA reduced AttHRV vaccine-associated diarrhea and substantially improved the protection against virus shedding conferred by the AttHRV vaccine against VirHRV challenge

Probiotic LA colonized the Gn pigs with similar counts in pigs with or without the AttHRV vaccine. From PID 5–28, the average daily fecal LA counts in the LA+AttHRV+ pigs ranged between 1.2×10^6 to 5.8×10^6 CFU/ml, and was similar to LA+AttHRV− pigs (5.2×10^5 to 4.7×10^6 CFU/ml).

After the first dose of AttHRV inoculation (Table 14.1), none of the LA+AttHRV+ and 50% of LA−AttHRV+ pigs shed virus nasally; and 29% of LA+AttHRV+ and 25% of LA−AttHRV+ pigs shed virus fecally as detected by ELISA and CCIF. The mean diarrhea scores were significantly higher in the LA−AttHRV+ than the LA+AttHRV+ pigs. No significant differences in proportions of fecal virus shedding and diarrhea, mean duration of virus fecal shedding, and mean peak titers of virus fecal shedding were observed between LA+AttHRV+ and LA−AttHRV+ pigs. These results indicate that 1) LA feeding prevented nasal AttHRV shedding, but not fecal shedding, thus LA did not reduce the intestinal replication of the AttHRV vaccine; and 2) LA reduced the diarrhea score after the first dose of AttHRV inoculation, suggesting that LA feeding may prevent side effects of the vaccine.

After challenge with VirHRV (Table 14.2), 40% of LA+AttHRV+ pigs, 73% of LA−AttHRV+ pigs and 100% of control (LA−AttHRV−) pigs shed virus fecally. The peak titers of HRV fecal shedding in the LA+AttHRV+ pigs were significantly lower (mean peak titer 1.3×10^2 versus 1.1×10^4 FFU/ml) than in the LA−AttHRV+ pigs. Figure 14.1 depicts the mean fecal shedding titers from PCD 1 to 6 among the two AttHRV+ groups and the controls determined by CCIF. The LA+AttHRV+ pigs shed virus only at PCD 2 and the mean shedding titer was significantly lower than the LA−AttHRV+ and control that time (1.3×10^2 versus 5.8×10^3 and 3.6×10^3 FFU/ml, respectively). There was a 95% and a 74% reduction in titers of virus shedding at PCD 2 and 3, respectively in the LA+AttHRV+ pigs compared to the LA−AttHRV+ pigs. Thus, LA feeding substantially improved the protection against virus shedding conferred by the AttHRV vaccine against the VirHRV challenge.

Furthermore, none of LA+AttHRV+ pigs, 100% of LA−AttHRV+ pigs, and 100% of control pigs had viremia/antigenemia at PCD 2 detected by antigen ELISA. The diarrhea scores in both the LA+AttHRV+ and LA−AttHRV+ groups were significantly lower than the controls. The protection rate against virus shedding and viremia in the LA+AttHRV+ group was higher (60% versus 27% and 100% versus 0%, respectively) than the LA−AttHRV+ group. The protection rate against diarrhea was similar in the LA+AttHRV+ and LA−AttHRV+ groups.

14.2.3 LA significantly enhanced the T cell, B cell, and antibody responses induced by the AttHRV vaccine

14.2.3.1 LA significantly enhanced HRV-specific IFN-γ producing CD8+ T cell responses induced by the AttHRV vaccine – MNCs from all pigs euthanized at PID 28 were subjected to flow cytometry for the detection of intestinal and systemic HRV-specific IFN-γ producing T cell responses. As shown in Figure 14.2, The frequencies of IFN-γ producing CD8+ T cells in the LA+AttHRV+ pigs were significantly higher in ileum (12-fold) and spleen (45-fold) and higher in blood (seven-fold) than the LA−AttHRV+ pigs. The frequencies of IFN-γ producing CD4+ T cells in the LA+AttHRV+ pigs were also higher (2–22-fold, but not significantly) than the LA−AttHRV+ pigs in all lymphoid tissues. Thus, LA feeding significantly increased the intestinal and systemic effector T cell responses induced by the AttHRV vaccine.

14.2.3.2 LA significantly enhanced intestinal HRV-specific IgA and IgG ASC responses induced by the AttHRV vaccine as well as the total intestinal IgA and IgG immunoglobulin-secreting cells (IgSC) – As shown in Figure 14.3A, the numbers of IgA ASC in the ileum of LA+AttHRV+ pigs were significantly higher (70 versus $4.6/5 \times 10^5$ MNC) than the LA−AttHRV+ pigs. Significantly higher numbers of IgG ASC were also detected in the ileum of LA+AttHRV+ pigs compared to the LA−AttHRV+ pigs. Thus, LA feeding significantly increased the intestinal HRV-specific IgA and IgG ASC responses induced by the AttHRV vaccine at PID28.

Table 14.1 Summary of AttHRV Shedding and Diarrhea in Pigs Inoculated with AttHRV Vaccines with or without LA

| | | Virus shedding after the first dose of AttHRV | | | | | | Diarrhea after the first dose of AttHRV | | | |
| | | Nasal | | | Fecal | | | | | | |
Treatment	n	% shed*	Mean duration days**	Mean peak titer shed (FFU/ml)b**	% shed*	Mean duration days**	Mean peak titer shed (FFU/ml)b**	% with diarrhea*	Mean days to onset**	Mean duration days**	Mean cumulative score**
LA+AttHRV+	7	0^a	0	0	29^a	1.0	3.9×10^3	14.3^a	6.0	1.0	4.2^b
						(0)	(1.8×10^3)				(0.5)
LA–AttHRV+	8	50^a	2.3	3×10^4	25^a	2.5	1×10^3	12.5^a	2.0	2.0	6.1^a
			(1)	(4.5×10^4)		(1)	(3.3×10^2)				(0.3)

Note:

* Proportions in the same column with the same superscript letters do not differ significantly (Fisher's exact test).

** Means in the same column with different superscript letters differ significantly; means with no letter indicate no significant difference (one-way ANOVA).

Numbers in parenthesis are the standard error of the mean.

a, b Means in the same column with different superscript letters (a, b) differ significantly; means with no letter indicate no significant difference (one-way ANOVA; $p < 0.05$).

Source: The table is modified from Zhang et al. (2008b). "Probiotic *Lactobacillus Acidophilus* Enhances the Immunogenicity of an Oral Rotavirus Vaccine in Gnotobiotic Pigs." *Vaccine* 2008. 26:3655–3661 with permission from Elsevier.

Table 14.2 Summary of HRV Shedding and Diarrhea in Pigs Inoculated with AttHRV Vaccine with or without LA and Challenged with VirHRV

Treatment	n	After VirHRV challenge								
		Virus shedding					Diarrhea			
		% shed	Mean days to onset	Mean duration days	Mean peak titer shed (FFU/ml)	% viremia	% with diarrhea	Mean duration days	Mean days to onset	Mean cumulative score
LA+AttHRV+	5	40^b	2.0^a	1.0^b	$1.3 \times 10^{2 \cdot b}$	0^b	80^a	2.5^a	1.3^a	7.1^b
			(0)	(0)	(0)			(1.2)	(0.3)	(2)
LA–AttHRV+	11	73^{ab}	2.0^a	2.3^{ab}	$1.1 \times 10^{4 \cdot a}$	100^a	64^a	3.1^a	2.1^a	7.2^b
			(0)	(0)	(5.5×10^4)			(0)	(0)	(1.1)
LA–AttHRV–	15	100^a	1.9^a	3.1^a	$1.2 \times 10^{4 \cdot a}$	100^a	100^a	3.0^a	2.1^a	10^a
			(0)	(0)	(1.4×10^5)			(0)	(0)	(0.5)

Note: See Table 14.1 note for legend and statistical analysis.
Source: Modified from (Zhang, 2007) MS Thesis "Effects of Probiotic Lactic Acid Bacteria on Innate and B Cell Responses to Rotavirus" with permission from the author.

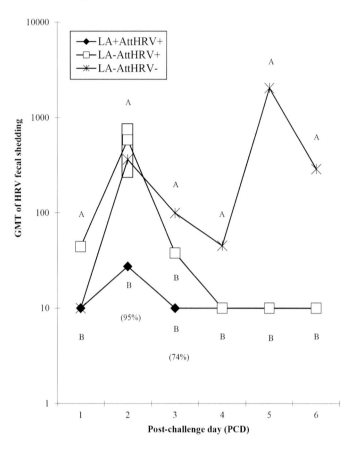

Figure 14.1   Geometric mean titers (GMT) of HRV fecal shedding from PCD 1 to 6 in Gn pigs colonized with LA or given 0.1% peptone water (LA-control) and infected with HRV or diluent (control). ELISA and CCIF were performed to determine HRV fecal shedding titers. Symbols with different letters differ significantly (one-way ANOVA followed by Duncan, $p < 0.05$). The percentages represent reductions of HRV fecal shedding titers in the LA+AttHRV+ group compared to the LA−AttHRV+ group. (Source: Figure 14.1 is reproduced from Zhang (2007), MS thesis, "Effects of Probiotic Lactic Acid Bacteria on Innate and B Cell Responses to Rotavirus" with permission from the author.)

As shown in Figure 14.3B, the numbers of IgG SC in the ileum and spleen of the LA+AttHRV+ pigs were significantly higher than the LA−AttHRV+ and LA+AttHRV− pigs. The numbers of IgA SC in ileum of the LA+AttHRV+ pigs were higher than the LA−AttHRV+ pigs and significantly higher than the LA+AttHRV− pigs.

The total IgSC consists of B cell responses to AttHRV, LA, and undefined background antigens (i.e., food antigens) in the Gn pigs. The IgSC were undetectable to very low numbers in all lymphoid tissues of LA−AttHRV− pigs and were significantly lower than other groups (except for IgM SC in spleen), indicating that the total IgSC responses were induced largely by LA and/or AttHRV. Both LA alone and AttHRV alone stimulated the development of total IgSC responses; however, AttHRV was more effective than LA in stimulating IgA SC in the ileum and significantly more effective in the spleen and blood (Figure 14.3B). LA plus AttHRV further enhanced the development of the total IgA and IgG SC responses in ileum.

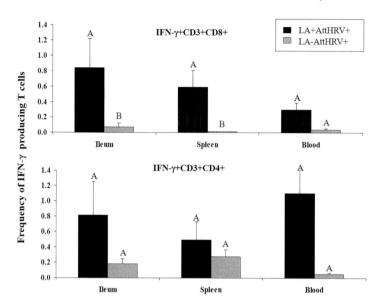

Figure 14.2   HRV-specific IFN-γ producing T cell responses in the Gn pigs vaccinated with AttHRV with or without LA. Mononuclear cells from the ileum, spleen, and peripheral blood of pigs were extracted and assayed on PID 28. Frequencies of IFN-γ+CD4+ and IFN-γ+CD8+ T cells among CD3+ MNC were determined by intracellular staining and flow cytometry assay after the MNC were stimulated with purified AttHRV antigen for 17 hours. Frequencies from the mock-stimulated MNC were subtracted from the frequencies of the AttHRV-stimulated MNC. Data represent the adjusted mean frequency (n = 4 for LA+AttHRV+; n = 7 for LA–AttHRV+) of HRV-specific IFN-γ producing T cells. Bars with different letters (A, B) on top differ significantly for the same tissue (Kruskal–Wallis Test, *p* < 0.05). (Source: Figure 14.2 is modified from Zhang et al. (2008b), "Probiotic *Lactobacillus Acidophilus* Enhances the Immunogenicity of an Oral Rotavirus Vaccine in Gnotobiotic Pigs." *Vaccine* 2008. 26:3655–3661 with permission from Elsevier.)

14.2.3.3 LA significantly enhanced serum IgM, IgA, IgG, and VN antibody responses induced by the AttHRV vaccine – As shown in Figure 14.4, the serum HRV-specific IgM, IgA, and IgG antibody titers were higher or significantly higher in the LA+AttHRV+ pigs than the LA–AttHRV+ pigs at PID 21 and PID 28. The fold differences were from four- (for IgM) to 18-fold (for IgG). The VN GMT in the LA+AttHRV+ pigs was six-fold higher than the LA–AttHRV+ pigs (832 ± 192 vs. 138 ± 55, *p* < 0.05). Thus, LA significantly increased the titers of HRV-specific IgM, IgA, IgG, and VN antibodies induced by the AttHRV vaccine.

14.2.4 Summary

This study provided clear evidence that *L. acidophilus* NCFM strain has a significant adjuvant effect on the intestinal and systemic HRV-specific T and B cell immune responses induced by the AttHRV vaccine. It is also important to note that LA intestinal colonization did not reduce the vaccine replication rate in the gut. Oral intake of this specific LA strain at the specified dosage during a period around vaccination is a safe and easy way for enhancing the effectiveness of oral rotavirus vaccines. Because neonatal Gn pigs are totally devoid of maternal antibodies, they are likely more susceptible than human infants to potential adverse effects of the *Lactobacilli* or vaccines. We confirmed that the LA feeding regimen did not induce any adverse effect in neonatal Gn pigs, yet was highly effective in enhancing the immunogenicity of the AttHRV vaccine.

A study by Olivares et al showed that oral intake of *L. fermentum* CECT5716 significantly enhanced serum Th1 type cytokine and influenza-specific IgA antibody responses to an intramuscular influenza vaccine in adults (Olivares et al., 2007). Another randomized double-blind placebo-controlled trial (Davidson et al., 2011) found that LGG taken twice daily for 28 days

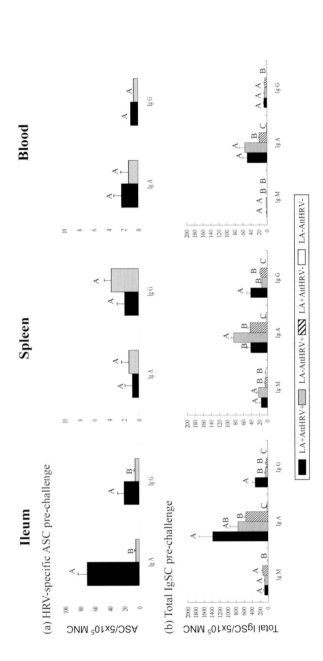

Figure 14.3 HRV-specific ASC and total Ig SC responses in Gn pigs vaccinated with AttHRV with or without LA. MNCs from the ileum, spleen, and peripheral blood of pigs were extracted and assayed on PID 28. ELISPOT assays for determining HRV-specific ASC and total IgSC numbers were performed on the day of MNC extraction. Data represent the mean numbers of HRV-specific ASC (a) and total IgSC (b) per 5 × 105 MNC, respectively (n = 4–7). Bars with different letters (A, B, C) on top differ significantly among groups for the same tissue and the same antibody isotype (Kruskal–Wallis Test, $p < 0.05$). (Source: Figure 14.3 is reproduced from Zhang et al. (2008b), "Probiotic *Lactobacillus Acidophilus* Enhances the Immunogenicity of an Oral Rotavirus Vaccine in Gnotobiotic Pigs." *Vaccine* 2008. 26:3655–3661 with permission from Elsevier.)

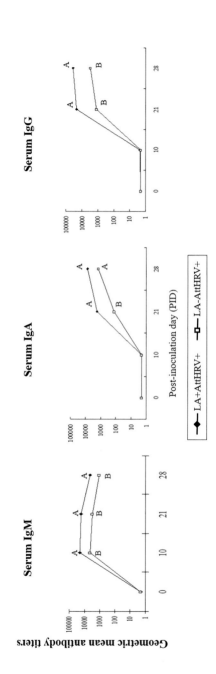

Figure 14.4 Geometric mean titers (GMT) of isotype-specific antibodies to HRV in the Gn pigs of the two AttHRV+ groups. Serum samples were collected on PID 0, 10, 21, and 28. Data represent the GMT of HRV-specific antibodies determined by ELISA at each time point (n = 7–9). Data points on lines marked with different letters (A, B, C, D) differ significantly (one-way ANOVA followed by Duncan's multiple range test on $\log_{10}$ transferred titers, $p < 0.05$). (Source: Figure 14.4 is reproduced from Zhang et al. (2008b), "Probiotic *Lactobacillus Acidophilus* Enhances the Immunogenicity of an Oral Rotavirus Vaccine in Gnotobiotic Pigs." *Vaccine* 2008. 26:3655–3661 with permission from Elsevier.)

significantly increased the protective antibody titers to H3N2 strain induced by a live-attenuated influenza vaccine in healthy adults. Therefore, these particular *Lactobacillus* strains may be used as oral adjuvants to improve the immunogenicity and protective efficacy of oral or parental vaccines, such as rotavirus and influenza vaccines. An increasing number of preclinical and clinical studies have demonstrated adjuvant effects of various specific probiotic bacterial strains for vaccines against different pathogens (Licciardi and Tang, 2011; Medina et al., 2010; Vitetta et al., 2017). The use of probiotic bacteria as oral adjuvants to enhance the immune responses induced by existing or new vaccines offers an exciting and promising new approach to improve the protective efficacy of vaccines against infectious diseases.

14.3 Dose effects of LAB on modulation of rotavirus vaccine-induced immune responses

14.3.1 Introduction

In subsequent studies, my laboratory further demonstrated that probiotic LA at the appropriate dose was effective both in reducing rotavirus diarrhea and enhancing the immunogenicity of oral rotavirus vaccines (Liu et al., 2014). We discovered that the dosage of probiotics plays a key role in the diverse effects observed in different studies. We reported that LA regulated rotavirus vaccine-induced immune responses in a dose-dependent manner in Gn pigs (Wen et al., 2012b). Low dose LA (five doses; up to 10^6 colony-forming units [CFU]/dose) significantly enhanced effector T cell responses and down-regulated Treg cell responses; high dose LA (14 doses; up to 10^9 CFU/dose) significantly down-regulated effector T cell responses and up-regulated Treg cell responses (Wen et al., 2012b). Intermediate dose LA (nine doses, up to 10^6 CFU/dose) significantly enhanced rotavirus-specific ASC and memory B cell responses induced by rotavirus vaccines in Gn pigs (Liu et al., 2014). These studies clearly demonstrated differential immune-modulating effects of high dose versus low dose LA on DC and T cell responses, and consequently different effects on the protection conferred by the AttHRV vaccine in Gn pigs challenged with virulent HRV. Low dose LA enhanced the protection against rotavirus diarrhea in AttHRV-vaccinated pigs whereas high dose LA had negative effects on the effectiveness of the vaccine. Thus, the same probiotic strains at different doses can exert qualitatively different modulating effects on immune responses induced by rotavirus vaccines and possibly other vaccines as well. The probiotic LGG strain enhanced the immunogenicity of rotavirus vaccines in Gn pigs, but the different dosages differentially modulated immune responses to favor either the mucosal IgA response (low dose) or the T cell response (intermediate dose) (Wen et al., 2015). The studies on probiotic dose effects were led by Ke Wen and Fangning Liu, who were my Ph.D. students. After graduation, they worked as post-doc research associates in my lab for a few more years. All these works were supported by the R01 award from NCCAM, NIH titled "Mechanisms of Immune Modulation by Probiotics" (R01AT004789. 2009–2014, PI: Lijuan Yuan).

Adjuvant effects of several probiotic lactic acid bacteria (LAB), mostly *Lactobacillus* strains, including *L. rhamnosus* GG, *L. acidophilus* NCFM, *L. acidophilus* CRL431, *L. acidophilus* La-14, *L. fermentum* CECT5716, *L. casei* DN-114 001, and *Bifidobacterium lactis* Bl-04 have been reported in studies of influenza, polio, rotavirus, and cholera vaccines and rotavirus and *Salmonella typhi* Ty21a infections (Boge et al., 2009; Davidson et al., 2011; Isolauri et al., 1995; Kaila et al., 1992; Mohamadzadeh et al., 2008; Olivares et al., 2007; Paineau et al., 2008; Winkler et al., 2005; Zhang et al., 2008b). The word adjuvant in the phrase "probiotic adjuvant" is not used in its traditional definition in which adjuvant implies a substance included in the vaccine formulation to aid the immune response to the vaccine antigen. Instead, probiotic adjuvants enhance the immunogenicity of vaccines when orally administered repeatedly around the time of vaccination and separately from the vaccine. By skewing the balance of pro- and anti-inflammatory innate immune responses and Th 1 and Treg cell adaptive immune responses in the context of vaccination, probiotic adjuvants act as "signal zero" to reduce Treg cell suppression and unleash effector T cell activation (Rowe et al., 2012).

Although the strain-specific effects of LAB in up- or down-regulating inflammatory immune responses have been well recognized, the dose effects of probiotics on innate and adaptive immune responses are not clearly understood. The same *Lactobacillus* strain is oftentimes reported by different research groups to have opposite immune-modulating effects. We hypothesized that the dose-effect is at least one of the reasons for the conflicting reports. Understanding the dose effects of probiotics has significant implications in their use as immunostimulatory (adjuvants) versus immunoregulatory agents.

In the following section, the findings from our serial studies of Gn pigs on the dose effects of the probiotic LA on innate and adaptive immune responses induced by the oral AttHRV vaccine are discussed (Yuan et al., 2013). We studied the effects of low dose (total 2.11×10^6 CFU) and high dose (total 2.22×10^9 CFU) LA on the intestinal and systemic 1) rotavirus-specific IFN-γ producing CD4+ and CD8+ T cell responses; 2) CD4+CD25+FoxP3+ and CD4+CD25−FoxP3+ Treg cell responses and the regulatory cytokine TGF-β and IL-10 production; 3) rotavirus-specific ASC and serum antibody responses; and 4) plasmacytoid dendritic cell (pDC) and conventional DC (cDC) frequencies, activation status, TLR expression, and cytokine production profile. The protective effect of the rotavirus vaccine against virus shedding and diarrhea was assessed in AttHRV-vaccinated Gn pigs fed with high, low, or no LA and challenged with the VirHRV.

14.3.2 Dose effects of LA on T cell responses

Probiotics have been reported to exert adjuvant properties by inducing pro-Th1 cytokines and promoting Th1 type immune responses. For example, *L. lactis* and *L. plantarum* induced production of IL-12 and IFN-γ by splenocytes when the LAB and an allergen were co-administered intranasally to mice (Repa et al., 2003). *L. fermentum* strain CECT5716 enhanced the Th1 responses induced by an influenza vaccine in addition to enhancing virus-neutralizing antibody responses (Olivares et al., 2007). Eleven different probiotic strains were tested for cytokine production in human peripheral blood MNC and each tested bacterium was shown to induce production of TNF-α and some strains also induced production of IL-12 and IFN-γ (Kekkonen et al., 2008). Previous studies of Gn pigs showed that a mixture of LA NCFM strain and *L. reuteri* strain enhanced both Th1 (IL-12, IFN-γ) and Th2 (IL-4 and IL-10) cytokine responses to virulent HRV infection (Azevedo et al., 2012). LA NCFM strain enhanced the HRV-specific IFN-γ producing CD8+ T cell response to a rotavirus vaccine in Gn pigs, indicating adjuvanticity of the LA strain (Zhang et al., 2008b).

Dose effects of probiotics on modulating T cell immune responses have not been well studied. To address this question, we examined the dose effects of LA NCFM (NCK56) strain on IFN-γ producing CD4+ and CD8+ T cell immune responses induced by an oral rotavirus vaccine in Gn pigs (Wen et al., 2012b). The animal treatment groups included 1) high dose LA plus AttHRV vaccine (HiLA+AttHRV), 2) low dose LA plus AttHRV (LoLA+AttHRV), 3) AttHRV only, 4) high dose LA only (HiLA), 5) low dose LA only (LoLA), and 6) mock-inoculated control (mock). Gn pigs were orally inoculated at five days of age with the AttHRV vaccine at 5×10^7 FFU per dose. A booster dose was given 10 days later at the same dose and route. Subsets of the pigs were euthanized at PID 28 to assess immune responses and the rest were challenged with the homotypic Wa strain VirHRV at a dose of 1×10^5 FFU to assess protection from PCD 1 to 7. Pigs in the high dose LA groups were fed daily with 10^3 to 10^9 CFU/dose of LA for 14 days with 10-fold incremental dose increases every other day from 3–16 days of age. The accumulative total LA dose was 2.22×10^9 CFU. Pigs in the low dose LA groups were fed with 10^3, 10^4, 10^5, 10^6, and 10^6 CFU/dose of LA every other day from 3–11 days of age. The accumulative total LA dose was 2.11×10^6 CFU. The accelerating LA dosing regimen was used to avoid diarrhea in newborn Gn pigs after initial colonization since the pigs do not have any maternal antibodies.

14.3.3 Low dose LA, but not high dose LA, enhanced HRV-specific IFN-γ producing T cell responses

The magnitude of HRV-specific IFN-γ producing T cell responses in pigs was differentially modulated by low versus high dose LA at both pre-challenge and post-challenge (PID 28 and PCD 7). AttHRV-vaccinated and low dose LA-fed pigs (LoLA+AttHRV) had significantly higher frequencies of HRV-specific IFN-γ+CD8+ T cells in the ileum (11- and five-fold higher pre- and post-challenge, respectively), spleen (3.8- and 2.1-fold higher pre- and post-challenge, respectively), and blood (three- and 20-fold higher pre- and post-challenge, respectively) compared to the AttHRV only pigs (Table 14.3). The LoLA+AttHRV pigs also had significantly higher frequencies of HRV-specific IFN-γ+CD4+ T cells in the blood (three-fold higher for both pre- and post-challenge) compared to the AttHRV only pigs. In contrast, high dose LA did not significantly

Table 14.3   Effect of Low Dose vs. High Dose LA on IFN-γ Producing CD8+ T Cell Responses

| | Frequencies of IFN-γ+CD8+ T cells among CD3+ cells | | | | | |
| | PID 28 | | | PCD 7 | | |
	Ileum	Spleen	Blood	Ileum	Spleen	Blood
HiLA+AttHRV	0.05	0.34	0.05	0.11	0.16	0.06
LoLA+AttHRV	1.21	0.46	0.24	0.56	0.46	0.98
AttHRV only	0.11	0.12	0.08	0.11	0.22	0.05

Source: Summarized from Wen et al. (2012b).

Table 14.4   Effect of Low Dose vs. High Dose LA on Frequencies of Treg Cells

| | Frequencies of CD4+CD25−FoxP3+Treg cells among total MNC | | | | | |
| | PID 28 | | | PCD 7 | | |
	Ileum	Spleen	Blood	Ileum	Spleen	Blood
HiLA+AttHRV	2.96	10.34	9.00	1.56	3.78	3.65
LoLA+AttHRV	0.24	0.54	0.10	0.08	0.22	0.05
AttHRV only	0.23	0.59	0.51	0.27	0.26	0.17

Source: Summarized from Wen et al. (2012b).

alter the HRV-specific IFN-γ producing CD4+ and CD8+ T cell responses in the HiLA+HRV pigs compared to AttHRV only pigs.

14.3.4 High dose LA significantly increased frequencies of intestinal and systemic CD4+CD25−FoxP3+ Treg cells whereas low dose LA decreased TGF-β and IL-10 producing Treg cell responses

Frequencies and cytokine production of Treg cells in pigs were differentially modulated by low versus high dose LA at both pre-challenge and post-challenge. HiLA+AttHRV pigs had significantly higher frequencies of CD4+CD25−FoxP3+ Treg cells (ranging from six- to 86-fold higher) in all the tissues compared to LoLA+AttHRV and AttHRV only pigs pre- and post-challenge (Table 14.4).

Because Treg cells exert regulatory functions through mechanisms involving TGF-β and IL-10, we also compared the effects of high and low dose LA on frequencies of the Treg cell subsets that produced TGF-β or IL-10 among the AttHRV-vaccinated pigs. Low dose LA reduced frequencies of TGF-β producing CD4+CD25+FoxP3+ and CD4+CD25−FoxP3+ Treg cells compared to high dose LA and AttHRV only pigs in all tissues pre- and post-challenge (Table 14.5; data for CD25+ Treg cells are not shown). Low dose LA also reduced pre- and post-challenge frequencies of IL-10 producing CD4+CD25+FoxP3+ and CD4+CD25−FoxP3+Treg cells compared to high dose LA and AttHRV− only (except for CD4+CD25−FoxP3+ Treg cells in ileum and spleen post-challenge) (Table 14.5). High dose LA induced 2.6-fold and 20-fold, respectively higher frequencies of IL-10 producing CD4+CD25−FoxP3+ Treg cells in ileum and spleen post-challenge compared to AttHRV only.

These data clearly demonstrated that low dose LA promoted IFN-γ producing T cell and downregulated Treg cell responses, whereas high dose LA induced a strong Treg cell response and promoted the regulatory cytokine production by tissue-residing Treg cells post-challenge in Gn pigs. Studies of other lactobacilli strains have reported similar findings. A mixture of *L. plantarum* CEC 7315 and CEC 7316 at high dose (5×10⁹ CFU/day) resulted in significant increases in the percentages of activated potential T-suppressor and NK cells, while at a low dose (5×10⁸ CFU/day) increased activated T-helper cells, B cells and antigen-presenting cells (APCs)

Table 14.5   Effect of Low Dose vs. High Dose LA on CD25–FoxP3+ Treg Cell Cytokine Production

	PID 28			PCD 7		
	Ileum	**Spleen**	**Blood**	**Ileum**	**Spleen**	**Blood**
	Frequencies of TGF-β+ cells among CD4+CD25–FoxP3+ Treg cells					
HiLA+AttHRV	4.51	1.13	0.92	4.87	10.05	1.39
LoLA+AttHRV	0.32	0.15	0.31	0.00	0.17	0.38
AttHRV only	4.19	1.19	1.55	4.34	0.79	1.53
	Frequencies of IL-10+ cells among CD4+CD25–FoxP3+ Treg cells					
HiLA+AttHRV	4.21	2.97	0.88	10.68	17.70	2.51
LoLA+AttHRV	0.92	0.15	0.00	5.22	6.79	0.00
AttHRV only	4.47	3.01	1.77	4.08	0.90	1.44

Source: Summarized from Wen et al. (2012b).

in institutionalized seniors (Mane et al., 2011). High concentration (≥1×10[6] colony-forming unit [CFU]/ml) of a combination containing LA and *Bifidobacterium* or *B. infantis* attenuated mitogen-induced immune responses by inhibiting cell proliferation and arresting the cell cycle at the G0/G1 stage in both mitogen-stimulated spleen cells and peripheral blood MNC. However, low concentration (≤1×10[6] CFU/ml) promoted a shift in the Th1/Th2 balance toward Th1-skewed immunity by enhancing IFN-γ and inhibiting the IL-4 response (Li et al., 2011). The differences between the "low dose" and "high dose" LAB in these studies are small, yet the immunomodulatory effects are qualitatively different.

Dose effects may explain some of the controversies that result from the same probiotic strain used by different research groups in animal studies showing opposite immunomodulatory functions. For example, administration of *L. casei* suppressed pro-inflammatory cytokine expression by CD4+ T cells and up-regulated IL-10 and TGF-β levels in rats (So et al., 2008a; So et al., 2008b). On the contrary, another study found that *L. casei* was a pure Th1 inducer in mice. In addition to the difference in animal species, the *L. casei* doses used by the different studies differed significantly, with much higher doses used in the studies of rats (So et al., 2008a; So et al., 2008b). In the studies of rats, the amount of *L. casei* was 5 × 10[9] or 2 x10[10] CFU/dose per rat, three times per week for 11-12 weeks. In the study of mice, the amount of *L. casei* was 2 × 10[8] CFU/dose per mouse, twice per week for 8 weeks (Van Overtvelt et al., 2010). Thus, different doses and frequency of administration of the same LAB strains may result in totally different *in vivo* effects.

The dose effects of LA on immune responses to the AttHRV vaccine in pigs may also partly explain why the efficacies of oral rotavirus vaccines are significantly reduced in low-income countries compared to developed countries. The two licensed rotavirus vaccines, RotaTeq and Rotarix have a protective efficacy of >85% against moderate to severe rotavirus gastroenteritis in middle and high-income countries (O'Ryan et al., 2009). However, the protective efficacy of the RotaTeq vaccine is only 39.3% against severe rotavirus gastroenteritis in sub-Saharan Africa (Armah et al., 2010) and 48.3% in developing countries in Asia (Zaman et al., 2010). Rotarix vaccine showed a similar disparity in efficacy in low-income countries in Africa (O'Ryan and Linhares, 2009). In addition to other factors that contribute to the reduction in rotavirus vaccine efficacy (e.g., higher titers of maternal antibodies, malnutrition), during the initial colonization of human infants, exposure to high doses of commensal bacteria (common in countries with lower hygiene standards) would have a suppressive effect on IFN-γ producing T cell responses and promote Treg cell responses, thus leading to the lowered protective immunity after rotavirus vaccination.

14.3.5 Dose effects of LA on antibody and B cell responses

Probiotics are known to modulate both humoral and cellular immune responses. Probiotics can induce antigen-specific and non-specific IgA antibody responses at mucosal surfaces (Perdigon et al., 2001; Wells and Mercenier, 2008) to prevent invasion by pathogenic microorganisms. Oral administration of *L. acidophilus* L-92 strain led to a significant increase of IgA production in

Peyer's patches in mice (Torii et al., 2007). *L. casei* CRL 431 strain increased induction of intestinal IgA secreting cells in mice (Galdeano and Perdigon, 2006). *L. acidophilus* La1 strain and bifidobacteria enhanced specific serum IgA titers to *S. typhi* strain Ty21a and also total serum IgA in humans (Link-Amster et al., 1994). *L. rhamnosus* GG enhanced rotavirus-specific IgA ASC responses in humans and promoted recovery from rotavirus diarrhea (Kaila et al., 1992). In our earlier study of Gn pigs, a mixture of LA strain and *L. reuteri* strain did not alter virus-specific intestinal and systemic antibody and ASC responses, but they significantly enhanced total intestinal IgA secreting cell responses and total serum IgM and intestinal IgM and IgG titers in HRV-infected pigs (Zhang et al., 2008a).

The first reported adjuvant effect of probiotic LAB in vaccination was from a human clinical trial in which *L. rhamnosus* GG was shown to enhance rotavirus-specific IgM secreting cells and rotavirus IgA seroconversion in infants receiving a live oral rhesus-human rotavirus reassortant vaccine (Isolauri et al., 1995). In recent years, an increasing number of human clinical trials have demonstrated adjuvant effects of probiotics in enhancing vaccine-induced antibody responses. In a double-blind randomized controlled trial, *L. rhamnosus* GG or *L. acidophilus* CRL 431 increased serum poliovirus neutralizing antibody titers and poliovirus-specific IgA and IgG titers 2- to 4-fold in adult human volunteers vaccinated with the live oral polio vaccine (de Vrese et al., 2005). In another human clinical trial, six out of the seven probiotic strains tested enhanced cholera-specific IgG antibody concentration in serum; for the *B. lactis* Bl-04 and *L. acidophilus* La-14 strains the increase was more significant (Paineau et al., 2008). Daily consumption of a fermented dairy drink (*L. casei* DN-114 001 and yogurt ferments, Actimel) was shown to increase virus-specific antibody responses to the intramuscular inactivated influenza vaccine in individuals over 70 years of age (Boge et al., 2009). In a randomized, double-blind placebo-controlled pilot study, *L. rhamnosus* GG significantly improved the development of serum antibody responses to the H3N2 strain influenza virus (84% receiving *L. rhamnosus* GG versus 55% receiving placebo had a protective titer 28 days after vaccination) in healthy adults receiving the live-attenuated influenza vaccine (FluMist, Medimmune Vaccines, Gaithersburg, MD, USA) (Davidson et al., 2011). Thus, specific strains of probiotics can act as adjuvants to enhance humoral immune responses following not only mucosal (oral or intranasal) but also parenteral vaccination. Yet, the dose effects of probiotics on antibodies responses have not been well studied.

14.3.6 High dose LA did not significantly alter the HRV-specific antibody responses whereas low dose LA had negative effects on the antibody responses

In our studies of Gn pigs, the effect of high and low dose LA NCFM strain on HRV-specific serum IgA and IgG antibody levels and HRV-specific ASC and memory B cell responses in the intestinal and systemic lymphoid tissues to rotavirus vaccination were examined. The animal treatment groups were the same as listed above in the studies of T cell responses. High dose LA did not significantly alter the HRV-specific antibody responses in serum and ASC responses in any tissue at PID 28 and PCD 7 in the AttHRV-vaccinated pigs (Figures 14.5 and 14.6), except to reduce the IgG ASC response in ileum of the mock-vaccinated pigs post-challenge (Figure 14.6c). In contrast, low dose LA significantly reduced the HRV-specific IgA antibody titers at PID 7 and 14 (Figure 14.5**a**) and reduced or significantly reduced IgG ASC responses in blood pre- and post-challenge as well as IgA ASC responses in spleen and blood post-challenge (Figure 14.6a and b). Low dose LA also significantly reduced the IgA and IgG ASC responses in ileum of the mock-vaccinated pigs post-challenge (Figure 14.6c).

The negative effects of low dose LA on the HRV-specific serum antibody responses and ASC responses induced by the AttHRV vaccine were undesirable for the vaccine's immunogenicity; however, it is consistent with the strong pro-Th1 effect of the low dose LA. The skewed balance toward a Th1 type immune response in the low dose LA group may have resulted in the weakened antibody responses. In the subsequent studies, we evaluated the effects of a low dose and an intermediate dose of *L. rhamnosus* GG on the effector T cell, antibody and ASC responses induced by the AttHRV vaccine and we found that *L. rhamnosus* GG enhanced the production of a balanced Th1 and Th2 immune responses to the AttHRV vaccine and significantly increased the virus-specific IFN-γ producing T cell responses, the antibody responses and the protection rate of the AttHRV vaccine (Wen et al., 2015).

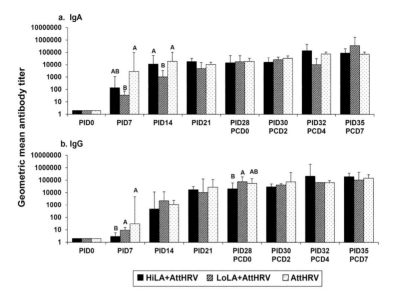

Figure 14.5 Rotavirus-specific serum IgA and IgG antibody responses in Gn pigs vaccinated with AttHRV with or without high or low dose LA feeding. Rotavirus-specific serum IgA (a) and IgG (b) antibodies were measured by an indirect isotype-specific antibody ELISA. Error bars indicate the standard error of the mean. Different capital letters (A, B) indicate significant differences among different pig groups at the same time point (Kruskal–Wallis Test, $p < 0.05$, n = 3–27), whereas shared letters or no letters on top indicate no significant difference.

14.3.7 Dose effects of LA on DC responses

The nature and consequences of a CD4+ T cell response (Th1, Th2, Th17, or Treg type) largely depend on the immune functions of DCs, which are the professional antigen-presenting cells that can prime and differentiate naive T cells. Both cDC and pDC are responsible for presenting microbial and dietary antigens to the adaptive immune systems, thereby influencing the polarization of the adaptive immune response (Konieczna et al., 2012). The pDC most effectively sense virus infections and are characterized by their capacity to produce large quantities of IFN-α and the pro-inflammatory cytokines IL-6 and TNF-α. These cytokines promote cDC maturation (Summerfield and McCullough, 2009). MHC II expression in professional APCs is tightly regulated. The MHC II of immature DCs are expressed at low levels at the plasma membrane, but abundantly in endocytic compartments. In the presence of inflammatory cytokines such as IFN-γ, DCs are activated; they stop capturing antigens and markedly increase MHC II expression on their plasma membrane. These MHC II are loaded with peptides derived from antigens captured at the site of inflammation. The mature DCs migrate to lymphoid tissues and up-regulate the co-stimulatory molecules (CD80/86) necessary to activate naive T cells (Villadangos et al., 2001).

It is known that probiotics can modify the distribution, phenotype, and function of DC subsets (Grangette, 2012). Both species-specific and strain-specific immunomodulatory effects of different LAB on DCs have been described in a large number of studies and were reviewed previously (Meijerink and Wells, 2010). Among the differential effects, several lactobacilli strains, including *L. acidophilus, L. gasseri, L. fermentum, L. casei, L. plantarum, L. johnsonii,* and *L. rhamnosus* have been reported to stimulate human or murine DCs to produce increased levels of pro-inflammatory cytokines (IL-2, IL-12, TNF-α) that favored Th1 and cytotoxic T cell polarization, and decreased levels of the regulatory cytokine TGF-β (Chiba et al., 2010; Christensen et al., 2002; Mohamadzadeh et al., 2005; Van Overtvelt et al., 2010; Vitini et al., 2000; Weiss et al., 2010; Yazdi et al., 2010). Such immune-stimulating effects are characteristics of adjuvants. However,

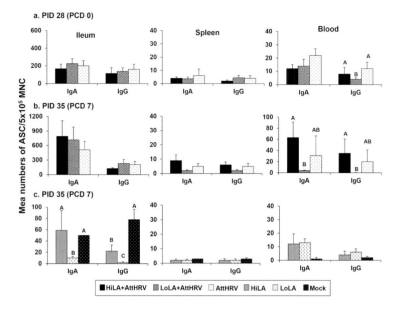

Figure 14.6 Rotavirus-specific IgA and IgG ASC responses in Gn pigs. Rotavirus-specific IgA and IgG ASC in the MNC isolated from the ileum, spleen, and blood of AttHRV-vaccinated pigs on PID 28 (PCD 0) (a) and PID 35 (PCD 7) (b) and of mock-vaccinated pigs on PID 35 (PCD 7) (c) were enumerated by using an ELISPOT assay and are presented as the mean numbers of virus-specific IgA and IgG ASC per 5 × 10⁵ MNC. Error bars indicate the standard error of the mean. Different capital letters (A, B, C) on top of the bars indicate significant differences among the treatment groups for the same isotype in the same tissue (Kruskal–Wallis test, $p < 0.05$, n = 3–14), whereas shared letters or no letters on top indicate no significant difference. Note the y axis scale difference (HiLA, high dose LA; LoLA, low dose LA).

studies of dose effects of lactobacilli on DC responses are scarce, with most consisting of *in vitro* experiments, and there is a dearth of comparative studies linking *in vitro* and *in vivo* results. *L. rhamnosus* Lcr35 was shown to induce dose-dependent immunomodulation of human DCs. Lcr35 at 10⁷ CFU/ml (10 multiplicity of infection), but not 10⁴ CFU/ml induced the semi-maturation of the DCs and a strong pro-inflammatory response (Evrard et al., 2011). LA NCFM induced a concentration-dependent production of IL-10, and low IL-12p70 in monocyte-derived DCs (Konstantinov et al., 2008). Immature DCs incubated with the LA NCFM at a bacterium to cell ratio of 1000:1 ("high dose") produced significantly higher IL-10 compared with the ratio of 10:1. In contrast, IL-12p70 was up-regulated at a lower concentration of the bacterium (10:1).

In our studies, dose effects of LA on pDC and cDC responses after rotavirus vaccination were examined in gnotobiotic pigs. The animal treatment groups were the same as listed earlier in the studies of T cell and B cell immune responses. Porcine pDC (CD172a+CD4+) and cDC (CD172a+CD11R1+) were defined as previously described (Jamin et al., 2006). The frequencies and tissue distribution, MHC II and co-stimulatory (CD80/86) molecular, TLR (2, 3, 9) and cytokine (IL-6, IL-10, IFN-α, TNF-α) expression by pDC and cDC in ileum, spleen, and blood of gnotobiotic pigs vaccinated with the AttHRV and fed with high dose, low dose or no LA were determined using multicolor flow cytometry.

The low dose LA group had significantly higher frequencies of pDC in ileum and spleen and cDC in spleen and blood compared to the high dose LA and AttHRV only groups (Figure 14.7a). The low dose LA group had overall lower MHC II expression on pDC and cDC in all tissues and lower CD80/86 expression in blood, but significantly higher CD80/86 expression on cDC in ileum, compared to the high dose LA and AttHRV only groups (Figure 14.7b). High dose LA did not have a significant effect on DC frequencies or activation marker MHC II and CD80/86

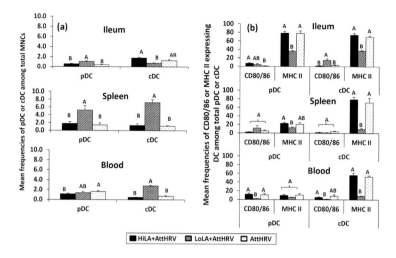

Figure 14.7   Frequencies of pDC and cDC (a) and the activation marker (CD80/86 and MHC II) expression (b) in intestinal and systemic lymphoid tissues of Gn pigs vaccinated with AttHRV vaccine with high dose, low dose, or no LA at PID 28. MNC were stained freshly without *in vitro* stimulation before being subjected to flow cytometry analyses. Data are presented as mean frequency ± standard error of the mean (n = 3–13). Different letters on top of bars indicate significant differences in frequencies among groups for the same cell type and tissue (Kruskal–Wallis test, *p* < 0.05), while shared letters indicate no significant difference.

expression, except for the significantly increased CD80/86 expression on pDC in ileum compared to the AttHRV only group (Figure 14.7b).

The low dose LA group had lower or significantly lower frequencies of TLR3 expression in both pDC and cDC in all tissues and significantly lower TLR2 expression on cDC in spleen compared to the high dose LA and AttHRV only groups (Figure 14.8). High dose LA did not have a significant effect on TLR expression in ileum and spleen. In blood, high dose LA group had significantly lower TLR3 expression in cDC (and lower in pDC) compared to the AttHRV only group.

The most striking dose-effect of LA on the cytokine production profile in DCs is the significantly increased IL-6 in the low dose LA group (Figure 14.9). The low dose LA group had significantly higher frequencies of IL-6 producing pDC and cDC in all tissues compared to the high dose LA and AttHRV only groups. Interestingly, the low dose LA reduced or significantly reduced the other cytokine TNF-α, IL-10, and IFN-α production in pDC in ileum and spleen. In contrast to ileum and spleen, the low dose LA increased or significantly increased TNF-α, IL-10, and IFN-α production in blood compared to the high dose LA or the AttHRV only group. High dose LA did not have a significant effect on IL-6, TNF-α, and IL-10 but lowered or significantly lowered IFN-α production in both pDC and cDC in all tissues compared to the AttHRV only group.

Therefore, the effects of high versus low dose LA on the frequencies, maturation status, and functions of DCs were strikingly different. Low dose LA significantly increased frequencies of both DC subsets, but these DCs were immature because they expressed lower frequencies of activation makers CD80/86 and MHC II and had reduced TNF-α, IL-10, and IFN-α production compared to the high dose LA and AttHRV only groups. Low dose LA promoted a strong IL-6 response in all tissues and increased all the other cytokine TNF-α, IL-10, and IFN-α production in blood for both pDC and cDC. High dose LA did not have such a significant modulating effect on the DC responses compared to the low dose (with a few exceptions). These findings are consistent with the differential effects of low dose versus high dose LA on the adaptive immune responses. The differential modulating effects of high versus low dose LA are intriguing. The biological and immunological implications of these effects and the underlying mechanisms require further investigation. From these data, it is clear that the same probiotic strain at different doses can exert qualitatively different modulating effects on DCs and consequently

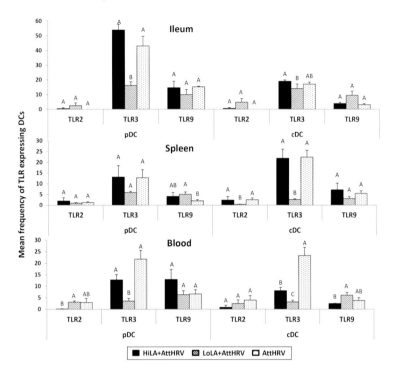

Figure 14.8   TLR expression patterns of pDC and cDC in intestinal and systemic lymphoid tissues of Gn pigs vaccinated with AttHRV vaccine with high dose, low dose, or no LA at PID 28. MNC were stained freshly without *in vitro* stimulation before flow cytometry analyses. Data are presented as mean frequency ± standard error of the mean (n = 3–5). Different letters on top of bars indicate significant differences in frequencies among groups for the same TLR in the same tissue (Kruskal–Wallis test, $p < 0.05$), while shared letters indicate no significant difference.

on adaptive immune responses induced by rotavirus vaccines. It has been reported that the effect of low dose microbe-associated pattern molecular (MAPM), such as lipopolysaccharide, was strikingly different as compared to that of high dose on macrophage cell functions: low dose lipopolysaccharide induced a strong inflammatory response in macrophages (Maitra et al., 2011). It is plausible that a similar interaction occurs between the MAPM from LA and DCs in the gut. Future studies are needed to identify the molecular mechanisms of the dose responses of different MAPM.

14.3.8   Dose effects of LA on protection conferred by the oral AttHRV vaccine against virulent HRV challenge

To examine the effects of low and high dose LA on improving the protection conferred by the AttHRV vaccine, subsets of Gn pigs from each treatment group were challenged with the virulent HRV Wa strain at PID 28. Clinical signs and virus shedding were monitored for 7 days post-challenge (Table 14.6).

After challenge, although the proportion of pigs that developed virus shedding and diarrhea did not differ significantly among the three AttHRV-vaccinated pig groups, the LoLA+AttHRV group had the shortest mean duration of fecal virus shedding and diarrhea and the lowest mean cumulative fecal consistency score among all the treatment groups. The durations of diarrhea in the LoLA+AttHRV pigs were significantly shorter compared to the AttHRV only and

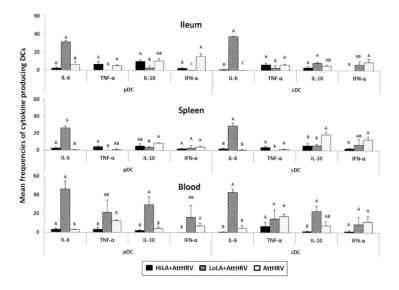

Figure 14.9   Cytokine production profiles of pDC and cDC in intestinal and systemic lymphoid tissues of Gn pigs vaccinated with AttHRV vaccine with high dose, low dose, or no LA at PID 28. MNC were stained freshly without *in vitro* stimulation before flow cytometry analyses. Data are presented as mean frequency ± standard error of the mean (n = 3–8). Different letters on top of bars indicate significant differences in frequencies among groups for the same cytokine in the same tissue (Kruskal–Wallis test, $p < 0.05$), while shared letters indicate no significant difference.

the mock-vaccinated control pigs. The durations of virus shedding in the LoLA+AttHRV pigs were significantly shorter compared to the HiLA+AttHRV and the mock control pigs. The mean cumulative fecal consistency scores in all the pigs in the LoLA+AttHRV and AttHRV only groups (8.4 and 9.0, respectively) were significantly lower than the control group, indicating significant protection against the severity of diarrhea. Thus, low dose LA slightly, but clearly improved the protection conferred by the AttHRV vaccine against rotavirus diarrhea. In contrast, high dose LA reduced the protection conferred by the AttHRV vaccine as indicated by the significantly longer mean duration of virus shedding (3.8 versus 1.3 days) and higher mean cumulative fecal scores compared to the AttHRV only pigs.

We reported previously that protection rates against rotavirus diarrhea are correlated with virus-specific intestinal IgA ASC and IFN-γ producing T cell responses at PID 28 in Gn pigs (Yuan et al., 1996; Yuan et al., 2008). A balanced Th1 and Th2 type response are needed for the optimal protective immunity against rotavirus. Although low dose LA further reduced the duration of diarrhea in the AttHRV-vaccinated pigs post-challenge, neither low nor high dose LA significantly altered the protection rate against rotavirus challenge (proportions of pigs that were infected and developed diarrhea after challenge). Because virus-specific intestinal IgA ASC responses probably play a more important role in rotavirus protective immunity than the IFN-γ producing CD8+ T cell responses (Yuan et al., 1996; Yuan et al., 2008), the effect of LA on virus-specific ASC responses also need to be taken into consideration regarding the differences in the protection conferred by the AttHRV vaccine with high and low dose LA. Although the low dose LA enhanced IFN-γ producing CD8+ T cell responses, it had negative effects on the serum antibody and ASC responses induced by the AttHRV vaccine. To improve the AttHRV vaccine efficacy, a different dose of LA (possibly an intermediate dose) or a different probiotic strain (i.e. LGG) may be optimal to promote a balanced Th1 and Th2 response without increasing Treg cell responses.

Table 14.6   Clinical Signs and Rotavirus Fecal Shedding in Gn Pigs after Virulent HRV Challenge[∆]

Treatments	n	Clinical signs			Fecal virus shedding (by CCIF and/or ELISA)		
		% with diarrhea[*]	Mean duration days[**]	Mean cumulative score[**]	% shedding virus[*]	Mean duration days[**]	Mean peak titer (FFU/ml)[**]
HiLA+AttHRV	13	92[a]	4.3 (0.7)[ab]	12.5 (1.4)[ab]	31[b]	3.8 (0.3)[a]	2.0[b]
LoLA+AttHRV	8	88[a]	2.4 (0.7)[b]	8.4 (1.3)[b]	36[b]	1.0 (0.0)[b]	5.6[b]
AttHRV only	12	67[a]	4.6 (0.5)[a]	9.8 (1.4)[b]	50[b]	1.3 (0.2)[b]	4.9[b]
Mock control	9	100[a]	5.6 (0.3)[a]	14.4 (1.0)[a]	100[a]	4.7 (0.7)[a]	4,558[a]

Note:

[∆] The data was partially presented previously (Wen et al., 2012b).

[*] Proportions in the same column with different superscript letters (a, b) differ significantly (Fisher's exact test, $p \leq 0.05$).

[**] Means in the same column with different superscript letters (a, b, c) differ significantly (Kruskal–Wallis test, $p \leq 0.05$).

14.3.9 Summary

Differential modulating effects on innate and adaptive immune responses by low dose versus high dose of the same LA NCFM strain were clearly demonstrated in Gn pigs. Low dose LA significantly enhanced the Th1 type effector T cell responses and decreased Treg cell functions in AttHRV-vaccinated pigs. Meanwhile, low dose LA resulted in a suppressed Th2 response, as evidenced by significantly reduced virus-specific ASC responses and serum antibody titers compared to the AttHRV only group. The dose effects of LA on IFN-γ-producing T cell and CD4+CD25−FoxP3+ Treg cell immune responses were similar between the intestinal and systemic lymphoid tissues. Thus, the same probiotic strain used in different doses can either increase or reduce mucosal and systemic immune responses induced by vaccines.

Lactobacillus rhamnosus GG dosage also affected its adjuvanticity in Gn pigs (Wen et al., 2015). The effects of two LGG dosages (five doses, LGG5X and nine doses, LGG9x) on intestinal and systemic T and B cell and antibody responses were evaluated in AttHRV-vaccinated Gn pigs. Only LGG5X significantly enhanced intestinal IgA ASC and memory B cell responses to AttHRV. However, both LGG5X and LGG9X significantly enhanced serum IgA antibody responses to AttHRV. Both regimens also enhanced rotavirus-specific IFN-γ-producing effector/memory T cell responses to the AttHRV vaccine, with LGG9X being more effective than LGG5X. Both regimens down-regulated CD4+CD25−FoxP3+ Treg cell responses in most lymphoid tissues examined pre-challenge and post-challenge and maintained the CD4+CD25+FoxP3+ Treg population in the ileum and intraepithelial lymphocyte post-challenge. Thus, LGG at both dosages functioned as an effective probiotic adjuvant for the AttHRV vaccine.

This series of studies of LA and LGG feeding in AttHRV-vaccinated Gn pigs revealed that the dose-dependent immunomodulating characteristics and the optimal dose for LA and LGG as vaccine adjuvants are different. The optimal dose for LA is the intermediate nine doses, whereas for LGG it depends on the desired type of the protective immune response, that is, LGG5X promotes better mucosal IgA response and LGG9X stronger systemic IgG and IFN-γ-producing T cell responses. These findings have significant implications in the use of probiotic *Lactobacilli* as immunostimulatory versus immunoregulatory agents. Probiotic products are increasingly used to improve health, alleviate disease symptoms, and enhance vaccine efficacy. Our findings suggest that probiotics can be ineffective if not used at the optimal dosage for the appropriate purposes, highlighting the importance of not only strain but also dose selection in probiotic studies.

The Gn pig model is a valuable animal model for the study of probiotic–virus host interaction because of the many similarities between human and porcine intestinal physiology and mucosal immune system (Meurens et al., 2012). The Gn status prevents confounding factors from commensal microflora that are present in conventionally reared animals or humans. Unlike Gn mice, Gn pigs are devoid of maternal antibodies, thus providing an immunologically naive background that allows clear identification of the immune responses to a single vaccine in hosts colonized with a qualitatively and quantitatively defined probiotic bacterial strain (Butler, 2009; Yuan and Saif, 2002). Although data from studies of Gn animal models may not be generalized directly to normal animals or humans, Gn animals provide a medium in which investigating the complex interrelationships of the host and its associated microbes become possible (Coates, 1975). Our findings provide a good starting point for the identification of the optimal dosage of a probiotic adjuvant. Nonetheless, the optimal dosage needs to be confirmed in conventionalized or humanized Gn pigs and human clinical trials to achieve the appropriate adjuvant effect for rotavirus and other vaccines.

15

Rice Bran as a Vaccine Adjuvant and as Prebiotics in Reducing Viral Diarrhea

This chapter describes our studies demonstrating that RB reduces rotavirus diarrhea and noro-virus infection and diarrhea, and the ability of RB to promote the development of intestinal and systemic T cell immune responses and improve the protective efficacy of oral AttHRV vaccine in Gn pigs. The studies were led by my Ph.D. students Xingdong Yang (for HRV) (Yang et al., 2015; Yang et al., 2014) and Shaohua Lei (for HuNoV) (Lei et al., 2016a). The projects were funded by an award titled "Dietary Rice Bran Supplementation for Gut Mucosal Immunity and Healthy Rice Crop Improvement" from the Bill and Melinda Gates Foundation Grand Challenges Exploration Grant in Global Health (OPP1043255 to PI Elizabeth P. Ryan) with a subcontract to Lijuan Yuan (G-6298-1. 6/15/2012–10/31/2015).

15.1 Rice bran as a vaccine adjuvant for AttHRV vaccines in Gn pigs

15.1.1 Introduction

Rice bran (RB), a globally accessible, abundant, and underutilized agricultural byproduct, has a distinct stoichiometry of bioactive compounds, phytochemicals, and minerals (Ryan, 2011). It has been studied for bioactive functions such as the prevention and treatment of chronic diseases, growth of beneficial intestinal microbes, induction of mucosal and systemic immune responses, and protection against enteric pathogens (Ghatak and Panchal, 2012; Henderson et al., 2012; Kumar et al., 2012; Sierra et al., 2005). Thus, this agricultural byproduct represents a promising and practical dietary-based solution for increasing innate resistance against enteric pathogens that cause diarrhea. In particular, its immune-stimulatory functions can be potentially used as a vaccine adjuvant for enteric pathogen infections. Previous studies have shown the immune-modulatory effects of RB on both innate and adaptive immunity *in vitro* and *in vivo* (Ghatak and Panchal, 2012). RB oil enhanced T and B lymphocytes proliferation, production of Th1 cytokines (IL-2, IFN-γ, and TNF-α) by lymphocytes, and reduced Th2 cytokines (both serum and lymphocyte-derived IL-4) as well as the level of serum IgE and IgG1 (Sierra et al., 2005). MGN-3, an arabinoxylan derived from rice bran, also increased levels of Th1 cytokines in human multiple myeloma patients (Cholujova et al., 2013). Importantly, γ-Oryzanol significantly promoted the development of antibody responses in rats stimulated with sheep red blood cells (Ghatak and Panchal, 2012). Total local (feces) and systemic (serum) IgA levels, and IgA expression on Peyer's patches B cells, were also enhanced in mice fed a 10% RB diet, suggesting RB promoted both mucosal and systemic B cell development (Henderson et al., 2012). These studies demonstrated the immune-stimulatory effects of RB on multiple components of the immune system. Given the RB-mediated protection against bacterial pathogens (Ghoneum, 1998; Ghosh et al., 2010; Ray et al., 2013) and stimulatory effects on both the innate and adaptive immune systems, RB represents a promising natural food product for modulating mucosal immunity and protecting against diarrhea and an adjuvant for enhancing protective efficacy of vaccines against major enteric pathogens, such as HRV and HuNoV.

15.1.2 Rice bran feeding, AttHRV inoculation, and VirHRV challenge of Gn pigs

Four treatment Gn pig groups were included: RB + AttHRV, AttHRV only, RB only, and mock control. Heat-stabilized, gamma-radiated RB (from the Neptune variety), provided by Anna

 DOI: 10.1201/b22816-15

McClung from the USDA-ARS Rice Research Center (Stuttgart, AK), was added to the Gn pigs' milk diet (ultra-high-temperature treated cow milk), starting at five days of age (PPD 5) until the end of the experiment. The amounts of RB were calculated to replace 10% of the pigs' daily caloric intake from milk. Two oral doses of the AttHRV vaccine were given at approximately 5×10^7 FFU/dose on PPD 5 and PPD 15. The mock group received neither RB nor AttHRV vaccine. A subset of pigs from each group was orally challenged with the Wa VirHRV at a dose of approximately 1×10^5 FFU on PID 28 and monitored from PCD 0–7 for assessment of diarrhea severity and virus shedding. Pigs were euthanized on PID 28 or PCD 7 for detection of total and HRV-specific T and B cell and antibody responses in intestinal and systemic lymphoid tissues.

15.1.3 Rice bran reduced HRV diarrhea but not virus shedding

The effects of RB on rotavirus infection and disease were determined by comparing the virus shedding and diarrhea parameters among the four treatment groups: RB + AttHRV; AttHRV only; RB only; and mock control. The results summarized in Table 15.1 showed that the RB-only group had significantly lower incidence (100% to 20%), shorter mean duration (5.6 to 0.2 days), and reduced severity of diarrhea (diarrhea score 14.4 to 4.4) compared to the mock control-diet group. When compared to the AttHRV group, RB + AttHRV vaccine-treated Gn pigs had significantly reduced incidence of diarrhea (67% to 0%), shorter mean duration (4.6 to 0 days), and reduced severity of diarrhea (diarrhea score 9.8 to 4.4). Importantly, RB only group had less diarrhea compared to the group with AttHRV vaccine alone, with reduced incidence (20% vs. 67%) of diarrhea, significantly shorter mean duration (0.2 vs. 4.6 days), and lower diarrhea scores (4.4 vs. 9.8).

No significant difference in virus shedding was observed between the RB-only group and mock controls. Compared to the AttHRV vaccine group, the RB + AttHRV group had increased virus shedding (50% vs. 100%), significantly earlier onset (6.0 vs. 2.0), and significantly longer mean duration of virus shedding (1.3 vs. 3.2 days). In addition, the RB alone group had a slightly longer (not significantly) mean duration of virus shedding (6.2 vs. 4.7 days) compared to the mock controls. These data suggest that RB protects against rotavirus-induced diarrhea through mechanisms that are independent of affecting rotavirus replication.

15.1.4 Rice bran enhanced IFN-γ+ CD4+ and CD8+ T cell responses to AttHRV vaccine

The effects of RB on effector T cells were assessed by the frequency of IFN-γ producing CD4+ and CD8+ T cell populations among total CD3+ mononuclear cells in both intestinal tissues (ileum and IEL) and systemic lymphoid tissues (spleen and blood). The results are shown in Figure 15.1. Compared to the mock control group, the RB-only group had significantly increased frequencies of IFN-γ+CD4+ T cell populations in the ileum, spleen, and blood on PID 28 and IFN-γ+CD8+ T cell populations in the ileum, spleen, and blood on both PID28 and PCD 7. Compared to the AttHRV vaccine group, RB + AttHRV group had significantly increased frequencies of both IFN-γ+CD4+ and IFN-γ+CD8+ T cell populations in the ileum, spleen, and blood on PID 28 and PCD 7, except for IFN-γ+CD4+ T cells in the ileum and spleen on PCD 7. These data demonstrate that RB has strong stimulating effects that favor Th1 type immune responses. RB did not influence the development of total Th and CTL cells (data not shown).

15.1.5 Rice bran promoted the development of intestinal and systemic IgSC

Total immunoglobulins (Ig) in the intestinal and systemic tissues, particularly intestinal IgA, play a significant role in non-specific mucosal protection against viral infections. The number of IgM, IgA, and IgG IgSCs in the ileum, spleen, and blood were measured by ELISPOT and compared between the RB-only group and mock group (Figure 15.2). Rice bran group alone showed significantly increased numbers of IgM IgSC in the ileum and spleen as well as numbers of IgA IgSC in the spleen and blood at PID 28. The numbers of IgA IgSC in the ileum and numbers of IgG IgSC in all tissues did not differ significantly between RB-only and mock control groups. These data indicated that dietary RB intake can promote the development of intestinal and systemic IgSCs (IgM in ileum and spleen, IgA in spleen and blood).

Table 15.1 Clinical Signs and Rotavirus Fecal Shedding in RB-Fed and AttHRV-Vaccinated Gn Pigs after VirHRV Challenge

Treatments	n	Clinical signs				Fecal virus shedding		
		% with diarrhea*	Mean days to onset**	Mean duration (days)**	Mean cumulative scores**	% shedding virus*	Mean days to onset**	Mean duration (days)**
RB+AttHRV	6	0[c]	8 (0)[a]	0 (0)[c]	4.4 (1.2)[c]	100[ab]	2.0 (0.5)[b]	3.2 (0.9)[b]
AttHRV only	12	67[ab]	4.4 (0.8)[b]	4.6 (0.5)[b]	9.8 (1.4)[b]	50[b]	6.0 (0.7)[a]	1.3 (0.2)[c]
RB only	5	20[bc]	7.2 (0.8)[ab]	0.2 (0.2)[c]	4.4 (1.6)[c]	100[ab]	1.6 (0.2)[b]	6.2 (0.2)[a]
Mock	9	100[a]	1.4 (0.2)[c]	5.6 (0.3)[a]	14.4 (1.0)[a]	100[a]	2.0 (0.3)[b]	4.7 (0.7)[ab]

Note:
* Fisher's exact test was used for comparisons. Different letters (a, b, c) indicate significant differences in protection rates among groups ($p < 0.05$), while shared letters indicate no significant difference.

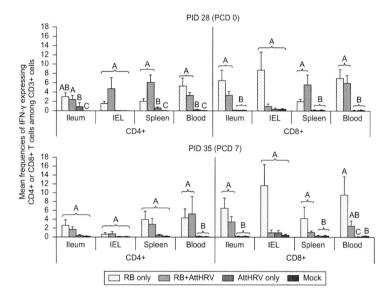

Figure 15.1   IFN-γ producing CD4+ and CD8+ T cell responses in Gn pigs fed with or without RB supplemented diet. MNCs from Gn pigs in each treatment group euthanized on PID 28 (left panel) or PCD 7 (right panel) were analyzed by flow cytometry after the MNCs were stimulated with semi-purified AttHRV antigen for 17 hrs. Frequencies of IFN-γ producing T cells among total CD3+ cells from each tissue were represented by frequencies of IFN-γ+ CD4+ subset (top panel), or IFN-γ+ CD8+ (bottom panel) T cell subset among CD3+ cells. Numbers on the y-axis indicate the percentage of IFN-γ producing CD4+ or CD8+ T cells among CD3+ cells in the respective tissues shown on the x-axis. Error bars indicate the standard error of the mean. Different capital letters (A, B, C) indicate significant differences between groups ($p < 0.05$), while shared letters indicate no significant difference (Kruskal–Wallis rank-sum test, $p < 0.05$; n = 3–6 for PID 28 and n = 4–12 for PCD 7).

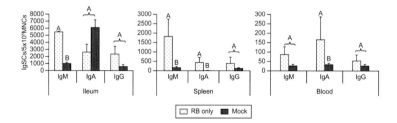

Figure 15.2   Mean numbers of total IgSCs in Gn pigs fed with or without RB supplemented diet for 28 days. MNCs from Gn pigs in RB-only and mock groups euthanized on PID 28 (without AttHRV vaccine and HRV challenge), were analyzed by total IgSC ELISPOT. Numbers on the y-axis indicate the number of total IgM−, IgA−, or IgG− immunoglobulin secreting cells per 5×10^5 MNCs in the respective tissues shown on the x-axis. Error bars indicate the standard error of the mean. Different capital letters (A, B) indicate significant differences between groups ($p < 0.05$), while shared letters indicate no significant difference (Kruskal–Wallis rank-sum test, $p < 0.05$; n = 3–4).

15.1.6 Rice bran stimulated the production of total IgM, IgA, and IgG in serum

Total serum IgM, IgA, and IgG antibody titers in Gn pigs fed with or without RB were determined using ELISA, and the results are shown in Figure 15.3. On PID 21, RB-only pigs had a significantly higher level of IgM and IgA titers compared to the controls. Additionally, RB-only pigs had significantly higher levels of serum IgM, IgA, and IgG antibody titers than the controls on PID 28. Post HRV challenge, RB-only pigs had a significantly higher level of IgA titer than the controls. In addition, RB + AttHRV pigs had significantly higher levels of IgM and IgA titers on PID 28 and IgM titers on PCD 7 than the AttHRV pigs. These data demonstrate that RB promoted the production of total serum IgM, IgA, and IgG antibody titer in both naïve and vaccinated Gn pigs.

15.1.7 Rice bran decreased the intestinal and systemic HRV-specific IgA and IgG ASC responses to AttHRV vaccination but not VirHRV challenge

HRV-specific serum IgA levels, as well as numbers of intestinal IgA and IgG ASCs, are associated with the protection against rotavirus infection and diarrhea (To et al., 1998; Yuan et al., 1996). HRV-specific ASC responses are shown in Figure 15.4. Compared to the AttHRV alone group, the RB + AttHRV group has significantly lower numbers of both IgA and IgG ASC in the ileum, spleen, and blood on PID 28. On PCD 7, compared to the non-vaccinated RB-only and mock groups, both AttHRV and RB + AttHRV groups had significantly higher numbers of HRV-specific IgA and IgG ASC in the ileum. The two vaccinated groups also had significantly higher numbers of IgA ASC in the blood than the mock group. The RB-only group had significantly higher numbers of IgA ASC in the ileum and blood in comparison to the mock group. Together, these results demonstrated that RB down-regulated virus-specific IgA and IgG effector responses induced by the AttHRV vaccine at PID 28, but not memory B cell responses upon VirHRV challenge.

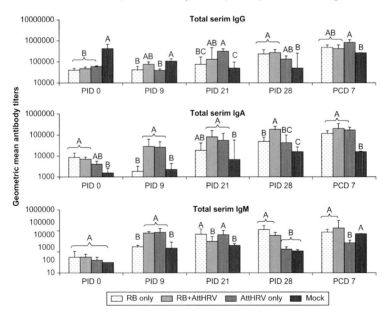

Figure 15.3   Total serum IgM, IgA, and IgG antibody responses in control or vaccinated Gn pigs fed with or without RB supplemented diet. Serum antibody titers were measured by ELISA. Data on PID 0, 9, 21, 28, and PCD 7 are shown. Different capital letters (A, B, C) indicate significant differences among different treatment groups for the same time point and same isotype while shared letters indicate no significant difference (ANOVA—general linear model [GLM], $p < 0.05$; n = 10–18).

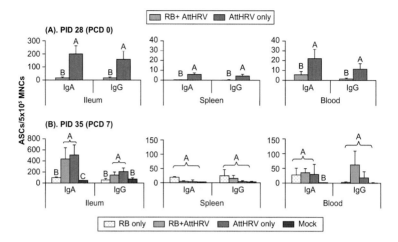

Figure 15.4 Mean numbers of HRV-specific ASCs in Gn pigs fed with or without RB supplemented diet at PID 28 and PCD 7. MNCs from Gn pigs in each treatment group euthanized on PID 28 (top figure) or PID 35/ PCD 7 (bottom figure), were analyzed by HRV-specific ELISPOT. Numbers on the y-axis indicate the number of HRV-specific IgA or IgG antibody-secreting cells per 5 × 10⁵ MNCs in the respective tissues shown on the x-axis. HRV-specific ASC responses were not detected in any tissue for both RB-only and mock groups on PID 28, therefore they are not presented. Error bars indicate the standard error of the mean. Different capital letters (A, B, C) indicate significant differences between groups, while shared letters indicate no significant difference (Kruskal–Wallis rank-sum test, $p < 0.05$; n = 4–7).

15.1.8 Rice bran reduced serum HRV-specific IgA and IgG antibody responses to AttHRV

To further confirm the results that RB down-regulated HRV-specific IgA and IgG ASC response at PID 28, HRV-specific serum IgA and IgG antibody titers were determined by ELISA (Figure 15.5). Consistent with HRV-specific IgA ASC data, serum IgA titer is significantly lower on both PID 21 and PID 28, but with no significant difference on PCD 7 in the RB + AttHRV group in comparison to the AttHRV vaccine group. For both RB + AttHRV and AttHRV groups, HRV-specific serum IgG antibody titers were not significantly different on PID 21, PID 28, and PCD 7, although RB + AttHRV group had a significantly higher IgG antibody titer on PID 10. The RB-only group had a significantly lower virus-specific IgG antibody titer than the mock group on PCD 7.

15.1.9 Rice bran increased HRV-specific IgA titer in the intestinal contents

HRV-specific antibody responses in the SIC and LIC were measured by ELISAs (Figure 15.6). HRV-specific IgA titers in both SIC and LIC of the RB + AttHRV pigs were higher at PID 28 and significantly higher at PCD 7 compared to the AttHRV pigs. These data demonstrated that RB can enhance the production of virus-specific IgA antibodies by intestinal memory B cells in the AttHRV-vaccinated pigs after VirHRV challenge, even though the numbers of virus-specific IgA ASC and the titers of virus-specific IgA antibody in serum before challenge were reduced.

15.1.10 Summary

In this study, we examined 1) the effects of RB supplementation on rotavirus infection and diarrhea, 2) the total and virus-specific T and B cell responses, and isotype-specific antibody responses induced by the AttHRV vaccine using neonatal Gn pigs as a model system. We

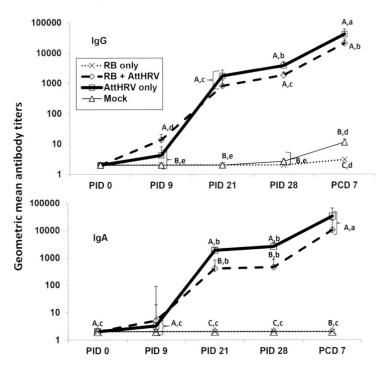

Figure 15.5 HRV-specific IgA and IgG antibody titers in serum of Gn pigs fed with or without RB supplemented diet. Antibody titers were measured by ELISA and are presented as geometric mean titers for each
treatment group. Error bars indicate the standard error of the mean. Different capital letters (A, B, C) indicate
significant differences among different treatment groups for the same time point, while different lower case
letters (a, b, c, d, e) indicate significant differences among different time points for the same treatment group.
Shared uppercase or lowercase letters indicate no significant difference (ANOVA—general linear model [GLM],
$p < 0.05$; n = 10–18).

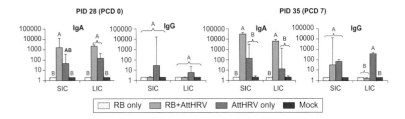

Figure 15.6 HRV-specific IgA and IgG antibody titers in small intestinal contents (SIC) and large intestinal
contents (LIC) of Gn pigs fed with or without RB supplemented diet. Antibody titers in intestinal contents were
measured by ELISA and are presented as geometric mean titers for each treatment group. Error bars indicate
the standard error of the mean. Different capital letters (A, B) indicate significant differences among treatment
groups for the same time point while shared letters indicate no significant difference (ANOVA—Tukey test,
$p < 0.05$; n = 3–6 for PID 28 and n = 4–12 for PCD 7).

observed that 10% dietary RB supplementation to milk significantly protected against rotavirus diarrhea, but did not reduce rotavirus replication. RB also strongly promoted the development of IFN-γ producing T cells, IgM and IgA producing IgSC, total serum IgM, IgA, and IgG antibody and HRV-specific intestinal IgA production, but significantly reduced HRV-specific IgA and IgG ASC in intestinal and systemic lymphoid tissues as well as HRV-specific serum IgA production at PID 28.

Rice bran alone reduced rotavirus diarrhea incidence and severity without reducing rotavirus shedding. Surprisingly, while RB and AttHRV vaccines synergistically and completely protected against rotavirus diarrhea, the protection of the AttHRV vaccine against rotavirus shedding was reduced by RB. These results strongly suggest that the mechanisms by which RB protects against rotavirus diarrhea are independent of rotavirus infection. The underlying mechanisms for rotavirus-induced diarrhea are currently not completely understood. The pathogenesis of rotavirus-induced diarrhea has been reviewed (Hagbom et al., 2012; Ramig, 2004). Four distinct but nonexclusive mechanisms have been implicated: 1) malabsorption due to the destruction of absorptive enterocytes in the villus, caused by rotavirus infection and increased intracellular [Ca^{2+}]; 2) NSP4 enterotoxin mediated increase in membrane permeability and tight junction disruption; 3) increased secretion from the crypt cells and intestinal motility via stimulation of enteric nervous system by rotavirus or NSP4 enterotoxin; 4) villus ischemia caused by unidentified vasoactive substances during rotavirus infection. RB could interfere with each of these four mechanisms. In fact, extracts from RB are effective in reducing diarrhea through inhibition of the intestinal mucosal Cl^- ion secretion by intestinal epithelial cells (Goldberg and Saltzman, 1996; Mathews et al., 1999). This mechanism is likely to have contributed to the protective effects of RB against rotavirus-induced diarrhea in the current study. It is also reported that zinc and enkephalinase inhibitors attenuate rotavirus-induced diarrhea (Hagbom et al., 2012). Certain RB phytochemicals might have functioned as such inhibitors. Further studies are underway to examine the effects of RB on the intestinal barrier integrity and permeability during rotavirus infection.

Both effector T and B cell responses play important roles during rotavirus infection and are associated with the protective efficacy of rotavirus vaccine against rotavirus infection and diarrhea (Blutt et al., 2012; To et al., 1998; Yuan et al., 2008). The significantly increased frequencies of IFN-γ producing CD4+ and CD8+ T cell responses in local (ileum) and systemic (spleen and blood) lymphoid tissues at both PID 28 and PCD 7 suggest that RB promoted the pathogenesis of effector T cell responses. However, this effect was not due to the enhanced expansion of total Th and CTL cells, as RB did not significantly increase their frequencies among lymphocytes in both intestinal and lymphoid tissues.

Rice bran also significantly enhanced the development of total IgM IgSCs in ileum and spleen and IgA IgSCs in spleen and blood pre- and post-challenge, as well as rotavirus-specific IgA antibody levels in intestinal contents after challenge, indicating the stimulatory effect of RB on the development of total and specific B cell responses to the AttHRV vaccine and VirHRV challenge. Similarly, a previous study showed that the number of peripheral blood lymphocytes was significantly increased in Wistar male rats fed a 10% hemicellulose extracted from RB fiber (RBF) diet for two weeks (Takenaka and Itoyama, 1993). However, the significantly reduced numbers of rotavirus-specific IgA and IgG ASCs in both local and systemic tissues, as well as the correspondingly lower levels of rotavirus-specific serum IgA and IgG antibody titers on PID 28, suggest that the immune-stimulatory effect (adjuvanticity) of RB are biased towards Th1 T cell responses before challenge and this effect is antigen-specific. Thus, RB functioned as a pro-Th1 type immune response "food adjuvant" for the AttHRV vaccine. This observation is consistent with previous studies showing that RB feeding in mice up-regulated Th1 cytokines and down-regulated Th2 and antibody responses. The reduction in rotavirus-specific B cell and serum antibody response at challenge may have contributed to the increased fecal rotavirus shedding in RB + AttHRV treatment group over the AttHRV only treatment group. However, RB did not negatively affect the rotavirus-specific memory B cell responses and enhanced rotavirus-specific intestinal IgA antibody responses at PCD 7, suggesting that RB increased priming of local virus-specific B cells even under the Th1 biased condition before challenge. The molecular mechanisms and kinetics by which RB modulates T and B cell responses warrant further studies.

In summary, results from this study demonstrated that RB significantly reduced the susceptibility to rotavirus diarrhea without reducing rotavirus shedding upon virulent HRV challenge in Gn pigs compared to the control diet. Furthermore, RB promoted the development of intestinal and systemic IFN-γ producing CD4+ and CD8+ T cell responses, total IgM IgSC in the ileum and spleen, total IgA IgSC in the spleen and blood, as well as total serum IgM, IgA, and IgG antibody production. Additionally, RB increased HRV-specific IgA titers in the intestinal contents post-challenge. RB alone also significantly increased the virus-specific IgA ASC response post-challenge in the ileum and blood. These results have significant clinical implications in the prevention and management of enteric pathogen-induced diarrhea using dietary RB in developing countries. Given that RB can support the colonization of gut probiotics (e.g., *Lactobacilli spp.*), enhance mucosal IgA production (Henderson et al., 2012), and significantly reduce the enteric burden of *Salmonella* infection in mice (Kumar et al., 2012) shown in other studies, continued investigation of dietary RB's mechanisms for protection against viral pathogens that cause significant global morbidity and mortality (e.g., rotavirus) is warranted.

15.2 Combination of probiotics and rice bran is highly effective in preventing HRV diarrhea in Gn pigs

Vaccines are believed to be the most cost-effective approaches to prevent rotavirus and norovirus gastroenteritis. In the last two sections of this book, however, two studies that demonstrate striking effectiveness of combined probiotics and rice bran in Gn pigs are described (Lei et al., 2016a; Yang et al., 2015). The protective efficacies of the probiotic LGG+EcN and RB combination against HRV (in section 15.2) and HuNoV (in section 15.3) diarrhea were higher than the candidate vaccines that we evaluated in the Gn pig models.

15.2.1 Introduction: probiotics and RB in HRV diarrhea

Rice bran contains prebiotic compounds (Komiyama et al., 2011) and a variety of bioactive components (i.e., polyphenols, fatty acids, and peptides) (Fabian and Ju, 2011; Ryan, 2011) that have been shown to have promising protective effects against diseases such as cancer (Norazalina et al., 2010), obesity (Gerhardt and Gallo, 1998), and diabetes (Chou et al., 2009) and immune-modulatory effects (Kataoka et al., 2008). Its therapeutic effects against enteric pathogen infections and diseases have also been studied in animal models (Kumar et al., 2012; Yang et al., 2014). By feeding mice a 10% RB diet, it was shown that RB reduced the colonization and invasion of *Salmonella enterica serovar Typhimurium* into enterocytes and intestinal mucosa (Kumar et al., 2012). In another mouse study, 10% dietary RB feeding for 28 days resulted in increased production of mucosal and systemic IgA (Henderson et al., 2012). In these mouse studies, modulation of gut microbiota, such as a 170-fold increase in the population of probiotic *Lactobacillus* spp. and decreased colonization of mucin degrading microbes (phylum Verrucomicrobia), was proposed as one of the possible mechanisms for RB to reduce the colonization and invasion of *Salmonella* bacteria (Kumar et al., 2012). We previously demonstrated that dietary RB feeding (10% RB diet) for 28 days significantly reduced HRV-induced diarrhea, without decreasing HRV replication and shedding in Gn pigs (Yang et al., 2014). Therefore, RB or its components protect against enteric pathogen infections and diseases probably through multiple mechanisms, including direct anti-microbial activities, prebiotic effects, and promoting intestinal epithelial health and mucosal immune responses. Identifying the RB-mediated gut barrier and mucosal immune mechanisms involved was the goal of the current study.

Lactobacillus rhamnosus GG is a gram-positive bacterium in the *L. rhamnosus* species that was first isolated in 1983 by Barry R. Goldin and Sherwood L. Gorbach (Silva et al., 1987). It is widely studied for treatment and prevention of gastrointestinal diseases and infections, and increasingly for extra-intestinal diseases as well, such as atopic dermatitis, allergic reactions, urogenital tract infections, and respiratory tract pathogens (Goldin and Gorbach, 2008). It has been shown to reduce the severity and duration of rotavirus diarrhea and persistent diarrhea in multiple clinical trials (Basu et al., 2007; Majamaa et al., 1995). LGG has also been found to reduce intestinal permeability in children with irritable bowel syndrome (Francavilla et al., 2010). *Escherichia coli* Nissle 1917 (EcN) is one of the best-characterized probiotics used to reduce acute and protracted diarrhea (Henker et al., 2007), and was shown to protect Gn pigs from lethal challenge by *Salmonella Typhimurium*. Given the above-discussed effects of RB, LGG, and EcN individually on rotavirus and *Salmonella* infection and diarrhea, studies to examine their combined therapeutic effects in the Gn pig model are warranted.

In this study, we hypothesized that RB can promote the growth of LGG and/or EcN, enhance gut health, reduce gut permeability, and together with LGG and EcN colonization, provide effective protection against HRV diarrhea and shedding. The objectives of this study were to identify: 1) the prebiotic effect of dietary RB on the growth of both LGG and EcN strains; 2) the protective effect of dietary RB supplementation against HRV diarrhea and shedding in LGG and EcN colonized Gn pigs; and 3) the effects of RB on intestinal health, permeability and innate immunity in LGG and EcN colonized Gn pigs.

15.2.2 RB feeding, probiotic colonization, and VirHRV challenge of Gn pigs

Neonatal Gn pigs were randomly divided into four treatment groups: probiotics plus RB feeding (RB+LGG+EcN), RB only (RB only), probiotics only (LGG+EcN), and mock control (mock). LGG (ATCC 53103) and EcN (a gift from Dr. Jun Sun, Rush University, Chicago, IL) used in the current study were propagated in *Lactobacilli* MRS broth (Weber, Hamilton, NJ) and Luria broth media, respectively. LGG and EcN 1:1 mixture at 10^4 CFU/dose each were administered orally to pigs at PPD 3, 5, and 7 to initiate colonization. The low dosage is purposely selected to be well below the therapeutic doses (10^9 to 10^{12} CFU). Heat-stabilized, gamma-irradiated RB (Calrose variety) was added to the Gn pigs' milk diet (ultra-high-temperature treated cow milk) replacing 10% daily calorie intake, starting at PPD 5 until the end of the experiment. A subset of pigs from RB+LGG+EcN and LGG+EcN groups were euthanized before challenge on PPD 33 (n = 6). The rest of the pigs were challenged orally with 10^5 FFU the Wa VirHRV on PPD 33 (PCD 0) and euthanized on PCD 3 (n = 6) or PCD 7 (n = 6). The pigs were weighed weekly starting on PPD 5 until euthanasia. Rectal swabs were collected daily for monitoring diarrhea and virus shedding by ELISA and CCIF from PCD 0 to 7.

15.2.3 RB completely protected against rotavirus diarrhea in LGG and EcN colonized Gn pigs

The effects of RB on HRV-induced diarrhea and virus shedding in LGG and EcN colonized Gn pigs were assessed (Table 15.2). Comparisons are all made between the treatment group to the mock group if not noted. RB+LGG+EcN completely protected against HRV diarrhea (100% protection rate); however, it did not significantly alter the percentage and duration of HRV shedding. RB alone significantly protected against HRV diarrhea (80% protection rate) but did not reduce HRV shedding. LGG+EcN alone also significantly reduced the incidence of HRV diarrhea (50% protection rate), but significantly prolonged HRV shedding. When comparing the LGG+EcN and the RB+LGG+ EcN pigs, the latter had no diarrhea and significantly reduced HRV shedding, with significantly delayed onset (1.2 versus 2.8 days), shortened mean duration (6.8 versus 5.2 days), and ~217-fold reduction in peak virus titer (1.3×10^5 versus 6×10^2 FFU/ml). Thus, although RB or LGG+EcN alone did not reduce virus shedding, RB supplementation prevented the increase of HRV shedding (duration and peak titers) in the LGG+EcN colonized pigs (Table 15.2).

15.2.4 RB significantly enhanced the growth and colonization of LGG and EcN and increased pig body weight gain in LGG+EcN colonized pigs

To determine whether dietary RB can promote the growth and colonization of probiotic bacteria in Gn pigs, similar to the effects observed in mice (Kumar et al., 2012), the titers of LGG and EcN were determined on specified time points following three low oral doses (10^4 CFU/dose) (Figure 15.7). RB significantly increased the load of both gram-positive LGG and gram-negative EcN in the gut of Gn pigs. The RB-fed pigs shed significantly higher counts of LGG (~10^4 increases) starting from post-feeding day (PFD) 7 through the entire monitored period until PFD 30. Similarly, RB also significantly enhanced the shedding of EcN (~10^5 increases) starting from PFD 2 through the entire monitored period until PFD 30. The peak titer for LGG shedding was 4×10^7 CFU and for EcN was 3×10^8 CFU. Together, these results showed that RB significantly enhanced the growth and colonization of the probiotic strains, with the effect on LGG growth manifested later and a slightly lower peak titer than that for EcN.

The growth rate is an important indication of the gut and the overall health of the host. To monitor the effect of RB on the growth rate of LGG+EcN colonized Gn pig, weekly body weight gains were compared between the LGG+EcN and the RB+LGG+ EcN pig groups (Table 15.3).

Table 15.2 RB Protects against HRV Diarrhea and Shedding in LGG and EcN Colonized Neonatal Gn Pigs

Treatments	N	Clinical signs				Fecal virus shedding			
		% with diarrhea*	Mean days to onset**§	Mean duration days**	Mean cumulative scores**	% shedding virus*	(ELISA) Mean days to onset**	Mean duration days**	(CCIF) Peak virus titer (FFU/ml)
RB+LGG+EcN	6	0[b]	N/A	0 (0)[b]	6.2 (0.5)[c]	100[a]	2.8 (0.3)[a]	5.2 (0.3)[c]	$6.0 \times 10^{2\ b}$
LGG+EcN	6	50[b]	5.2 (1.3)[a]	0.7 (0.3)[b]	8.9 (0.6)[b]	100[a]	1.2 (0.2)[b]	6.8 (0.2)[a]	$1.3 \times 10^{5\ a}$
RB only	5	20[b]	7.2 (0.8)[a]	0.2 (0.2)[b]	4.4 (1.6)[c]	100[a]	1.6 (0.2)[b]	6.2 (0.2)[b]	ND
Mock	9	100[a]	1.4 (0.2)[b]	5.6 (0.3)[a]	14.4 (1.0)[a]	100[a]	2.0 (0.3)[ab]	4.7 (0.7)[bc]	ND

Note:

* Fisher's exact test was used for comparisons. Different letters indicate significant differences in protection rates among groups ($n = 5–9$; $p < 0.05$), while shared letters indicate no significant difference.

** Kruskal–Wallis rank-sum test was used for comparisons. Different letters indicate significant differences in protection rates among groups ($p < 0.05$), while shared letters indicate no significant difference. ELISA, enzyme-linked immunosorbent assay; CCIF, cell culture immunofluorescent assay; FFU, fluorescence forming unit; N/A, not applicable; ND, not determined.

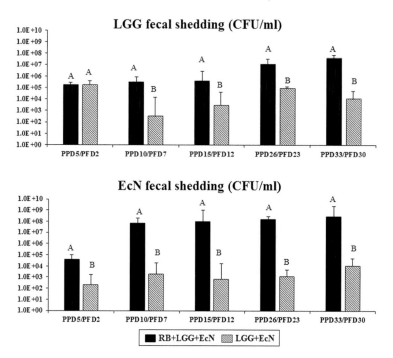

Figure 15.7 RB promotes the growth and colonization of the probiotics LGG and EcN in Gn pigs. Rectal swabs were washed in 4 ml 10% peptone water (ten-fold dilution) and additional ten-fold series dilutions from 10^2 to 10^4 were prepared and plated on LGG agar plates (LGG counting) or LB agar plates (EcN counting). The plates were incubated at 37° C incubator for three days. Colonies on each plate are then counted and titers calculated. Geometric means of the counts in each group at the specified time points are presented. Error bars are standard errors of the mean. Filled bars, RB+LGG+EcN group; hatched bars, LGG+EcN group. PPD, post-partum day; PFD, post–probiotic feeding starting day. Kruskal–Wallis rank-sum test was used for comparisons. Different letters indicate significant differences between groups (n =10–18; $p < 0.05$), while shared letters indicate no significant difference.

RB increased the weekly body weight gain in the pigs on the fourth and fifth week after RB feeding started, as shown by the mean body weight gain of 0.75 versus 0.51 kg on the fourth week, and 0.80 versus 0.42 kg on the fifth week for the RB+LGG+EcN versus LGG+EcN group, respectively. These results suggest that RB can promote the growth of LGG+EcN colonized Gn pigs after four weeks of RB feeding, rendering the hosts more resistant to HRV infections and diarrhea.

15.2.5 The combination of RB and LGG+EcN prevented epithelial damage from HRV challenge

To examine the effects of RB on the health of intestinal epithelium of Gn pigs colonized with LGG and EcN, H&E stained slides of the distal ileum were evaluated and scored via light micros-copy by Erica Twitchell who was my Ph.D. student and a board-certified veterinary pathologist. After euthanasia, sections of ileum from the LGG+EcN colonized groups were collected for histopathologic examination. Samples were fixed in 4% paraformaldehyde, routinely processed for H&E staining. The pathologist was blinded to the identification of the animal until after microscopic analysis of all samples was complete. A histopathological scoring system for ileal samples was developed using guidance from previous publications (Crouch and Woode, 1978; Gibson-Corley et al., 2013; Kim et al., 2011; Pearson and McNulty, 1977; Shackelford et al., 2002). The mitotic index was obtained by dividing the total number of mitotic figures in 50

Table 15.3   RB Enhanced the Growth of Gn Pigs from PPD 27 to PPD 33

Treatment group	n	PPD 13–19	PPD 20–26	PPD 27–33 (PCD 0)	PPD 34–40 (PCD 7)
RB+LGG+EcN	12	0.58 (0.09)[a]	0.90 (0.08)[a]	0.75 (0.11)[a]	0.80 (0.08)[a]
LGG+EcN	16	0.54 (0.06)[a]	0.96 (0.05)[a]	0.51 (0.05)[b]	0.42 (0.17)[a]

Note: RB (replacing 10% of total daily caloric intake) was added to the Gn pigs' milk diet (ultra-high-temperature treated cow milk) daily, starting at PPD 5 until the end of the experiment. Weight gain over a specific period was calculated by subtracting the weight (in kilograms) at the beginning of a period from the weight at the end of the period. Mean weekly body weight gain of each treatment group is presented. The number in the parenthesis is the standard error of the mean. Kruskal–Wallis rank-sum test was used for comparisons. Different letters indicate significant differences in weight changes between groups on the same time point (n = 3–16; $p < 0.05$), while shared letters indicate no significant difference. PPD, post-partum day; PCD, post-challenge day.

randomly selected crypts not associated with Peyer's patches by 50. Ten randomly selected ileal villi and crypts, not overlying Peyer's patches were measured for each sample to provide an average villus length and crypt depth. Villus length was measured from the tip of the villus to the junction with the crypt and crypt depth was measured from the junction with the villus to the crypt base. Villus length to crypt depth ratio (V:C) was obtained by dividing the mean villus length by the mean crypt depth for each sample. V:C score was assigned as follows: 0 (normal), > or = 6:1; 1 (mild), 5.0–5.9:1; 2 (moderate), 4.0–4.9:1; 3 (marked), 3–3.9:1; 4 (severe), < 3:1. The mid-villus widths of 10 random ileal villi, not overlying Peyer's patches, were measured and averaged. The number of cells within the ileal lamina propria was given a subjective score ranging from 1+ to 4+.

Out of all seven parameters assessed, RB feeding significantly changed three parameters: mitotic index, villus width, and the abundance of lamina propria cells (Figure 15.8). Mitosis of intestinal crypt cells is increased to replace the damaged intestinal epithelial cells following HRV infection in Gn pigs. Dietary RB maintained the mitotic index in LGG and EcN colonized Gn pigs during HRV infection on PCD 3 and PCD 7, whereas LGG and EcN colonized Gn pigs without dietary RB feeding had significantly increased mitotic index on PCD 3 and PCD 7. Villus width is increased due to the influx of immune cells and edema during rotavirus infection and inflammation. Dietary RB maintained the width of a villus in the ileum of LGG and EcN colonized Gn pigs during HRV infection on PCD 3 and PCD 7, whereas in the non-RB-fed pigs, villus width increased significantly from PCD 0 to PCD 7. Additionally, RB maintained the abundance of lamina propria cells in LGG and EcN colonized Gn pigs during HRV infection, whereas non-RB-fed pigs had significantly reduced lamina propria cells on PCD 3 and PCD 7. Furthermore, RB prevented the decrease in villus length during HRV infection from PCD 0 to PCD 3 and PCD 7, although these differences were not statistically significant.

When we compared the RB+LGG+EcN to the LGG+EcN pigs post HRV challenge at PCD 3 and 7 between the two treatment groups, RB-fed pigs had a significantly lower mitotic index and narrower villus width, but a higher or significantly higher abundance of lamina propria cells and longer villus length, further supporting its protective effects against HRV-induced ileum epithelial damage and inflammation, and its mucosal immune-stimulatory effects. Therefore, RB not only protected against damage to the intestinal epithelium but also maintained intestinal homeostasis (a balance of inflammation and immune response) in the face of HRV challenge.

15.2.6 RB enhanced the innate immune response during HRV infection

To evaluate the effects of RB plus LGG and EcN on innate immunity during HRV infection, we measured the intestinal IFN-γ concentration and total IgA and IgG titers in small and large intestinal contents using ELISA. RB+LGG+EcN group had significantly higher intestinal IFN-γ concentrations compared to the LGG+EcN group on PCD 3 in both SIC and LIC and on PCD 7 in SIC (Figure 15.9A).

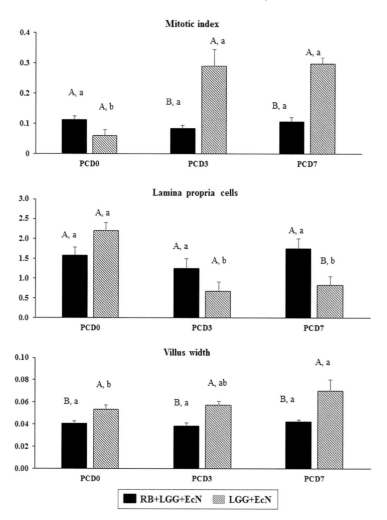

Figure 15.8 RB promoted intestinal epithelial health and maintained intestinal homeostasis during HRV infection in Gn pigs colonized with LGG and EcN. Kruskal–Wallis rank-sum test was used for comparisons between different groups at the same time point (upper case letters) and between different time points for the same treatment group (lower case letters). Different letters indicate significant differences, while shared letters indicate no significant difference (n = 4–6; $p < 0.05$). For data and error bar descriptions, see Figure 15.7 legend.

Compared to the LGG+EcN group, the RB+LGG+EcN group had significantly higher total IgA titers (nine-fold; GMT 2580 versus 23171) in LIC on PCD 3 (Figure 15.9B). No significant difference was observed for total IgA titers between the two groups on PCD 0; however, RB+LGG+EcN group had five-fold higher IgA levels (GMT 2,702 versus 13,004) in LIC than the LGG+EcN group. Together, these data suggest that RB strongly enhanced the innate protective immunity in LGG+EcN colonized pigs during HRV infection.

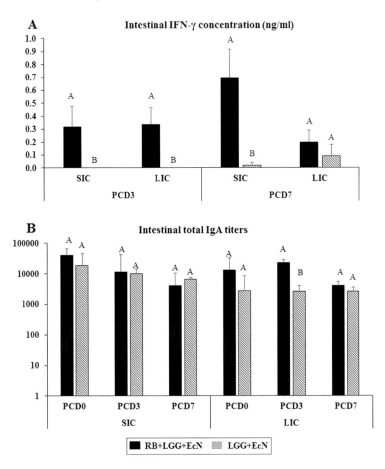

Figure 15.9 RB increased IFN-γ and total IgA levels in intestinal contents during HRV infection in Gn pigs colonized with LGG and EcN. SIC and LIC samples were collected upon euthanasia and stored at –20° C until further analysis. Samples were diluted two-fold before determination of the IFN-γ concentration with a commercial ELISA kit. The average value was calculated from duplicate for each sample and the means for all pigs in the same group at a specific time point were calculated and are presented (15.9A). Kruskal–Wallis rank-sum test was used for comparisons. Total IgA titers in intestinal contents were measured by ELISA and are presented as geometric mean titers for each treatment group (15.9B). ANOVA—Tukey test was used for comparisons. Different capital letters (A, B) indicate a significant difference between the treatment groups for the same time point (n = 4–6; $p < 0.05$), while shared letters indicate no significant difference. For data and error bar descriptions, see Figure 15.7 legend.

15.2.7 Summary

In this study, we used the Gn pig model of HRV infection and diarrhea and demonstrated that dietary RB provided complete protection against HRV diarrhea in LGG and EcN colonized Gn pigs. The results also showed that dietary RB significantly enhanced the growth and colonization of both LGG and EcN in the intestine of Gn pigs, promoted body weight gain, protected against damage to intestinal epithelium while maintaining intestinal homeostasis (a balance of inflammation and immune response), maintained intestine permeability and enhanced the innate IFN-γ and IgA protective immunity during HRV infection. Together, these results demonstrated that the combination of RB and LGG and EcN can provide complete protection against HRV diarrhea and pointed to its potential mechanisms.

This is the first study that has tested the therapeutic effects of combined RB and probiotic LGG and EcN against HRV diarrhea. While RB, LGG, or EcN individually can confer varying degrees of protection against HRV diarrhea, this study showed that combining RB and initial colonization of the two diarrhea-reducing probiotics can achieve complete protection. We have previously shown that RB together with the oral AttHRV vaccine provided complete protection against HRV diarrhea in Gn pigs (Yang et al., 2014). The complete protection against rotavirus diarrhea is significant as no rotavirus vaccines or antiviral drugs have shown such complete effective-ness, regardless of their specific mechanisms for reducing HRV diarrhea. RB's effect on reduc-ing diarrhea is not closely related to reducing HRV shedding. RB strongly enhanced the innate immunity as indicated by the increased intestinal IFN-γ and total IgA responses. Thus, this non-specific therapeutic effect against diarrhea suggests that the combination of RB and LGG+EcN could potentially be used to provide broad-spectrum protection against diarrhea caused by other enteric pathogens and diarrhea of unknown etiology. Despite potential differences in the pathogenesis and molecular mechanisms of a certain type of diarrhea between humans and pigs, it is expected that RB will enhance innate immunity and provide similar protection against diarrhea in humans. The optimal timing and dosage of RB and LGG and EcN may need to be further determined for each pathogen or type of diarrhea. Additionally, specific molecular mechanisms underlying this protection require further studies. Further studies on metabolomes altered by RB and probiotics were conducted to address some of these questions (Nealon et al., 2017; Zambrana et al., 2019; Zambrana et al., 2021a).

Despite their adjuvant effects on rotavirus vaccine and protection against rotavirus diarrhea (Liu et al., 2013; Yang et al., 2014), neither RB alone nor LGG+EcN alone (Table 15.2) reduced HRV shedding in Gn pigs. LGG at both high dose and low dose did not reduce HRV shedding in our previous study (Wen et al., 2015). Thus, RB and LGG+EcN seem to have similar effects on HRV pathogenesis, reducing diarrhea without impacting virus replication and shedding. Interestingly, RB significantly reduced HRV shedding in the LGG and EcN colonized pigs, as shown by sig-nificantly delayed onset of shedding, mean duration days, and lower peak shedding titers. The mechanism for this phenomenon is unclear, but may possibly be due to the synergistic effects of RB with LGG and EcN. LGG is normally found in the gut microbiota of human infants and young children (Zhang et al., 2014). Therefore, RB, when used in humans who have already been colonized with LGG, is likely to be more effective in reducing HRV shedding than in germ-free pigs. LGG was found to significantly reduce rotavirus shedding in a conventional mouse model (Zhang et al., 2013c). These results indicate that RB is promising to provide more effective pro-tection against HRV diarrhea and a significant reduction in rotavirus shedding in young children if their gut is colonized with LGG+EcN. This approach will be more effective, affordable, and safer than probiotic therapy (reducing the potential risk of septicemia in immunocompromised children), especially for children in developing countries. Further preclinical studies in Gn pigs transplanted with human gut microbiota (Wen et al., 2014a) and human clinical trials of this novel therapeutic combination against HRV diarrhea and shedding are warranted.

In this study, dietary RB intake in Gn pigs increased the growth and colonization of both pro-biotic strains LGG and EcN up to 5 logs. This result supports previous findings that RB feeding can increase the abundance of the beneficial gut bacteria *Lactobacillus* spp. in mice (Henderson et al., 2012) and *Bifidobacterium* in adults (Sheflin et al., 2015) and reduce the colonization and invasion of pathogenic *Salmonella enterica* in mice (Henderson et al., 2012). As LGG is a gram-positive bacterium in the Firmicutes phylum and EcN is a gram-negative bacterium in the Proteobacteria phylum, these results suggest that RB can promote the growth of a variety of probiotic strains. This is not surprising given the complex composition of RB (Friedman, 2013; Ryan, 2011). However, the growth rate and abundance achieved by different probiotic strains with RB may be different. It is important to consider these differences when using RB and probiotics clinically.

Rice bran enhanced the growth of Gn pigs, indicating increased nutrient absorption via the gut and overall health. It also promoted gut health by preventing epithelial damage (intestinal crypt cell mitosis) while maintaining the homeostasis of the mucosal immune system (maintained the number of lamina propria cells and villus width) during HRV infection in Gn pigs. During inflammation induced by HRV infection, intestinal permeability is increased, resulting in edema and diarrhea. Preventing the changes in gut permeability by RB may have contributed to its remarkable effects in preventing HRV-induced diarrhea. Meanwhile, RB maintained the number of lamina propria cells, which are mainly lymphocytes (CD2+ and CD4+ T lymphocytes and sIgA secreting plasma cells) (Rothkotter et al., 1994), suggesting its ability to stimulate the intestinal mucosal immune system during HRV infection in LGG and EcN colonized pigs. Previous studies in mice (Henderson et al., 2012) and Gn pigs (Yang et al., 2014) indicated that RB alone or together with probiotic bacteria such as *Lactobacillus spp* can increase the production of mucosal and systemic total IgA by plasma cells. Consistent with these findings, our result also showed RB significantly increased the intestinal total IgA titers during HRV infection. Thus, the increased lamina propria cells in the RB group pigs may be due to its immune-stimulatory effects on the mucosal immune system. Together, the anti-inflammatory and immune-modulatory effects of RB and LGG and EcN promoted intestinal epithelial health and homeostasis, contributing to an intact intestinal barrier that is resistant to HRV diarrhea.

It is not known which components or specific compounds of RB contributed to the HRV diarrhea-reducing activities. However, heat-resistant amylase, protease, and hemicellulose-treated rice fiber, which has significantly lower contents of protein, lipids, and carbohydrates have been shown to be able to prevent diarrhea in dextran sodium sulfate (DSS)-induced experimental colitis mouse models (Komiyama et al., 2011). This result suggests that the dietary fiber portion of RB, such as cellulose, hemicellulose, and lignin, may also play important roles in decreasing diarrhea during inflammatory bowel disease. In fact, arabinoxylan, a dietary fiber from RB, significantly decreased the diarrhea score in irritable bowel syndrome adult patients through its anti-inflammatory and immune-modulating activities (Kamiya et al., 2014). RB components promoting probiotic bacteria growth and colonization are likely to vary depending on the specific bacterial species. However, heat-resistant amylase, protease, and hemicellulase-treated dietary fiber were unable to increase the shedding of *Lactobacillus* spp and *Bifidobacterium* (Komiyama et al., 2011), suggesting that carbohydrate or lipid components of RB could be the main prebiotics for LGG and EcN in this study. A recent study in mice found that a 10% RB oil diet significantly increased the occupation ratios of *Lactobacillales* group of bacteria in the gut microbiota (Tamura et al., 2012). Further studies are underway to identify the RB components that are responsible for its HRV diarrhea fighting properties and prebiotic properties.

Both RB and probiotics are natural products and have been demonstrated to have various health benefits for disease prevention and treatment in humans and animals. LGG is safe in all age and health groups, even in immune-compromised individuals (Snydman, 2008). Given its natural colonization in the gastrointestinal systems, LGG has been studied extensively for its activities in treating gastrointestinal diseases and infections, such as diarrhea and enteric pathogens. However, this is the first study that showed the combined effects of RB and LGG+EcN in treating enteric pathogen infection and diseases. The results here indicated the synergistic effects of RB and LGG+EcN in preventing HRV diarrhea. Further preclinical studies in Gn pigs transplanted with human gut microbiota (Wen et al., 2014a) and human clinical studies are necessary to determine the optimal dosage and formulation for maximal safety and efficacy. In conclusion, the combination of RB and LGG+EcN may represent a novel, safe, and highly effective therapeutic against diarrhea and infection caused by HRV, and potentially diarrhea caused by other enteric pathogens and etiologies in young children.

15.3 Combination of probiotics and rice bran is highly effective in preventing HuNoV diarrhea in Gn pigs

Aimed to develop an effective and ready-to-use anti-HuNoV therapeutic strategy, we first screened a group of probiotics to identify the virus-binding bacteria using HuNoV P particles and native virions. Subsequently, probiotics and RB were evaluated individually or combined as cocktail regimens for their effects on HuNoV infection and diseases in the well-established Gn pig model (Bui et al., 2013). The mechanisms of antiviral and diarrhea-reducing activities from those treatments were also explored (Lei et al., 2016a). In this section, we only describe the *in vivo* study of the combination of RB and probiotics in reducing HuNoV infection and diarrhea in Gn pigs.

15.3.1 Gnotobiotic pig treatment groups

Neonatal Gn pigs (male and female) were randomly assigned to the five treatment groups upon derivation: cocktail−7d ($n = 5$), cocktail+1d ($n = 5$), RB−7d ($n = 4$), LGG+EcN ($n = 5$), and control ($n = 9$). To initiate the colonization of LGG and EcN, 10^4 CFU of each were mixed in 5 ml of minimal essential medium and administered orally to pigs on PPD 3, 5, and 7. The low dosage was chosen on purpose to be well below the therapeutic practice (10^9 to 10^{12} CFU). LGG and EcN fecal shedding was determined by rectal swab sampling of pig feces and enumeration of colonies grown on media agar plates as described previously (Yang et al., 2015). For RB feeding of pigs, heat-stabilized and gamma-irradiated RB (Calrose variety) was added to pigs' milk diet by replacing 10% daily calorie intake (Yang et al., 2015). Daily feeding started seven days prior to or one day after HuNoV inoculation until euthanasia. All pigs were orally inoculated on PPD33 with 6.43×10^5 viral genome copies of HuNoV GII.4/2006b variant 092895. To reduce gastric acidity, 4 ml 200 mM sodium bicarbonate was given to pigs 15 min prior to inoculation. Fecal consistency and virus shedding were assessed daily until euthanasia on PPD40 when blood, tissues, and intestinal contents were collected. Fecal consistency scores were obtained based on the previous scaling system (Bui et al., 2013), and fecal virus shedding was measured by a one-step TaqMan qRT-PCR as described previously (Lei et al., 2016b). Jejunum tissues collected after pig euthanasia were fixed in 4% paraformaldehyde for 12–16 h, paraffin-embedded, sectioned into 5 μm slices, and placed on positively charged slides for routine H&E staining. A pathologist who was blinded to the sample identifications evaluated the villus length using an ocular micrometer under a light microscope.

15.3.2 LGG+EcN inhibited HuNoV shedding and RB reduced diarrhea in Gn pigs

To develop a ready-to-use anti-HuNoV therapeutic strategy, LGG and EcN were chosen since they could bind HuNoVs *in vitro* and are commercially available as diarrhea-reducing probiotics. The previous study showed that RB protected against HRV-induced diarrhea in the presence of LGG and EcN in Gn pigs (Yang et al., 2015), we tested RB feeding and/or LGG+EcN co-colonization in five treatment groups in this study: cocktail–7d ($n = 5$), pigs were pre-colonized with LGG and EcN, RB feeding started seven days prior to HuNoV inoculation; cocktail+1d ($n = 5$), pigs were pre-colonized with LGG and EcN, RB feeding started one day after virus inoculation; RB–7d ($n = 4$), RB feeding started seven days prior to inoculation; LGG+EcN ($n = 5$), pigs were colonized with LGG and EcN only; control ($n = 9$), non-RB-fed and non-LGG+EcN-colonized. All pigs were inoculated with a HuNoV GII.4/2006b variant 092895 on PPD 33/PID 0 and euthanized on PID 7 (Figure 15.10A).

Fecal consistency and virus shedding were assessed daily after the HuNoV inoculation. The results summarized in Table 15.4 showed that compared to the control group, LGG+EcN group had similar rates of HuNoV diarrhea (89% versus 60%), yet undetectable HuNoV shedding. RB–7d group had a slightly shorter mean duration of diarrhea (2.2 versus 1.3 days) and significantly delayed shedding onset (2.8 versus 6.3 days). More importantly, cocktail–7d and cocktail+1d groups had a significantly lower incidence (20%), delayed onset (3.9 versus 7.0 and 7.2 days, respectively), shorter mean duration of diarrhea (2.2 versus 0.2 days), and shorter mean duration of virus shedding (3.2 versus 1.0 days). In both cocktail groups, the diarrhea reduction rates were 78% (Table 15.4), and the reduced severity of diarrhea was also shown by the significantly lower cumulative fecal scores (Figure 15.10B). Interestingly, only the LGG+EcN group had significantly reduced cumulative and peak virus shedding compared to the control group. RB feeding with or without LGG+EcN colonization did not significantly alter virus shedding pattern, except that shedding in the cocktail+1d group trended lower when compared to the other RB-fed groups and the controls (Figures 15.10C and D).

15.3.3 LGG+EcN and RB stimulated the production of total IFN-γ^+ T cells

To elucidate the mechanisms of the inhibitory effects of LGG+EcN and RB on HuNoV infection and diarrhea, their immunomodulatory roles were first assessed regarding effector T cells. After euthanasia on PID7, MNCs were isolated from both intestinal and systemic lymphoid tissues, and the frequencies of IFN-γ^+ CD4$^+$ and CD8$^+$ T cells were determined by flow cytometry (Figure 15.11A). MNCs were stimulated with P particles to detect HuNoV-specific IFN-γ^+ T cells, which was the increased frequency compared to the mock stimulated sample. For pigs in the control, LGG+EcN, and RB–7d groups, no significant increase of IFN-γ^+ T cells was observed in P particle stimulated MNCs, suggesting low or short-term HuNoV-specific IFN-γ^+ T cell responses. However, compared with control pigs, both LGG+EcN colonization and RB feeding significantly increased frequencies of non-specific total IFN-γ^+ T cells (Figure 15.11B). In addition, compared

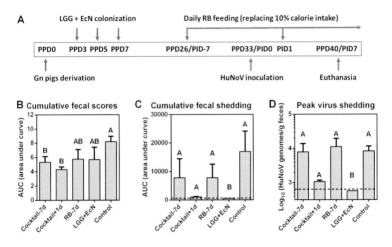

Figure 15.10 Design and summary of Gn pig study. (A) Experimental timeline. Cumulative fecal scores (B) and shedding (C) are shown as areas under the curve from daily measurements of individual pigs. (D) Mean peak virus shedding titers from PID 1 to PID 7 in individual pigs. Sample sizes are shown in Table 15.4. Data are presented as mean ± SEM. Dashed lines indicate the limit of detection. Statistics were determined by Kruskal–Wallis test. Different letters indicate significant differences among groups ($p < 0.05$), while shared letters indicate no significant difference.

with the LGG+EcN group, the RB-7 group had significantly higher frequencies of IFN-γ^+ CD8$^+$ T cell population in the ileum and IFN-γ^+ CD4$^+$ T cell population in all assayed lymphoid tissues (duodenum, ileum, spleen, and blood) (Figure 15.11B), indicating that RB has strong stimulatory effects on total IFN-γ^+ T cell responses, which may contribute to the reduction of HuNoV diarrhea in Gn pigs.

15.3.4 Probiotics plus RB cocktail regimens enhanced gut immunity

The immunomodulatory roles of LGG+EcN and RB on gut immunity were evaluated by testing total intestinal IgA, IgG, and IFN-γ levels since PID 7 is too early to detect virus-specific IgA and IgG antibody responses. Compared with the control group, the cocktail–7d, cocktail+1d, and LGG+EcN groups had significantly higher IgA titers in both small and large intestinal contents (SIC and LIC), but the increase was not observed in the RB–7d group (Figure 15.12A), indicating that LGG+EcN but not RB enhanced the production of IgA. The cocktail–7d and cocktail+1d groups had significantly higher IgG titers in both SIC and LIC, whereas no differences were observed in either the LGG+EcN or RB groups (Figure 15.12B). Consistent with the strong stimulation of RB on total IFN-γ^+ T cells (Figure 15.11B), significantly higher IFN-γ concentrations were detected in LIC from the cocktail–7d, cocktail+1d, and RB–7d groups (Figure 15.12C). In all, cocktail regimens remarkably enhanced gut immunity in Gn pigs by secretion of intestinal immunoglobulins and interferon, which might protect against HuNoV infection.

15.3.5 Probiotics plus RB cocktail regimens increased jejunal villi length

Villus blunting is a major manifestation of impaired intestinal health, such as in Crohn's disease (Cadwell et al., 2010), celiac disease (Chand and Mihas, 2006), and virus-induced gastroenteritis (Hodges and Gill, 2010). To examine the beneficial effects of LGG+EcN and RB on the health of the small intestine in Gn pigs, sections of the jejunum were stained with H&E and evaluated for all the treatment groups after euthanasia. Compared with control, both LGG+EcN colonization and RB feeding were associated with significantly longer jejunal villus length. Their stimulatory roles might be additive as the two cocktail groups displayed greater villus length than either single treatment (Figure 15.13). These data indicate that the cocktail regimens promote the

Table 15.4 Incidence of Diarrhea and Fecal Virus Shedding in Gn Pigs after HuNoV GII.4 Challenge[a]

| Group | n | Diarrhea[b] | | | | Virus shedding[b] | | |
		Pigs with diarrhea (%)*	Mean days to onset (SEM)**	Mean duration days (SEM)**	Rate of Reduction[c]	Pigs shedding virus (%)*	Mean days to onset (SEM)**	Mean duration days (SEM)**
Cocktail–7d	5	1 (20%)[a]	7.0 (1.0)[a]	0.2 (0.2)[a]	78%	4 (80%)[a]	4.0 (1.2)[ab]	1.0 (0.3)[ab]
Cocktail+1d	5	1 (20%)[a]	7.2 (0.8)[a]	0.2 (0.2)[a]	78%	5 (100%)[a]	4.8 (0.2)[ab]	1.0 (0)[c]
RB–7d	4	3 (75%)[ab]	3.3 (1.6)[ab]	1.3 (0.5)[ab]	16%	3 (75%)[a]	6.3 (0.8)[a]	1.5 (0.6)[ab]
LGG+EcN	5	3 (60%)[ab]	4.0 (1.6)[ab]	1.8 (1.1)[ab]	33%	0[b]	8.0 (0)[c]	0[c]
Control	9	8 (89%)[b]	3.9 (0.7)[b]	2.2 (0.4)[b]	n/a	8 (89%)[b]	2.8 (0.8)[b]	3.2 (0.9)[b]

Note:

[a] Gn pigs were inoculated with a HuNoV GII.4 2006b variant 092895 at 33 days of age. Diarrhea and virus shedding was monitored by daily rectal swab sampling and RT-qPCR after inoculation. Calculations included all pigs in each group from PID 1 to PID 7.

[b] If diarrhea or virus shedding was not observed, the days to onset were recorded as 8 and the duration days were recorded as 0 for statistical purposes.

[c] Calculated as (1 – [% of treated pigs with diarrhea / % of control pigs with diarrhea] × 100%. N/A, not applicable.

* Fisher's exact test or **Kruskal–Wallis test was used for comparisons. Different letters indicate significant differences among treatment groups ($p < 0.05$), while shared letters indicate no significant difference.

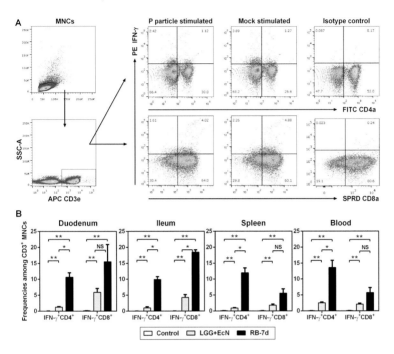

Figure 15.11   LGG+EcN and RB stimulated IFN-γ+ T cell responses. (A) Gating strategies for IFN-γ+ CD3+CD4+ (Th) cells and CD3+CD8+ (CTL) cells. Representative dot plots showing frequencies of HuNoV-specific (P particle stimulated) and non-specific (mock stimulated) IFN-γ+ T cells in ileum isolated from LGG+EcN colonized Gn pigs. SSC-A, side scatter area; APC, allophycocyanin; FITC, fluorescein isothiocyanate; SPRD, spectral red; PE, phycoerythrin. (B) Non-specific IFN-γ+ T cells in intestinal (duodenum, ileum) and systemic (spleen, blood) tissues on PID 7. Sample sizes are shown in Table 15.1. Data are presented as means ± SEM. Statistics were determined by the Kruskal–Wallis test. NS, not significant, *p < 0.05, **p < 0.01.

growth and health of intestinal epithelium, which might contribute to the protection of HuNoV-induced disease.

15.3.6 Summary

In this study, after HuNoV inoculation in Gn pigs colonized with LGG+EcN, virus fecal shedding was below the limit of detection, indicating significant inhibition on HuNoV infection by their colonization. Similar to the reduced virus shedding but the unaffected incidence of diarrhea observed in *Enterobacter cloacae*–colonized Gn pigs in our previous study (Lei et al., 2016c), LGG+EcN colonization did not alter the occurrence of diarrhea, suggesting that HuNoV gastroenteritis could be induced by extremely low viral loads and that anti-HuNoV agents inhibiting viral replication may have insufficient efficacy in reducing the disease. Given that bacterial anti-HuNoV capacity might depend on the extent of viral retention ability, it is likely that LGG plays a more important role in the inhibition of HuNoV infectivity than EcN since LGG has a greater HuNoV-binding ability (Lei et al., 2016a), but further investigations will be required to clarify the effects of LGG or EcN mono-colonization on HuNoV infection. Nevertheless, cocktail regimens containing LGG and EcN offer great promise to simultaneously protect against HuNoV and HRV infection (Kandasamy et al., 2016).

Although RB was shown to promote the colonization of *Lactobacilli* in mice (Henderson et al., 2012), LGG fecal shedding was lower after RB feeding in cocktail groups in this study. When colonized together with EcN in Gn pigs, LGG fecal shedding and concentration in intestinal tissues trended toward lower than those of single colonization (Kandasamy et al., 2016), suggesting that the presence of EcN might inhibit the growth of LGG. Thus, it is likely that higher

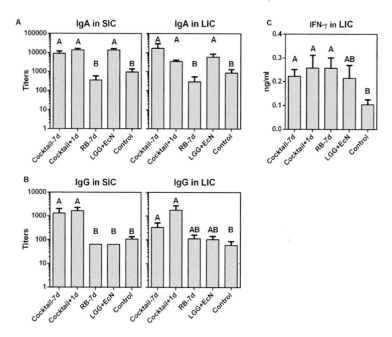

Figure 15.12 IgA, IgG, and IFN-γ levels in intestinal contents after HuNoV infection. Total IgA (A) and IgG (B) titers in small and large intestinal contents (SIC and LIC) were measured by ELISA. (C) IFN-γ concentration in LIC was measured by ELISA. Data are presented as means ± SEM. Statistics were determined by the Kruskal–Wallis test. Different letters indicate significant differences among groups ($p < 0.05$), while shared letters indicate no significant difference.

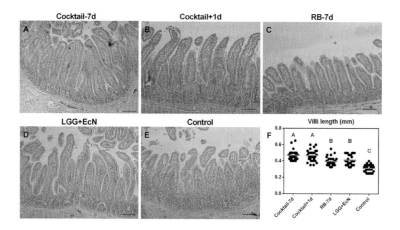

Figure 15.13 LGG+EcN and RB are associated with longer villi. (A–E) Representative images of H&E stained jejunum showing the villi length in the five groups. Scale bar, 0.25 mm. (F) 30 random villi including all pigs in each group were measured to quantify the villi length. Data are presented as means with individual points. Statistics were determined by the Kruskal–Wallis test. Different letters indicate significant differences among groups ($p < 0.05$), while shared letters indicate no significant difference.

growth of EcN led to lower growth of LGG after RB feeding, and underlying mechanisms utilized by EcN need to be identified, such as competing for the nutrients and colonization sites, improving intestinal barrier, and modulating immune responses (Hering et al., 2014; Splichalova et al., 2011). In all, higher protective efficacy against HuNoV shedding and diarrhea might be achieved if only RB and LGG are given.

Effector T cells are a crucial immune component to eliminate virus-infected cells, and their responses in the small intestine are associated with protective immunity against HRV (Yuan et al., 2008). However, HuNoV infection or P particle vaccination did not significantly stimulate virus-specific IFN-γ+ CD4+ or CD8+ T cell responses (Kocher et al., 2014), and neither did LGG+EcN colonization nor RB feeding in this study. Still, significantly increased frequencies of non-specific IFN-γ+ T cells were observed especially after RB feeding, which might be correlated with the diarrhea-reducing property of RB, but a significant reduction in virus shedding was not observed along with enhanced T cell responses in the RB–7d group, which was similar to our previous study on HRV (Yang et al., 2014). For the cocktail regimens, the intestinal IgA was increased by LGG+EcN alone, while the increased intestinal IgG might be induced by the synergism between LGG+EcN and RB. The additive effects of the probiotics and RB appeared to be associated with longer jejunal villus length.

In this study, the cocktail–7d group displayed a 78% reduction of diarrhea, as well as significantly shortened duration of diarrhea and virus shedding after HuNoV challenge, indicating the regimen is an effective preventive measure. In addition, similar effects in reducing diarrhea and virus shedding were observed in the cocktail+1d group, in which RB feeding started one day after HuNoV challenge, thus this regimen could be considered as a therapeutic strategy to treat HuNoV gastroenteritis. The first HuNoV vaccine candidate evaluated in clinical trials was an adjuvanted monovalent GI.1 VLP, which provided 47% and 26% protection against Norwalk virus gastroenteritis and infection compared with the placebo group, respectively (Atmar et al., 2011). A bivalent VLP-based vaccine containing both GI.1 and GII.4 components is under development as well, and human clinical trials showed a 52% reduction in vomiting and/or diarrhea compared with the control after challenge (Debbink et al., 2014). Our previous evaluations of adjuvanted GII.4 VLP and P particle vaccines in Gn pigs demonstrated reductions of diarrhea by 60% and 47%, respectively (Kocher et al., 2014). Therefore, the probiotics plus RB cocktail regimens may provide an alternative strategy with better anti-HuNoV effects than the current vaccine candidates.

In summary, colonization with LGG+EcN completely inhibited HuNoV fecal shedding in Gn pigs. The two cocktail regimens had RB feeding started either seven days prior to or one day after viral inoculation in the LGG+EcN colonized Gn pigs, and both regimens exhibited dramatic anti-HuNoV effects, including reduced incidence and shorter duration of diarrhea, as well as shorter duration of virus fecal shedding. The anti-HuNoV effects of the cocktail regimens were associated with the stimulated IFN-γ+ T cell responses, increased production of intestinal IgA and IgG, and longer villus length. Considering the natural source and commercial availability of probiotics and RB, the cocktail regimens may represent a novel, safe, and ready-to-use strategy against diarrhea and infection caused by HuNoV infection and other enteric pathogens.

15.4 Impacts of studies of RB in Gn pigs

Our article reporting the high protective efficacy of probiotics and RB against HuNoV infection and diarrhea in Gn pigs (Lei et al., 2016a) is part of the research topic "New Frontiers in the Search of Antimicrobials from Natural Products" in *Frontiers Microbiology* (https://doi.org/10.3389/fmicb.2016.01699). This article was viewed more than 10,444 times from its online publication in November 2016 to December 2021. Diarrhea-reducing effects of RB are also summarized in a book chapter written by Shaohua Lei and me, "Rice Bran Usage in Diarrhea" in *Dietary Interventions in Gastrointestinal Disease: Food, Nutrients and Dietary Supplements* published by Academic Press, Elsevier Sciences (Lei and Yuan, 2019). Our studies of RB in Gn pigs strongly indicate that RB plus diarrhea-reducing probiotic colonization represent an effective prophylactic and/or therapeutic approach against HRV, HuNoV, and potentially a variety of other diarrhea-inducing enteric viral and bacterial pathogens.

Based on these encouraging findings, Dr. Elizabeth Ryan, the PI of the Gates Foundation Phase II project, decided that RB's effect should be evaluated directly in human clinical trials, instead of performing another study in a mouse model of *Salmonella* infection as originally proposed. Dr. Ryan organized the collaborations with Dr. Sylvia Becker-Dreps at the University of North Carolina at Chapel Hill, Dr. Samuel Vilchez at the National Autonomous University of Nicaragua, and Dr. Ousmane Koita at the University of Sciences, Techniques, and Technologies of Bamako, Mali for the clinical trials in Nicaragua and Mali. Two pilot feasibility trials were conducted in a total of 95 infants from 6 to 12 months old (NCT02615886 and NCT0255737315). Daily consumption of RB was safe and feasible (Zambrana et al., 2021b), and improved infants' health as evidenced by increased length for age z-scores in Nicaragua and Mali infants when compared to control (Zambrana et al., 2019). Diarrheal disease incidence was 33% in the RB intervention group compared to 79% in the control group over the six-month study period in Mali infants. Because Nicaragua infants had a low rate of diarrhea in both the control and RB treated group, no significant effect on diarrhea reduction was observed there. This was the first randomized controlled trial of RB supplementation in LMIC infants and provides a compelling rationale for continued follow-up larger-scale investigation of RB supplementation for reducing the risk of malnutrition, as well as for protect against enteric pathogens and diarrheal diseases in infants and young children (Zambrana et al., 2019).

15.5 Closing remarks

This book on vaccine evaluation in Gn pig models ends with a chapter that is not all about vaccines, but rather one about probiotics and RB. The decision for doing so has some sentimental reasons for me. Although my Ph.D. dissertation (Yuan, 2000) was on vaccine evaluation, my first NIH R21 grant in 2005 and first NIH R01 grant in 2009 as a PI were for studying probiotics' effects on rotavirus immunity. Thus, in this sense, studying probiotics established me as an independent researcher. The article "High Protective Efficacy of Probiotics and Rice Bran against Human Norovirus Infection and Diarrhea in Gnotobiotic Pigs" (Lei et al., 2016a) may be the last paper on studies of probiotics from Yuan Lab. It is a milestone and the end of the pursuit of a research direction. The authors' list of this paper also brings a sentiment of gratitude to me because it includes all the Ph.D. students who graduated from my lab at Virginia Tech so far, Shaohua Lei (2013–2018), Ashwin Ramesh (2014–2019), Erica Twitchell (2014–2018), Ke Wen (2007–2011), Tammy Bui (2008–2017, Ph.D./DVM dual degree), Xingdong Yang (2009–2015), and Jacob Kocher (2009–2014). Mariah Weiss was the first laboratory and research specialist and lab manager I was able to afford to hire for 3 and half years (2012-2016) with my research grants. She was an exceptional lab manager and a talented technician. From 2016-2018, she was the project and lab manager at the Teaching and Research Animal Care Support Service (TRACSS), Virginia-Maryland College of Veterinary Medicine and provided excellent leadership and technical support for our Gn pig program. With Peter Jobst, the former manager of TRACSS, we three co-authored the book chapter "Gnotobiotic pigs: from establishing facility to modeling human infectious diseases" in Gnotobiotics (Trenton Schoeb, Kathryn Eaton edit, Academic Press) (Yuan et al., 2017). I missed her very much after she moved to North Carolina following her husband's new job posting in 2018. Also, on this list is the first research scientist in my lab, Guohua Li who is a rotavirus molecular virologist trained by my former supervisor Dr. Yuan Qian in China, at the Capital Institute of Pediatrics, and the first Fulbright post-doctoral scholar in my lab, Ernawati Giri-Rachman from Indonesia. Erna and I continued collaboration on the development of a dual hepatitis B virus and norovirus mucosal vaccine after she finished her Fulbright scholarship and went back to her home university Institut Teknologi Bandung in 2013. Another collaborator, from Vietnam, Nguyen Van Trang is also on the list, who was the first Ph.D. student I co-mentored with Dr. Linda Saif when I was an adjunct assistant professor at The Ohio State University from 2002–2007. Dr. Xi (Jason) Jiang and Dr. Elizabeth Ryan are my collaborators and I want them to know how deeply grateful I am to them for their enormous generosity and kindness in trusting me to be their research partner. They first reached out to me to initiate collaborations and these collaborations are ongoing and highly productive. With Dr. Jiang, we have co-authored eight peer-reviewed journal articles from 2013–2021; with Dr. Ryan, we have co-authored six peer-reviewed journal articles from 2014–2021.

Now the focus of Yuan Lab's research has switched back to vaccines. I have never stopped vaccine studies in the past 28 years since I came to the United States as a visiting scholar in Saif Lab in 1993. My persistence on this single topic led to the perfection of Gn pig models and their recognition by vaccine developers and researchers as the best animal models for human rotavirus and norovirus preclinical studies. They are used as the gatekeeper of most, if not all, candidate rotavirus and norovirus vaccines, and prophylactics and therapeutics before entering human clinical trials. Not mentioned in this book, the vaccines we evaluated in Gn pigs also include a thermostable oral film live attenuated tetravalent rhesus-human reassortant rotavirus vaccine from Universal Stabilization Technologies, San Diego, CA, (Hensley et al., 2021), Lanzhou trivalent human-lamb reassortant rotavirus vaccine (LLR3), and Anhui Zhifei quadrivalent VLP norovirus vaccine from China. The LLR3 vaccine was highly immunogenic in Gn pigs and it has been evaluated in Chinese infants in a randomized, double-blind, placebo-controlled trial and was shown to confer up to 70.3% protection against severe rotavirus gastroenteritis (RVGE) and 56.6% against RVGE of any severity (Xia et al., 2020). The quadrivalent VLP vaccine contains recombinant HuNoV GI.1, GII.3, GII.4, and GII.17 VP1 proteins and is under a human clinical trial phase I/IIa in China (NCT04563533). Manuscripts on the studies of these two vaccines in Gn pigs are under preparation.

Currently, mRNA-based trivalent (P2-P[8], P[6], P[4]) rotavirus vaccines that are generated by CureVac and S60-VP8* nanoparticle-based quadrivalent (P[8], P[6], P[4], P[11]) rotavirus vaccines that are developed by Dr. Ming Tan at Cincinnati Children's Hospital Medical Center are being evaluated in the Gn pig model of Wa HRV infection and diarrhea in Yuan Lab. We are also developing a new Gn pig challenge model for the G4P[6] Arg HRV strain which is zoonotic between pigs and humans (Degiuseppe et al., 2013). Arg HRV infected Gn pigs revealed diarrhea with similar duration and severity as those caused by the virulent Wa HRV. The Arg HRV strain has been passaged in neonate Gn pigs multiple times and a large inoculum pool has been generated from the intestinal contents of the infected neonatal Gn pigs. A dose-response study will determine the ID_{50} and DD_{50} of this Arg HRV strain in Gn pigs at four to five weeks of age. Hence, two Gn pig models are ready in our laboratory to assess the protective efficacy of rotavirus vaccines against infection and diarrhea of both P[8] and P[6] HRVs.

And last, but certainly not least, a Gn pig model of *Clostridium difficile* infection and disease has been established in Yuan Lab. Following oral inoculation with 1×10^5 spores from *C. diff* UK1 strain, Gn pigs exhibited typical clinical signs (moderate to severe diarrhea) and histo-pathological changes in the large intestine at 5–10 days post-infection as described previously (Steele et al., 2010). This pig model together with the HuNoV Gn pig challenge model will be used for the evaluation of engineered antibodies against HuNoV and *C. diff* expressed in probiotic yeast, *Saccharomyces boulardii* generated by Dr. Hanping Feng at the University of Maryland, Baltimore. This project is funded by an R01 from NIAID, NIH with Dr. Feng and me as PIs (2020–2024).

The usefulness of Gn pig models and their roles in the development of vaccines and therapeutics against diarrheal diseases have been well demonstrated in this book. However, to date, only six Gn pig facilities are operating in North America for research of infectious diseases. There are many challenges in the use of Gn pig models that we have discussed in the book chapter "Gnotobiotic Pigs: From Establishing Facility to Modeling Human Infectious Diseases" (Yuan et al., 2017). Among the challenges, the biggest one is that the establishment of a Gn pig facility and program requires strong institutional commitment and financial support. I was extremely fortunate to have been provided such support from our college and university from the time I started my lab at Virginia Tech in 2007 till now.

References

Abe, F., Ishibashi, N., Shimamura, S., 1995. Effect of administration of bifidobacteria and lactic acid bacteria to newborn calves and piglets. *J Dairy Sci* 78(12), 2838–2846.

Agarwal, S., Hickey, J.M., McAdams, D., White, J.A., Sitrin, R., Khandke, L., Cryz, S., Joshi, S.B., Volkin, D.B., 2020a. Effect of aluminum adjuvant and preservatives on structural integrity and physicochemical stability profiles of three recombinant subunit rotavirus vaccine antigens. *J Pharm Sci* 109(1), 476–487.

Agarwal, S., Hickey, J.M., Sahni, N., Toth, R.T.t., Robertson, G.A., Sitrin, R., Cryz, S., Joshi, S.B., Volkin, D.B., 2020b. Recombinant subunit rotavirus trivalent vaccine candidate: Physicochemical comparisons and stability evaluations of three protein antigens. *J Pharm Sci* 109(1), 380–393.

Agarwal, S., Sahni, N., Hickey, J.M., Robertson, G.A., Sitrin, R., Cryz, S., Joshi, S.B., Volkin, D.B., 2020c. Characterizing and minimizing aggregation and particle formation of three recombinant fusion-protein bulk antigens for use in a candidate trivalent rotavirus vaccine. *J Pharm Sci* 109(1), 394–406.

Ahlroos, T., Tynkkynen, S., 2009. Quantitative strain-specific detection of *Lactobacillus rhamnosus* GG in human faecal samples by real-time PCR. *J Appl Microbiol* 106(2), 506–514.

Akira, S., Uematsu, S., Takeuchi, O., 2006. Pathogen recognition and innate immunity. *Cell* 124(4), 783–801.

Ali, A., Iqbal, N.T., Sadiq, K., 2016. Environmental enteropathy. *Curr Opin Gastroenterol* 32(1), 12–17.

Aliabadi, N., Lopman, B.A., Parashar, U.D., Hall, A.J., 2015. Progress toward Norovirus vaccines: Considerations for further development and implementation in potential target populations. *Expert Rev Vaccines*, 14(9), 1241–1253.

Anderson, M.J., 2008. A new method for non-parametric multivariate analysis of variance. *Austral Ecol* 26(1), 32–46.

Ang, L., Arboleya, S., Lihua, G., Chuihui, Y., Nan, Q., Suarez, M., Solis, G., de los Reyes-Gavilan, C.G., Gueimonde, M., 2014. The establishment of the infant intestinal microbiome is not affected by rotavirus vaccination. *Sci Rep* 4, 7417.

Armah, G.E., Sow, S.O., Breiman, R.F., Dallas, M.J., Tapia, M.D., Feikin, D.R., Binka, F.N., Steele, A.D., Laserson, K.F., Ansah, N.A., Levine, M.M., Lewis, K., Coia, M.L., Attah-Poku, M., Ojwando, J., Rivers, S.B., Victor, J.C., Nyambane, G., Hodgson, A., Schodel, F., Ciarlet, M., Neuzil, K.M., 2010. Efficacy of pentavalent rotavirus vaccine against severe rotavirus gastroenteritis in infants in developing countries in sub-Saharan Africa: A randomised, double-blind, placebo-controlled trial. *Lancet* 376(9741), 606–614.

Atarashi, K., Tanoue, T., Shima, T., Imaoka, A., Kuwahara, T., Momose, Y., Cheng, G., Yamasaki, S., Saito, T., Ohba, Y., Taniguchi, T., Takeda, K., Hori, S., Ivanov, I.I., Umesaki, Y., Itoh, K., Honda, K., 2011. Induction of colonic regulatory T cells by indigenous *Clostridium* species. *Science* 331(6015), 337–341.

Atmar, R.L., Bernstein, D.I., Harro, C.D., Al-Ibrahim, M.S., Chen, W.H., Ferreira, J., Estes, M.K., Graham, D.Y., Opekun, A.R., Richardson, C., Mendelman, P.M., 2011. Norovirus vaccine against experimental human Norwalk virus illness. *N Engl J Med* 365(23), 2178–2187.

Atmar, R.L., Opekun, A.R., Gilger, M.A., Estes, M.K., Crawford, S.E., Neill, F.H., Ramani, S., Hill, H., Ferreira, J., Graham, D.Y., 2014. Determination of the 50% human infectious dose for Norwalk virus. *J Infect Dis* 209(7), 1016–1022.

Avershina, E., Storro, O., Oien, T., Johnsen, R., Pope, P., Rudi, K., 2014. Major faecal microbiota shifts in composition and diversity with age in a geographically restricted cohort of mothers and their children. *FEMS Microbiol Ecol* 87(1), 280–290.

Azevedo, M.P., Vlasova, A.N., Saif, L.J., 2013. Human rotavirus virus-like particle vaccines evaluated in a neonatal gnotobiotic pig model of human rotavirus disease. *Expert Rev Vaccines* 12(2), 169–181.

Azevedo, M.S., Gonzalez, A.M., Yuan, L., Jeong, K.I., Iosef, C., Van Nguyen, T., Lovgren-Bengtsson, K., Morein, B., Saif, L.J., 2010. An oral versus intranasal prime/boost regimen using attenuated human rotavirus or VP2 and VP6 virus-like particles with immunostimulating complexes influences protection and antibody-secreting cell responses to rotavirus in a neonatal gnotobiotic pig model. *Clin Vaccin Immunol* 17(3), 420–428.

Azevedo, M.S., Yuan, L., Iosef, C., Chang, K.O., Kim, Y., Nguyen, T.V., Saif, L.J., 2004. Magnitude of serum and intestinal antibody responses induced by sequential replicating and non-replicating rotavirus vaccines in gnotobiotic pigs and correlation with protection. *Clin Diagn Lab Immunol* 11(1), 12–20.

Azevedo, M.S., Yuan, L., Jeong, K.I., Gonzalez, A., Nguyen, T.V., Pouly, S., Gochnauer, M., Zhang, W., Azevedo, A., Saif, L.J., 2005. Viremia and nasal and rectal shedding of rotavirus in gnotobiotic pigs inoculated with Wa human rotavirus. *J Virol* 79(9), 5428–5436.

Azevedo, M.S., Yuan, L., Pouly, S., Gonzales, A.M., Jeong, K.I., Nguyen, T.V., Saif, L.J., 2006. Cytokine responses in gnotobiotic pigs after infection with virulent or attenuated human rotavirus. *J Virol* 80(1), 372–382.

Azevedo, M.S., Zhang, W., Wen, K., Gonzalez, A.M., Saif, L.J., Yousef, A.E., Yuan, L., 2012. *Lactobacillus acidophilus* and *Lactobacillus reuteri* modulate cytokine responses in gnotobiotic pigs infected with human rotavirus. *Benef Microbes* 3(1), 33–42.

Baeuerle, P.A., Baltimore, D., 1996. NF-kappa B: Ten years after. *Cell* 87(1), 13–20.

Baldwin, A.S., Jr., 1996. The NF-kappa B and I kappa B proteins: New discoveries and insights. *Annu Rev Immunol* 14, 649–683.

Basu, S., Chatterjee, M., Ganguly, S., Chandra, P.K., 2007. Effect of *Lactobacillus rhamnosus* GG in persistent diarrhea in Indian children: A randomized controlled trial. *J Clin Gastroenterol* 41(8), 756–760.

Becker-Dreps, S., Allali, I., Monteagudo, A., Vilchez, S., Hudgens, M.G., Rogawski, E.T., Carroll, I.M., Zambrana, L.E., Espinoza, F., Azcarate-Peril, M.A., 2015a. Gut microbiome composition in young Nicaraguan children during diarrhea episodes and recovery. *Am J Trop Med Hyg* 93(6), 1187–1193.

Becker-Dreps, S., Vilchez, S., Bucardo, F., Twitchell, E., Choi, W.S., Hudgens, M.G., Perez, J., Yuan, L., 2017. The association between fecal biomarkers of environmental enteropathy and rotavirus vaccine response in Nicaraguan infants. *Pediatr Infect Dis J* 36(4), 412–416.

Becker-Dreps, S., Vilchez, S., Velasquez, D., Moon, S.S., Hudgens, M.G., Zambrana, L.E., Jiang, B., 2015b. Rotavirus-specific IgG antibodies from mothers' serum may inhibit infant immune responses to the pentavalent rotavirus vaccine. *Pediatr Infect Dis J* 34(1), 115–116.

Bernstein, D.I., Atmar, R.L., Lyon, G.M., Treanor, J.J., Chen, W.H., Jiang, X., Vinje, J., Gregoricus, N., Frenck, R.W., Jr., Moe, C.L., Al-Ibrahim, M.S., Barrett, J., Ferreira, J., Estes, M.K., Graham, D.Y., Goodwin, R., Borkowski, A., Clemens, R., Mendelman, P.M., 2015. Norovirus vaccine against experimental human GII.4 virus illness: A challenge study in healthy adults. *J Infect Dis* 211(6), 870–878.

Bertolotti-Ciarlet, A., Ciarlet, M., Crawford, S.E., Conner, M.E., Estes, M.K., 2003. Immunogenicity and protective efficacy of rotavirus 2/6-virus-like particles produced by a dual baculovirus expression vector and administered intramuscularly, intranasally, or orally to mice. *Vaccine* 21(25–26), 3885–3900.

Bhattacharjee, A., Burr, A.H.P., Overacre-Delgoffe, A.E., Tometich, J.T., Yang, D., Huckestein, B.R., Linehan, J.L., Spencer, S.P., Hall, J.A., Harrison, O.J., Morais da Fonseca, D., Norton, E.B., Belkaid, Y., Hand, T.W., 2021. Environmental enteric dysfunction induces regulatory T cells that inhibit local CD4+ T cell responses and impair oral vaccine efficacy. *Immunity* 54(8), 1745–1757 e1747.

Bianchi, A.T., Zwart, R.J., Jeurissen, S.H., Moonen-Leusen, H.W., 1992. Development of the B- and T-cell compartments in porcine lymphoid organs from birth to adult life: An immunohistological approach. *Vet Immunol Immunopathol* 33(3), 201–221.

Biasucci, G., Benenati, B., Morelli, L., Bessi, E., Boehm, G., 2008. Cesarean delivery may affect the early biodiversity of intestinal bacteria. *J Nutr* 138(9), 1796S–1800S.

Blanchette, F., Rivard, N., Rudd, P., Grondin, F., Attisano, L., Dubois, C., 2001. Cross-talk between the p42/p44 MAP kinase and Smad pathways in transforming growth factor beta 1-induced furin gene transactivation. *J Biol Chem* 276(36), 33986–33994.

Blutt, S.E., Miller, A.D., Salmon, S.L., Metzger, D.W., Conner, M.E., 2012. IgA is important for clearance and critical for protection from rotavirus infection. *Mucosal Immunol* 5(6), 712–719.

Blutt, S.E., Warfield, K.L., Estes, M.K., Conner, M.E., 2008. Differential requirements for T cells in viruslike particle- and rotavirus-induced protective immunity. *J Virol* 82(6), 3135–3138.

Boge, T., Remigy, M., Vaudaine, S., Tanguy, J., Bourdet-Sicard, R., van der Werf, S., 2009. A probiotic fermented dairy drink improves antibody response to influenza vaccination in the elderly in two randomised controlled trials. *Vaccine* 27(41), 5677–5684.

Bomba, A., Nemcova, R., Gancarcikova, S., Herich, R., Kastel, R., 1999. Potentiation of the effectiveness of *Lactobacillus casei* in the prevention of *E. coli* induced diarrhea in conventional and gnotobiotic pigs. *Adv Exp Med Biol* 473, 185–190.

Boshuizen, J.A., Reimerink, J.H., Korteland-van Male, A.M., van Ham, V.J., Koopmans, M.P., Buller, H.A., Dekker, J., Einerhand, A.W., 2003. Changes in small intestinal homeostasis, morphology, and gene expression during rotavirus infection of infant mice. *J Virol* 77(24), 13005–13016.

Brisse, S., Verhoef, J., 2001. Phylogenetic diversity of *Klebsiella pneumoniae* and *Klebsiella oxytoca* clinical isolates revealed by randomly amplified polymorphic DNA, gyrA and parC genes sequencing and automated ribotyping. *Int J Syst Evol Microbiol* 51(3), 915–924.

Brito, L.A., Singh, M., 2011. Acceptable levels of endotoxin in vaccine formulations during preclinical research. *J Pharm Sci* 100(1), 34–37.

Brown, E.M., Wlodarska, M., Willing, B.P., Vonaesch, P., Han, J., Reynolds, L.A., Arrieta, M.C., Uhrig, M., Scholz, R., Partida, O., Borchers, C.H., Sansonetti, P.J., Finlay, B.B., 2015. Diet and specific microbial exposure trigger features of environmental enteropathy in a novel murine model. *Nat Commun* 6, 7806.

Bui, T., Kocher, J., Li, Y., Wen, K., Li, G., Liu, F., Yang, X., LeRoith, T., Tan, M., Xia, M., Zhong, W., Jiang, X., Yuan, L., 2013. Median infectious dose of human Norovirus GII.4 in gnotobiotic pigs is decreased by simvastatin treatment and increased by age. *J Gen Virol* 94(9), 2005–2016.

Butler, J.E., 2009. Isolator and other neonatal piglet models in developmental immunology and identification of virulence factors. *Anim Health Res Rev* 10(1), 35–52.

Butler, J.E., Lager, K.M., Splichal, I., Francis, D., Kacskovics, I., Sinkora, M., Wertz, N., Sun, J., Zhao, Y., Brown, W.R., DeWald, R., Dierks, S., Muyldermans, S., Lunney, J.K., McCray, P.B., Rogers, C.S., Welsh, M.J., Navarro, P., Klobasa, F., Habe, F., Ramsoondar, J., 2009a. The piglet as a model for B cell and immune system development. *Vet Immunol Immunopathol* 128(1–3), 147–170.

Butler, J.E., Wertz, N., Deschacht, N., Kacskovics, I., 2009b. Porcine IgG: Structure, genetics, and evolution. *Immunogenetics* 61(3), 209–230.

Butler, J.E., Zhao, Y., Sinkora, M., Wertz, N., Kacskovics, I., 2009c. Immunoglobulins, antibody repertoire and B cell development. *Dev Comp Immunol* 33(3), 321–333.

Cadwell, K., Patel, K.K., Maloney, N.S., Liu, T.C., Ng, A.C., Storer, C.E., Head, R.D., Xavier, R., Stappenbeck, T.S., Virgin, H.W., 2010. Virus-plus-susceptibility gene interaction determines Crohn's disease gene Atg16L1 phenotypes in intestine. *Cell* 141(7), 1135–1145.

Caporaso, J.G., Kuczynski, J., Stombaugh, J., Bittinger, K., Bushman, F.D., Costello, E.K., Fierer, N., Pena, A.G., Goodrich, J.K., Gordon, J.I., Huttley, G.A., Kelley, S.T., Knights, D., Koenig, J.E., Ley, R.E., Lozupone, C.A., McDonald, D., Muegge, B.D., Pirrung, M., Reeder, J., Sevinsky, J.R., Turnbaugh, P.J., Walters, W.A., Widmann, J., Yatsunenko, T., Zaneveld, J., Knight, R., 2010. QIIME allows analysis of high-throughput community sequencing data. *Nat Methods* 7(5), 335–336.

Caporaso, J.G., Lauber, C.L., Walters, W.A., Berg-Lyons, D., Huntley, J., Fierer, N., Owens, S.M., Betley, J., Fraser, L., Bauer, M., Gormley, N., Gilbert, J.A., Smith, G., Knight, R., 2012. Ultra-high-throughput microbial community analysis on the Illumina HiSeq and MiSeq platforms. *ISME J* 6(8), 1621–1624.

Casadevall, A., Pirofski, L.A., 2003. Exploiting the redundancy in the immune system: Vaccines can mediate protection by eliciting 'unnatural' immunity. *J Exp Med* 197(11), 1401–1404.

Chachu, K.A., LoBue, A.D., Strong, D.W., Baric, R.S., Virgin, H.W., 2008. Immune mechanisms responsible for vaccination against and clearance of mucosal and lymphatic Norovirus infection. *PLOS Pathog* 4(12), e1000236.

Chand, N., Mihas, A.A., 2006. Celiac disease: Current concepts in diagnosis and treatment. *J Clin Gastroenterol* 40(1), 3–14.

Chander, Y., Ramakrishnan, M.A., Jindal, N., Hanson, K., Goyal, S., 2011. Differentiation of *Klebsiella pneumoniae* and *K. oxytoca* by multiplex polymerase chain reaction. *Intern J Appl Res Vet Med* 9, 138–142.

Chang, K.O., 2009. Role of cholesterol pathways in Norovirus replication. *J Virol* 83(17), 8587–8595.

Chang, K.O., Vandal, O.H., Yuan, L., Hodgins, D.C., Saif, L.J., 2001. Antibody-secreting cell responses to rotavirus proteins in gnotobiotic pigs inoculated with attenuated or virulent human rotavirus. *J Clin Microbiol* 39(8), 2807–2813.

Chattha, K.S., Vlasova, A.N., Kandasamy, S., Esseili, M.A., Siegismund, C., Rajashekara, G., Saif, L.J., 2013. Probiotics and colostrum/milk differentially affect neonatal humoral immune responses to oral rotavirus vaccine. *Vaccine* 31(15), 1916–1923.

Che, C., Pang, X., Hua, X., Zhang, B., Shen, J., Zhu, J., Wei, H., Sun, L., Chen, P., Cui, L., Zhao, L., Yang, Q., 2009. Effects of human fecal flora on intestinal morphology and mucosal immunity in human flora-associated piglet. *Scand J Immunol* 69(3), 223–233.

Cheetham, S., Souza, M., McGregor, R., Meulia, T., Wang, Q., Saif, L.J., 2007. Binding patterns of human norovirus-like particles to buccal and intestinal tissues of gnotobiotic pigs in relation to A/H histo-blood group antigen expression. *J Virol* 81(7), 3535–3544.

Cheetham, S., Souza, M., Meulia, T., Grimes, S., Han, M.G., Saif, L.J., 2006. Pathogenesis of a genogroup II human Norovirus in gnotobiotic pigs. *J Virol* 80(21), 10372–10381.

Chen, S.C., Fynan, E.F., Greenberg, H.B., Herrmann, J.E., 1999. Immunity obtained by gene-gun inoculation of a rotavirus DNA vaccine to the abdominal epidermis or anorectal epithelium. *Vaccine* 17(23–24), 3171–3176.

Chen, S.C., Fynan, E.F., Robinson, H.L., Lu, S., Greenberg, H.B., Santoro, J.C., Herrmann, J.E., 1997. Protective immunity induced by rotavirus DNA vaccines. *Vaccine* 15(8), 899–902.

Chen, S.C., Jones, D.H., Fynan, E.F., Farrar, G.H., Clegg, J.C., Greenberg, H.B., Herrmann, J.E., 1998. Protective immunity induced by oral immunization with a rotavirus DNA vaccine encapsulated in microparticles. *J Virol* 72(7), 5757–5761.

Chiba, Y., Shida, K., Nagata, S., Wada, M., Bian, L., Wang, C., Shimizu, T., Yamashiro, Y., Kiyoshima-Shibata, J., Nanno, M., Nomoto, K., 2010. Well-controlled proinflammatory cytokine responses of Peyer's patch cells to probiotic *Lactobacillus casei. Immunology* 130(3), 352–362.

Cholesterol Treatment Trialists Collaborators, Mihaylova, B., Emberson, J., Blackwell, L., Keech, A., Simes, J., Barnes, E.H., Voysey, M., Gray, A., Collins, R., Baigent, C., 2012. The effects of lowering LDL cholesterol with statin therapy in people at low risk of vascular disease: Meta-analysis of individual data from 27 randomised trials. *Lancet* 380(9841), 581–590.

Cholujova, D., Jakubikova, J., Czako, B., Martisova, M., Hunakova, L., Duraj, J., Mistrik, M., Sedlak, J., 2013. MGN-3 arabinoxylan rice bran modulates innate immunity in multiple myeloma patients. *Cancer Immunol Immunother CII* 62(3), 437–445.

Chon, H., Choi, B., Jeong, G., Lee, E., Lee, S., 2010. Suppression of proinflammatory cytokine production by specific metabolites of *Lactobacillus plantarum* 10hk2 via inhibiting NF-kappaB and p38 MAPK expressions. *Comp Immunol Microbiol Infect Dis* 33(6), e41–49.

Chou, T.W., Ma, C.Y., Cheng, H.H., Chen, Y.Y., Lai, M.H., 2009. A rice bran oil diet improves lipid abnormalities and suppress hyperinsulinemic responses in rats with streptozotocin/nicotinamide-induced type 2 diabetes. *J Clin Biochem Nutr* 45(1), 29–36.

Christensen, H.R., Frokiaer, H., Pestka, J.J., 2002. Lactobacilli differentially modulate expression of cytokines and maturation surface markers in murine dendritic cells. *J Immunol* 168(1), 171–178.

Ciarlet, M., 2007. Update on clinical programs for pentavalent rotavirus vaccine. *Vaccines for Enteric Disease* 2007. The 4th International Conference on Vaccines for Enteric Diseases 25–27 April 2007, Lisbon, Portugal.

Coates, M.E., 1975. Gnotobiotic animals in research: Their uses and limitations. *Lab Anim* 9(4), 275–282.

Cojocaru, L., Rusali, A.C., Suta, C., Radulescu, A.M., Suta, M., Craiu, E., 2013. The role of simvastatin in the therapeutic approach of rheumatoid arthritis. *Autoimmune Dis* 2013, 326258.

Cox, M.J., Huang, Y.J., Fujimura, K.E., Liu, J.T., McKean, M., Boushey, H.A., Segal, M.R., Brodie, E.L., Cabana, M.D., Lynch, S.V., 2010. *Lactobacillus casei* abundance is associated with profound shifts in the infant gut microbiome. *PLOS ONE* 5(1), e8745.

Crane, R.J., Jones, K.D., Berkley, J.A., 2015. Environmental enteric dysfunction: An overview. *Food Nutr Bull* 36(1) Suppl, S76–87.

Crawford, S.E., Labbe, M., Cohen, J., Burroughs, M.H., Zhou, Y.J., Estes, M.K., 1994. Characterization of virus-like particles produced by the expression of rotavirus capsid proteins in insect cells. *J Virol* 68(9), 5945–5952.

Crouch, C.F., Woode, G.N., 1978. Serial studies of virus multiplication and intestinal damage in gnotobiotic piglets infected with rotavirus. *J Med Microbiol* 11(3), 325–334.

Dai, Y., Vaught, T.D., Boone, J., Chen, S.H., Phelps, C.J., Ball, S., Monahan, J.A., Jobst, P.M., McCreath, K.J., Lamborn, A.E., Cowell-Lucero, J.L., Wells, K.D., Colman, A., Polejaeva, I.A., Ayares, D.L., 2002. Targeted disruption of the alpha1,3-galactosyltransferase gene in cloned pigs. *Nat Biotechnol* 20(3), 251–255.

Dalcico, R., de Menezes, A.M., Deocleciano, O.B., Oria, R.B., Vale, M.L., Ribeiro, R.A., Brito, G.A., 2013. Protective mechanisms of simvastatin in experimental periodontal disease. *J Periodontol* 84(8), 1145–1157.

Davidson, L.E., Fiorino, A.M., Snydman, D.R., Hibberd, P.L., 2011. Lactobacillus GG as an immune adjuvant for live-attenuated influenza vaccine in healthy adults: A randomized double-blind placebo-controlled trial. *Eur J Clin Nutr* 65(4), 501–507.

Davidson, M.K., Lindsey, J.R., Davis, J.K., 1987. Requirements and selection of an animal model. *Isr J Med Sci* 23(6), 551–555.

de Vrese, M., Rautenberg, P., Laue, C., Koopmans, M., Herremans, T., Schrezenmeir, J., 2005. Probiotic bacteria stimulate virus-specific neutralizing antibodies following a booster polio vaccination. *Eur J Nutr* 44(7), 406–413.

Debbink, K., Lindesmith, L.C., Baric, R.S., 2014. The state of Norovirus vaccines. *Clin Infect Dis Off Publ Infect Dis Soc Am* 58(12), 1746–1752.

Degiuseppe, J.I., Beltramino, J.C., Millan, A., Stupka, J.A., Parra, G.I., 2013. Complete genome analyses of G4P[6] rotavirus detected in Argentinean children with diarrhoea provides evidence of interspecies transmission from swine. *Clin Microbiol Infect* 19(8), E367–371.

Dharakul, T., Labbe, M., Cohen, J., Bellamy, A.R., Street, J.E., Mackow, E.R., Fiore, L., Rott, L., Greenberg, H.B., 1991. Immunization with baculovirus-expressed recombinant rotavirus proteins VP1, VP4, VP6, and VP7 induces CD8+ T lymphocytes that mediate clearance of chronic rotavirus infection in SCID mice. *J Virol* 65(11), 5928–5932.

Ding, S., Yu, B., van Vuuren, A.J., 2021. Statins significantly repress rotavirus replication through downregulation of cholesterol synthesis. *Gut Microbes* 13(1), 1955643.

Dominguez-Bello, M.G., Costello, E.K., Contreras, M., Magris, M., Hidalgo, G., Fierer, N., Knight, R., 2010. Delivery mode shapes the acquisition and structure of the initial microbiota across multiple body habitats in newborns. *Proc Natl Acad Sci U S A* 107(26), 11971–11975.

Donowitz, J.R., Haque, R., Kirkpatrick, B.D., Alam, M., Lu, M., Kabir, M., Kakon, S.H., Islam, B.Z., Afreen, S., Musa, A., Khan, S.S., Colgate, E.R., Carmolli, M.P., Ma, J.Z., Petri, W.A., Jr., 2016. Small intestine bacterial overgrowth and environmental enteropathy in Bangladeshi children. *mBio* 7(1), e02102–02115.

Doria-Rose, N.A., Ohlen, C., Polacino, P., Pierce, C.C., Hensel, M.T., Kuller, L., Mulvania, T., Anderson, D., Greenberg, P.D., Hu, S.L., Haigwood, N.L., 2003. Multigene DNA priming-boosting vaccines protect macaques from acute CD4+-T-cell depletion after simian-human immunodeficiency virus SHIV89.6P mucosal challenge. *J Virol* 77(21), 11563–11577.

Edgar, R.C., 2010. Search and clustering orders of magnitude faster than BLAST. *Bioinformatics* 26(19), 2460–2461.

El-Kamary, S.S., Pasetti, M.F., Mendelman, P.M., Frey, S.E., Bernstein, D.I., Treanor, J.J., Ferreira, J., Chen, W.H., Sublett, R., Richardson, C., Bargatze, R.F., Sztein, M.B., Tacket, C.O., 2010. Adjuvanted intranasal Norwalk virus-like particle vaccine elicits antibodies and antibody-secreting cells that express homing receptors for mucosal and peripheral lymphoid tissues. *J Infect Dis* 202(11), 1649–1658.

Eo, S.K., Gierynska, M., Kamar, A.A., Rouse, B.T., 2001. Prime-boost immunization with DNA vaccine: Mucosal route of administration changes the rules. *J Immunol* 166(9), 5473–5479.

Esona, M.D., Foytich, K., Wang, Y., Shin, G., Wei, G., Gentsch, J.R., Glass, R.I., Jiang, B., 2010. Molecular characterization of human rotavirus vaccine strain CDC-9 during sequential passages in Vero cells. *Hum Vaccin* 6(3), 10409.

Esquivel, F.R., Lopez, S., Guitierrez, X.L., Arias, C., 2000. The internal rotavirus protein VP6 primes for an enhanced neutralizing antibody response. *Arch Virol* 145(4), 813–825.

Evrard, B., Coudeyras, S., Dosgilbert, A., Charbonnel, N., Alame, J., Tridon, A., Forestier, C., 2011. Dose-dependent immunomodulation of human dendritic cells by the probiotic *Lactobacillus rhamnosus* Lcr35. *PLOS ONE* 6(4), e18735.

Fabian, C., Ju, Y.H., 2011. A review on rice bran protein: Its properties and extraction methods. *Crit Rev Food Sci Nutr* 51(9), 816–827.

Fang, H., Tan, M., Xia, M., Wang, L., Jiang, X., 2013. Norovirus P particle efficiently elicits innate, humoral and cellular immunity. *PLOS ONE* 8(4), e63269.

Favier, C.F., Vaughan, E.E., De Vos, W.M., Akkermans, A.D., 2002. Molecular monitoring of succession of bacterial communities in human neonates. *Appl Environ Microbiol* 68(1), 219–226.

Fei, N., Zhao, L., 2013. An opportunistic pathogen isolated from the gut of an obese human causes obesity in germfree mice. *ISME J* 7(4), 880–884.

Fix, A., Kirkwood, C.D., Steele, D., Flores, J., 2020. Next-generation rotavirus vaccine developers meeting: Summary of a meeting sponsored by PATH and the Bill & Melinda Gates Foundation (19–20 June 2019, Geneva). *Vaccine* 38(52), 8247–8254.

Fix, A.D., Harro, C., McNeal, M., Dally, L., Flores, J., Robertson, G., Boslego, J.W., Cryz, S., 2015. Safety and immunogenicity of a parenterally administered rotavirus VP8 subunit vaccine in healthy adults. *Vaccine* 33(31), 3766–3772.

Francavilla, R., Miniello, V., Magista, A.M., De Canio, A., Bucci, N., Gagliardi, F., Lionetti, E., Castellaneta, S., Polimeno, L., Peccarisi, L., Indrio, F., Cavallo, L., 2010. A randomized controlled trial of Lactobacillus GG in children with functional abdominal pain. *Pediatrics* 126(6), e1445–1452.

Franco, M.A., Greenberg, H.B., 1997. Immunity to rotavirus in T cell deficient mice. *Virology* 238(2), 169–179.

Franco, M.A., Greenberg, H.B., 1999. Immunity to rotavirus infection in mice. *J Infect Dis* 179(Suppl 3), S466–469.

Franco, M.A., Greenberg, H.B., 2000. Immunity to homologous rotavirus infection in adult mice. *Trends Microbiol* 8(2), 50–52.

Franke, E.D., Corradin, G., Hoffman, S.L., 1997. Induction of protective CTL responses against the *Plasmodium yoelii* circumsporozoite protein by immunization with peptides. *J Immunol* 159(7), 3424–3433.

Frenck, R., Bernstein, D.I., Xia, M., Huang, P., Zhong, W., Parker, S., Dickey, M., McNeal, M., Jiang, X., 2012. Predicting susceptibility to Norovirus GII.4 by use of a challenge model involving humans. *J Infect Dis* 206(9), 1386–1393.

Friedman, M., 2013. Rice brans, rice bran oils, and rice hulls: Composition, food and industrial uses, and bioactivities in humans, animals, and cells. *J Agric Food Chem* 61(45), 10626–10641.

Fu, J., Liang, J., Kang, H., Lin, J., Yu, Q., Yang, Q., 2014. The stimulatory effect of different CpG oligonucleotides on the maturation of chicken bone marrow-derived dendritic cells. *Poult Sci* 93(1), 63–69.

Gaboriau-Routhiau, V., Rakotobe, S., Lecuyer, E., Mulder, I., Lan, A., Bridonneau, C., Rochet, V., Pisi, A., De Paepe, M., Brandi, G., Eberl, G., Snel, J., Kelly, D., Cerf-Bensussan, N., 2009. The key role of segmented filamentous bacteria in the coordinated maturation of gut helper T cell responses. *Immunity* 31(4), 677–689.

Galdeano, C.M., Perdigon, G., 2006. The probiotic bacterium *Lactobacillus casei* induces activation of the gut mucosal immune system through innate immunity. *Clin Vaccin Immunol* 13(2), 219–226.

Garcia-Lopez, R., Perez-Brocal, V., Diez-Domingo, J., Moya, A., 2012. Gut microbiota in children vaccinated with rotavirus vaccine. *Pediatr Infect Dis J* 31(12), 1300–1302.

George, C.M., Oldja, L., Biswas, S., Perin, J., Lee, G.O., Kosek, M., Sack, R.B., Ahmed, S., Haque, R., Parvin, T., Azmi, I.J., Bhuyian, S.I., Talukder, K.A., Mohammad, S., Faruque, A.G., 2015. Geophagy is associated with environmental enteropathy and stunting in children in rural Bangladesh. *Am J Trop Med Hyg* 92(6), 1117–1124.

Gerdts, V., Wilson, H.L., Meurens, F., van Drunen Littel-van den Hurk, S., Wilson, D., Walker, S., Wheler, C., Townsend, H., Potter, A.A., 2015. Large animal models for vaccine development and testing. *ILAR J* 56(1), 53–62.

Gerhardt, A.L., Gallo, N.B., 1998. Full-fat rice bran and oat bran similarly reduce hypercholesterolemia in humans. *J Nutr* 128(5), 865–869.

Gerner, W., Kaser, T., Saalmuller, A., 2009. Porcine T lymphocytes and NK cells--An update. *Dev Comp Immunol* 33(3), 310–320.

Gerondakis, S., Fulford, T.S., Messina, N.L., Grumont, R.J., 2014. NF-kappaB control of T cell development. *Nat Immunol* 15(1), 15–25.

Gerondopoulos, A., Jackson, T., Monaghan, P., Doyle, N., Roberts, L.O., 2010. Murine norovirus-1 cell entry is mediated through a non-clathrin-, non-caveolae-, dynamin- and cholesterol-dependent pathway. *J Gen Virol* 91(6), 1428–1438.

Ghadimi, D., Vrese, M., Heller, K.J., Schrezenmeir, J., 2010. Effect of natural commensal-origin DNA on toll-like receptor 9 (TLR9) signaling cascade, chemokine IL-8 expression, and barrier integritiy of polarized intestinal epithelial cells. *Inflamm Bowel Dis* 16(3), 410–427.

Ghatak, S.B., Panchal, S.J., 2012. Investigation of the immunomodulatory potential of oryzanol isolated from crude rice bran oil in experimental animal models. *Phytother Res PTR* 26(11), 1701–1708.

Ghoneum, M., 1998. Anti-HIV activity in vitro of MGN-3, an activated arabinoxylane from rice bran. *Biochem Biophys Res Commun* 243(1), 25–29.

Ghosh, T., Auerochs, S., Saha, S., Ray, B., Marschall, M., 2010. Anti-cytomegalovirus activity of sulfated glucans generated from a commercial preparation of rice bran. *Antivir Chem Chemother* 21(2), 85–95.

Giahi, L., Aumueller, E., Elmadfa, I., Haslberger, A.G., 2012. Regulation of TLR4, p38 MAPkinase, IkappaB and miRNAs by inactivated strains of lactobacilli in human dendritic cells. *Benef Microbes* 3(2), 91–98.

Gibson-Corley, K.N., Olivier, A.K., Meyerholz, D.K., 2013. Principles for valid histopathologic scoring in research. *Vet Pathol* 50(6), 1007–1015.

Gilmartin, A.A., Petri, W.A., Jr., 2015. Exploring the role of environmental enteropathy in malnutrition, infant development and oral vaccine response. *Philos Trans R Soc Lond B* 370(1671), 20140143.

Glass, R.I., Parashar, U., Patel, M., Gentsch, J., Jiang, B., 2014. Rotavirus vaccines: Successes and challenges. *J Infect* 68 Suppl 1, S9–18.

Glitscher, M., Martin, D.H., Woytinek, K., Schmidt, B., Tabari, D., Scholl, C., Stingl, J.C., Seelow, E., Choi, M., Hildt, E., 2021. Targeting cholesterol metabolism as efficient antiviral strategy against the hepatitis E virus. *Cell Mol Gastroenterol Hepatol* 12(1), 159–180.

Goenka, R., Parent, M.A., Elzer, P.H., Baldwin, C.L., 2011. B cell-deficient mice display markedly enhanced resistance to the intracellular bacterium Brucella abortus. *J Infect Dis* 203(8), 1136–1146.

Goldberg, E.D., Saltzman, J.R., 1996. Rice inhibits intestinal secretions. *Nutr Rev* 54(1 Pt 1), 36–37.

Goldin, B.R., Gorbach, S.L., 2008. Clinical indications for probiotics: An overview. *Clin Infect Dis Off Publ Infect Dis Soc Am* 46(Suppl 2), S96–100; discussion S144–151.

Goldstein, J.L., Brown, M.S., 2009. The LDL receptor. *Arterioscler Thromb Vasc Biol* 29(4), 431–438.

Gonzalez, A.M., Azevedo, M.S., Jung, K., Vlasova, A., Zhang, W., Saif, L.J., 2010. Innate immune responses to human rotavirus in the neonatal gnotobiotic piglet disease model. *Immunology* 131(2), 242–256.

Gonzalez, A.M., Nguyen, T.V., Azevedo, M.S., Jeong, K., Agarib, F., Iosef, C., Chang, K., Lovgren-Bengtsson, K., Morein, B., Saif, L.J., 2004. Antibody responses to human rotavirus (HRV) in gnotobiotic pigs following a new prime/boost vaccine strategy using oral attenuated HRV priming and intranasal VP2/6 rotavirus-like particle (VLP) boosting with ISCOM. *Clin Exp Immunol* 135(3), 361–372.

Grangette, C., 2012. Bifidobacteria and subsets of dendritic cells: Friendly players in immune regulation! *Gut* 61(3), 331–332.

Groenen, M.A., 2016. A decade of pig genome sequencing: A window on pig domestication and evolution. *Genet Sel Evol* 48, 23.

Groenen, M.A., Archibald, A.L., Uenishi, H., Tuggle, C.K., Takeuchi, Y., Rothschild, M.F., Rogel-Gaillard, C., Park, C., Milan, D., Megens, H.J., Li, S., Larkin, D.M., Kim, H., Frantz, L.A., Caccamo, M., Ahn, H., Aken, B.L., Anselmo, A., Anthon, C., Auvil, L., Badaoui, B., Beattie, C.W., Bendixen, C., Berman, D., Blecha, F., Blomberg, J., Bolund, L., Bosse, M., Botti, S., Bujie, Z., Bystrom, M., Capitanu, B., Carvalho-Silva, D., Chardon, P., Chen, C., Cheng, R., Choi, S.H., Chow, W., Clark, R.C., Clee, C., Crooijmans, R.P., Dawson, H.D., Dehais, P., De Sapio, F., Dibbits, B., Drou, N., Du, Z.Q., Eversole, K., Fadista, J., Fairley, S., Faraut, T., Faulkner, G.J., Fowler, K.E., Fredholm, M., Fritz, E., Gilbert, J.G., Giuffra, E., Gorodkin, J., Griffin, D.K., Harrow, J.L., Hayward, A., Howe, K., Hu, Z.L., Humphray, S.J., Hunt, T., Hornshoj, H., Jeon, J.T., Jern, P., Jones, M., Jurka, J., Kanamori, H., Kapetanovic, R., Kim, J., Kim, J.H., Kim, K.W., Kim, T.H., Larson, G., Lee, K., Lee, K.T., Leggett, R., Lewin, H.A., Li, Y., Liu, W., Loveland, J.E., Lu, Y., Lunney, J.K., Ma, J., Madsen, O., Mann, K., Matthews, L., McLaren, S., Morozumi, T., Murtaugh, M.P., Narayan, J., Nguyen, D.T., Ni, P., Oh, S.J., Onteru, S., Panitz, F., Park, E.W., Park, H.S., Pascal, G., Paudel, Y., Perez-Enciso, M., Ramirez-Gonzalez, R., Reecy, J.M., Rodriguez-Zas, S., Rohrer, G.A., Rund, L., Sang, Y., Schachtschneider, K., Schraiber, J.G., Schwartz, J., Scobie, L., Scott, C., Searle, S., Servin, B., Southey, B.R., Sperber, G., Stadler, P., Sweedler, J.V., Tafer, H., Thomsen, B., Wali, R., Wang, J., Wang, J., White, S., Xu, X., Yerle, M., Zhang, G., Zhang, J., Zhang, J., Zhao, S., Rogers, J., Churcher, C., Schook, L.B., 2012. Analyses of pig genomes provide insight into porcine demography and evolution. *Nature* 491(7424), 393–398.

Groome, M.J., Fairlie, L., Morrison, J., Fix, A., Koen, A., Masenya, M., Jose, L., Madhi, S.A., Page, N., McNeal, M., Dally, L., Cho, I., Power, M., Flores, J., Cryz, S., 2020. Safety and immunogenicity of a parenteral trivalent P2-VP8 subunit rotavirus vaccine: A multisite, randomised, double-blind, placebo-controlled trial. *Lancet Infect Dis* 20(7), 851–863.

Groome, M.J., Koen, A., Fix, A., Page, N., Jose, L., Madhi, S.A., McNeal, M., Dally, L., Cho, I., Power, M., Flores, J., Cryz, S., 2017. Safety and immunogenicity of a parenteral P2-VP8-P[8] subunit rotavirus vaccine in toddlers and infants in South Africa: A randomised, double-blind, placebo-controlled trial. *Lancet Infect Dis* 17(8), 843–853.

Grzeskowiak, L., Collado, M.C., Mangani, C., Maleta, K., Laitinen, K., Ashorn, P., Isolauri, E., Salminen, S., 2012. Distinct gut microbiota in southeastern African and northern European infants. *J Pediatr Gastroenterol Nutr* 54(6), 812–816.

Gu, Z., Zhu, H., Rodriguez, A., Mhaissen, M., Schultz-Cherry, S., Adderson, E., Hayden, R.T., 2015. Comparative evaluation of broad-panel PCR assays for the detection of gastrointestinal pathogens in pediatric oncology patients. *J Mol Diagn* 17(6), 715–721.

Haas, B.J., Gevers, D., Earl, A.M., Feldgarden, M., Ward, D.V., Giannoukos, G., Ciulla, D., Tabbaa, D., Highlander, S.K., Sodergren, E., Methe, B., DeSantis, T.Z., Human Microbiome Consortium, Petrosino, J.F., Knight, R., Birren, B.W., 2011. Chimeric 16S rRNA sequence formation and detection in Sanger and 454-pyrosequenced PCR amplicons. *Genome Res* 21(3), 494–504.

Hagbom, M., Sharma, S., Lundgren, O., Svensson, L., 2012. Towards a human rotavirus disease model. *Curr Opin Virol* 2(4), 408–418.

Hansen, C.H., Nielsen, D.S., Kverka, M., Zakostelska, Z., Klimesova, K., Hudcovic, T., Tlaskalova-Hogenova, H., Hansen, A.K., 2012. Patterns of early gut colonization shape future immune responses of the host. *PLOS ONE* 7(3), e34043.

Harrington, P.R., Vinje, J., Moe, C.L., Baric, R.S., 2004. Norovirus capture with histo-blood group antigens reveals novel virus-ligand interactions. *J Virol* 78(6), 3035–3045.

Heart Protection Study Collaborative Group, Bulbulia, R., Bowman, L., Wallendszus, K., Parish, S., Armitage, J., Peto, R., Collins, R., 2011. Effects on 11-year mortality and morbidity of lowering LDL cholesterol with simvastatin for about 5 years in 20,536 high-risk individuals: A randomised controlled trial. *Lancet* 378(9808), 2013–2020.

Hemmi, H., Takeuchi, O., Kawai, T., Kaisho, T., Sato, S., Sanjo, H., Matsumoto, M., Hoshino, K., Wagner, H., Takeda, K., Akira, S., 2000. A toll-like receptor recognizes bacterial DNA. *Nature* 408(6813), 740–745.

Henderson, A.J., Kumar, A., Barnett, B., Dow, S.W., Ryan, E.P., 2012. Consumption of rice bran increases mucosal immunoglobulin A concentrations and numbers of intestinal *Lactobacillus* spp. *J Med Food* 15(5), 469–475.

Henker, J., Laass, M., Blokhin, B.M., Bolbot, Y.K., Maydannik, V.G., Elze, M., Wolff, C., Schulze, J., 2007. The probiotic *Escherichia coli* strain Nissle 1917 (EcN) stops acute diarrhoea in infants and toddlers. *Eur J Pediatr* 166(4), 311–318.

Hensley, C., Zhou, P., Schnur, S., Mahsoub, H.M., Liang, Y., Wang, M.-X., Page, C., Yuan, L., Bronshtein, V., 2021. Thermostable, dissolvable buccal film rotavirus vaccine is highly effective in neonatal gnotobiotic pig challenge model. *Vaccines* 9(437). https://doi.org/10.3390/vaccines9050437.

Hering, N.A., Richter, J.F., Fromm, A., Wieser, A., Hartmann, S., Gunzel, D., Bucker, R., Fromm, M., Schulzke, J.D., Troeger, H., 2014. TcpC protein from *E. coli* Nissle improves epithelial barrier function involving PKCzeta and ERK1/2 signaling in HT-29/B6 cells. *Mucosal Immunol* 7(2), 369–378.

Herrmann, J.E., Chen, S.C., Jones, D.H., Tinsley-Bown, A., Fynan, E.F., Greenberg, H.B., Farrar, G.H., 1999. Immune responses and protection obtained by oral immunization with rotavirus VP4 and VP7 DNA vaccines encapsulated in microparticles. *Virology* 259(1), 148–153.

Hillyard, D.Z., Cameron, A.J., McDonald, K.J., Thomson, J., MacIntyre, A., Shiels, P.G., Panarelli, M., Jardine, A.G., 2004. Simvastatin inhibits lymphocyte function in normal subjects and patients with cardiovascular disease. *Atherosclerosis* 175(2), 305–313.

Hillyard, D.Z., Nutt, C.D., Thomson, J., McDonald, K.J., Wan, R.K., Cameron, A.J., Mark, P.B., Jardine, A.G., 2007. Statins inhibit NK cell cytotoxicity by membrane raft depletion rather than inhibition of isoprenylation. *Atherosclerosis* 191(2), 319–325.

Hirayama, K., 1999. Ex-germfree mice harboring intestinal microbiota derived from other animal species as an experimental model for ecology and metabolism of intestinal bacteria. *Exp Anim* 48(4), 219–227.

Hodges, K., Gill, R., 2010. Infectious diarrhea: Cellular and molecular mechanisms. *Gut Microbes* 1(1), 4–21.

Hodgins, D.C., Kang, S.Y., deArriba, L., Parreno, V., Ward, L.A., Yuan, L., To, T., Saif, L.J., 1999. Effects of maternal antibodies on protection and development of antibody responses to human rotavirus in gnotobiotic pigs. *J Virol* 73(1), 186–197.

Hoshino, Y., Saif, L.J., Kang, S.Y., Sereno, M.M., Chen, W.K., Kapikian, A.Z., 1995. Identification of group A rotavirus genes associated with virulence of a porcine rotavirus and host range restriction of a human rotavirus in the gnotobiotic piglet model. *Virology* 209(1), 274–280.

Huang, P., Farkas, T., Marionneau, S., Zhong, W., Ruvoen-Clouet, N., Morrow, A.L., Altaye, M., Pickering, L.K., Newburg, D.S., LePendu, J., Jiang, X., 2003. Noroviruses bind to human ABO, Lewis, and secretor histo-blood group antigens: Identification of 4 distinct strain-specific patterns. *J Infect Dis* 188(1), 19–31.

Huang, P., Farkas, T., Zhong, W., Tan, M., Thornton, S., Morrow, A.L., Jiang, X., 2005. Norovirus and histo-blood group antigens: Demonstration of a wide spectrum of strain specificities and classification of two major binding groups among multiple binding patterns. *J Virol* 79(11), 6714–6722.

Huda, M.N., Lewis, Z., Kalanetra, K.M., Rashid, M., Ahmad, S.M., Raqib, R., Qadri, F., Underwood, M.A., Mills, D.A., Stephensen, C.B., 2014. Stool microbiota and vaccine responses of infants. *Pediatrics* 134(2), e362–372.

Humphray, S.J., Scott, C.E., Clark, R., et al. 2007. A high utility integrated map of the pig genome. *Genome Biol* 8(7), R139.

Huo, Y., Wan, X., Ling, T., Wu, J., Wang, W., Shen, S., 2018. Expression and purification of Norovirus virus like particles in *Escherichia coli* and their immunogenicity in mice. *Mol Immunol* 93, 278–284.

Hutson, A.M., Atmar, R.L., Graham, D.Y., Estes, M.K., 2002. Norwalk virus infection and disease is associated with ABO histo-blood group type. *J Infect Dis* 185(9), 1335–1337.

Iosef, C., Chang, K.O., Azevedo, M.S., Saif, L.J., 2002a. Systemic and intestinal antibody responses to NSP4 enterotoxin of Wa human rotavirus in a gnotobiotic pig model of human rotavirus disease. *J Med Virol* 68(1), 119–128.

Iosef, C., Van Nguyen, T., Jeong, K., Bengtsson, K., Morein, B., Kim, Y., Chang, K.O., Azevedo, M.S., Yuan, L., Nielsen, P., Saif, L.J., 2002b. Systemic and intestinal antibody secreting cell responses and protection in gnotobiotic pigs immunized orally with attenuated Wa human rotavirus and Wa 2/6-rotavirus-like-particles associated with immunostimulating complexes. *Vaccine* 20(13–14), 1741–1753.

Isolauri, E., Joensuu, J., Suomalainen, H., Luomala, M., Vesikari, T., 1995. Improved immuno-genicity of oral D x RRV reassortant rotavirus vaccine by *Lactobacillus casei* GG. *Vaccine* 13(3), 310–312.

Isolauri, E., Juntunen, M., Rautanen, T., Sillanaukee, P., Koivula, T., 1991. A human Lactobacillus strain (*Lactobacillus casei* sp strain GG) promotes recovery from acute diarrhea in children. *Pediatrics* 88(1), 90–97.

Jamin, A., Gorin, S., Le Potier, M.F., Kuntz-Simon, G., 2006. Characterization of conventional and plasmacytoid dendritic cells in swine secondary lymphoid organs and blood. *Vet Immunol Immunopathol* 114(3–4), 224–237.

Jan, R.H., Lin, Y.L., Chen, C.J., Lin, T.Y., Hsu, Y.C., Chen, L.K., Chiang, B.L., 2012. Hepatitis B virus surface antigen can activate human monocyte-derived dendritic cells by nuclear factor kappa B and p38 mitogen-activated protein kinase mediated signaling. *Microbiol Immunol* 56(10), 719–727.

Jiang, B., Gentsch, J.R., Glass, R.I., 2002. The role of serum antibodies in the protection against rotavirus disease: An overview. *Clin Infect Dis* 34(10), 1351–1361.

Jiang, B., Wang, Y., Glass, R.I., 2013. Does a monovalent inactivated human rotavirus vaccine induce heterotypic immunity? Evidence from animal studies. *Hum Vaccin Immunother* 9(8), 1634–1637.

Jiang, X., Wang, M., Wang, K., Estes, M.K., 1993. Sequence and genomic organization of Norwalk virus. *Virology* 195(1), 51–61.

Jiang, Y., Lu, X., Man, C., Han, L., Shan, Y., Qu, X., Liu, Y., Yang, S., Xue, Y., Zhang, Y., 2012. *Lactobacillus acidophilus* induces cytokine and chemokine production via NF-kappaB and p38 mitogen-activated protein kinase signaling pathways in intestinal epithelial cells. *Clin Vaccin Immunol CVI* 19(4), 603–608.

Jothikumar, N., Cromeans, T.L., Hill, V.R., Lu, X., Sobsey, M.D., Erdman, D.D., 2005. Quantitative real-time PCR assays for detection of human adenoviruses and identification of serotypes 40 and 41. *Appl Environ Microbiol* 71(6), 3131–3136.

Jung, K., Wang, Q., Kim, Y., Scheuer, K., Zhang, Z., Shen, Q., Chang, K.O., Saif, L.J., 2012. The effects of simvastatin or interferon-alpha on infectivity of human Norovirus using a gnotobiotic pig model for the study of antivirals. *PLOS ONE* 7(7), e41619.

Kaila, M., Isolauri, E., Soppi, E., Virtanen, E., Laine, S., Arvilommi, H., 1992. Enhancement of the circulating antibody secreting cell response in human diarrhea by a human Lactobacillus strain. *Pediatr Res* 32(2), 141–144.

Kamiya, T., Shikano, M., Tanaka, M., Ozeki, K., Ebi, M., Katano, T., Hamano, S., Nishiwaki, H., Tsukamoto, H., Mizoshita, T., Mori, Y., Kubota, E., Tanida, S., Kataoka, H., Okuda, N., Joh, T., 2014. Therapeutic effects of biobran, modified arabinoxylan rice bran, in improving symptoms of diarrhea predominant or mixed type irritable bowel syndrome: A pilot, randomized controlled study. *Evid Based Complement Alternat Med eCAM* 2014, 828137.

Kanda, H., Yokota, K., Kohno, C., Sawada, T., Sato, K., Yamaguchi, M., Komagata, Y., Shimada, K., Yamamoto, K., Mimura, T., 2007. Effects of low-dosage simvastatin on rheumatoid arthritis through reduction of Th1/Th2 and CD4/CD8 ratios. *Mod Rheumatol* 17(5), 364–368.

Kandasamy, S., Chattha, K.S., Vlasova, A.N., Rajashekara, G., Saif, L.J., 2014. Lactobacilli and bifidobacteria enhance mucosal B cell responses and differentially modulate systemic antibody responses to an oral human rotavirus vaccine in a neonatal gnotobiotic pig disease model. *Gut Microbes* 5(5), 639–651.

Kandasamy, S., Vlasova, A.N., Fischer, D., Kumar, A., Chattha, K.S., Rauf, A., Shao, L., Langel, S.N., Rajashekara, G., Saif, L.J., 2016. Differential effects of *Escherichia coli* Nissle and *Lactobacillus rhamnosus* strain GG on human rotavirus binding, infection, and B cell immunity. *J Immunol* 196(4), 1780–1789.

Kane, M., Case, L.K., Kopaskie, K., Kozlova, A., MacDearmid, C., Chervonsky, A.V., Golovkina, T.V., 2011. Successful transmission of a retrovirus depends on the commensal microbiota. *Science* 334(6053), 245–249.

Kaplon, J., Cros, G., Ambert-Balay, K., Leruez-Ville, M., Chomton, M., Fremy, C., Pothier, P., Blanche, S., 2015. Rotavirus vaccine virus shedding, viremia and clearance in infants with severe combined immune deficiency. *Pediatr Infect Dis J* 34(3), 326–328.

Karst, S.M., Wobus, C.E., Lay, M., Davidson, J., Virgin, H.W.t., 2003. STAT1-dependent innate immunity to a Norwalk-like virus. *Science* 299(5612), 1575–1578.

Kataoka, K., Ogasa, S., Kuwahara, T., Bando, Y., Hagiwara, M., Arimochi, H., Nakanishi, S., Iwasaki, T., Ohnishi, Y., 2008. Inhibitory effects of fermented brown rice on induction of acute colitis by dextran sulfate sodium in rats. *Dig Dis Sci* 53(6), 1601–1608.

Kaumaya, P.T., Kobs-Conrad, S., Seo, Y.H., Lee, H., VanBuskirk, A.M., Feng, N., Sheridan, J.F., Stevens, V., 1993. Peptide vaccines incorporating a 'promiscuous' T-cell epitope bypass certain haplotype restricted immune responses and provide broad spectrum immunogenicity. *J Mol Recognit* 6(2), 81–94.

Kekkonen, R.A., Kajasto, E., Miettinen, M., Veckman, V., Korpela, R., Julkunen, I., 2008. Probiotic Leuconostoc mesenteroides ssp. cremoris and *Streptococcus thermophilus* induce IL-12 and IFN-gamma production. *World J Gastroenterol* 14(8), 1192–1203.

Kim, H.J., Park, J.G., Matthijnssens, J., Lee, J.H., Bae, Y.C., Alfajaro, M.M., Park, S.I., Kang, M.I., Cho, K.O., 2011. Intestinal and extra-intestinal pathogenicity of a bovine reassortant rotavirus in calves and piglets. *Vet Microbiol* 152(3–4), 291–303.

Kim, Y.B., 2007. Gnotobiotic miniature swine for immunobiology and medicine. *Microb Ecol Health Dis* 19, 251–252.

Kim, Y.C., Kim, K.K., Shevach, E.M., 2010. Simvastatin induces Foxp3+ T regulatory cells by modulation of transforming growth factor-beta signal transduction. *Immunology* 130(4), 484–493.

Kim, Y.G., Ohta, T., Takahashi, T., Kushiro, A., Nomoto, K., Yokokura, T., Okada, N., Danbara, H., 2006. Probiotic *Lactobacillus casei* activates innate immunity via NF-kappaB and p38 MAP kinase signaling pathways. *Microbes Infect* 8(4), 994–1005.

Kitazawa, H., Villena, J., 2014. Modulation of respiratory TLR3-anti-viral response by probiotic microorganisms: Lessons learned from *Lactobacillus rhamnosus* CRL1505. *Front Immunol* 5, 201.

Kocher, J., 2014. *Evaluation of the Novel P Particle Vaccine Candidate against Human Norovirus Using the Gnotobiotic Pig Challenge Model*. Dissertation. Virginia Polytechnic Institute and State University, Blacksburg, VA.

Kocher, J., Bui, T., Giri-Rachman, E., Wen, K., Li, G., Yang, X., Liu, F., Tan, M., Xia, M., Zhong, W., Jiang, X., Yuan, L., 2014. Intranasal P particle vaccine provided partial cross-variant protection against human GII.4 norovirus diarrhea in gnotobiotic pigs. *J Virol* 88(17), 9728–9743.

Kocher, J., Castellucci, T.B., Wen, K., Li, G., Yang, X., Lei, S., Jiang, X., Yuan, L., 2021. Simvastatin reduces protection and intestinal T cell responses induced by a Norovirus P particle vaccine in gnotobiotic pigs. *Pathogens* 10(7), 829.

Koenig, J.E., Spor, A., Scalfone, N., Fricker, A.D., Stombaugh, J., Knight, R., Angenent, L.T., Ley, R.E., 2011. Succession of microbial consortia in the developing infant gut microbiome. *Proc Natl Acad Sci U S A* 108(Suppl 1), 4578–4585.

Kohno, M., Pouyssegur, J., 2006. Targeting the ERK signaling pathway in cancer therapy. *Ann Med* 38(3), 200–211.

Komiyama, Y., Andoh, A., Fujiwara, D., Ohmae, H., Araki, Y., Fujiyama, Y., Mitsuyama, K., Kanauchi, O., 2011. New prebiotics from rice bran ameliorate inflammation in murine colitis models through the modulation of intestinal homeostasis and the mucosal immune system. *Scand J Gastroenterol* 46(1), 40–52.

Konieczna, P., Groeger, D., Ziegler, M., Frei, R., Ferstl, R., Shanahan, F., Quigley, E.M., Kiely, B., Akdis, C.A., O'Mahony, L., 2012. Bifidobacterium infantis 35624 administration induces Foxp3 T regulatory cells in human peripheral blood: Potential role for myeloid and plasmacytoid dendritic cells. *Gut* 61(3), 354–366.

Konstantinov, S.R., Smidt, H., de Vos, W.M., Bruijns, S.C., Singh, S.K., Valence, F., Molle, D., Lortal, S., Altermann, E., Klaenhammer, T.R., van Kooyk, Y., 2008. S layer protein A of *Lactobacillus acidophilus* NCFM regulates immature dendritic cell and T cell functions. *Proc Natl Acad Sci U S A* 105(49), 19474–19479.

Konstantinov, S.R., Zhu, W.Y., Williams, B.A., Tamminga, S., Vos, W.M., Akkermans, A.D., 2003. Effect of fermentable carbohydrates on piglet faecal bacterial communities as revealed by denaturing gradient gel electrophoresis analysis of 16S ribosomal DNA. *FEMS Microbiol Ecol* 43(2), 225–235.

Kosek, M., Haque, R., Lima, A., Babji, S., Shrestha, S., Qureshi, S., Amidou, S., Mduma, E., Lee, G., Yori, P.P., Guerrant, R.L., Bhutta, Z., Mason, C., Kang, G., Kabir, M., Amour, C., Bessong, P., Turab, A., Seidman, J., Olortegui, M.P., Quetz, J., Lang, D., Gratz, J., Miller, M., Gottlieb, M., and for the MAL-ED Network, 2013. Fecal markers of intestinal inflammation and permeability associated with the subsequent acquisition of linear growth deficits in infants. *Am J Trop Med Hyg* 88(2), 390–396.

Kovacs-Nolan, J., Mine, Y., 2006. Tandem copies of a human rotavirus VP8 epitope can induce specific neutralizing antibodies in BALB/c mice. *Biochim Biophys Acta* 1760(12), 1884–1893.

Kukkonen, K., Nieminen, T., Poussa, T., Savilahti, E., Kuitunen, M., 2006. Effect of probiotics on vaccine antibody responses in infancy--A randomized placebo-controlled double-blind trial. *Pediatr Allergy Immunol* 17(6), 416–421.

Kumar, A., Arora, R., Kaur, P., Chauhan, V.S., Sharma, P., 1992. "Universal" T helper cell determinants enhance immunogenicity of a Plasmodium falciparum merozoite surface antigen peptide. *J Immunol* 148(5), 1499–1505.

Kumar, A., Henderson, A., Forster, G.M., Goodyear, A.W., Weir, T.L., Leach, J.E., Dow, S.W., Ryan, E.P., 2012. Dietary rice bran promotes resistance to *Salmonella enterica* serovar Typhimurium colonization in mice. *BMC Microbiol* 12, 71.

Kunisawa, J., Kiyono, H., 2011. Peaceful mutualism in the gut: Revealing key commensal bacteria for the creation and maintenance of immunological homeostasis. *Cell Host Microbe* 9(2), 83–84.

Kurokawa, N., Lavoie, P.O., D'Aoust, M.A., Couture, M.M., Dargis, M., Trepanier, S., Hoshino, S., Koike, T., Arai, M., Tsutsui, N., 2021. Development and characterization of a plant-derived rotavirus-like particle vaccine. *Vaccine* 39(35), 4979–4987.

Kuss, S.K., Best, G.T., Etheredge, C.A., Pruijssers, A.J., Frierson, J.M., Hooper, L.V., Dermody, T.S., Pfeiffer, J.K., 2011. Intestinal microbiota promote enteric virus replication and systemic pathogenesis. *Science* 334(6053), 249–252.

Kwak, B., Mulhaupt, F., Myit, S., Mach, F., 2000. Statins as a newly recognized type of immunomodulator. *Nat Med* 6(12), 1399–1402.

Kwak, B., Mulhaupt, F., Veillard, N., Pelli, G., Mach, F., 2001. The HMG-CoA reductase inhibitor simvastatin inhibits IFN-gamma induced MHC class II expression in human vascular endothelial cells. *Swiss Med Wkly* 131(3–4), 41–46.

Kyriakis, S.C., Tsiloyiannis, V.K., Vlemmas, J., Sarris, K., Tsinas, A.C., Alexopoulos, C., Jansegers, L., 1999. The effect of probiotic LSP 122 on the control of post-weaning diarrhoea syndrome of piglets. *Res Vet Sci* 67(3), 223–228.

Lakatos, K., McAdams, D., White, J.A., Chen, D., 2020. Formulation and preclinical studies with a trivalent rotavirus P2-VP8 subunit vaccine. *Hum Vaccin Immunother* 16(8), 1957–1968.

Laycock, G., Sait, L., Inman, C., Lewis, M., Smidt, H., van Diemen, P., Jorgensen, F., Stevens, M., Bailey, M., 2012. A defined intestinal colonization microbiota for gnotobiotic pigs. *Vet Immunol Immunopathol* 149(3–4), 216–224.

Lee, K., Kwon, D.N., Ezashi, T., Choi, Y.J., Park, C., Ericsson, A.C., Brown, A.N., Samuel, M.S., Park, K.W., Walters, E.M., Kim, D.Y., Kim, J.H., Franklin, C.L., Murphy, C.N., Roberts, R.M., Prather, R.S., Kim, J.H., 2014. Engraftment of human iPS cells and allogeneic porcine cells into pigs with inactivated RAG2 and accompanying severe combined immunodeficiency. *Proc Natl Acad Sci U S A* 111(20), 7260–7265.

Lee, K.J., Moon, J.Y., Choi, H.K., Hur, H.O., Jung, K.H., Lee, S.Y., Kim, J.H., Shin, C., Shim, J.J., In, K.H., Yoo, S.H., Kang, K.H., Lee, S.Y., 2010. Immune regulatory effects of simvastatin on regulatory T cell-mediated tumour immune tolerance. *Clin Exp Immunol* 161(2), 298–305.

Lefrancois, L., Goodman, T., 1989. In vivo modulation of cytolytic activity and Thy-1 expression in TCR-gamma delta+ intraepithelial lymphocytes. *Science* 243(4899), 1716–1718.

Lei, S., Ramesh, A., Twitchell, E., Wen, K., Bui, T., Weiss, M., Yang, X., Kocher, J., Li, G., Giri-Rachman, E., Trang, N.V., Jiang, X., Ryan, E.P., Yuan, L., 2016a. High protective efficacy of probiotics and rice bran against human Norovirus infection and diarrhea in gnotobiotic pigs. *Front Microbiol* 7, 1699.

Lei, S., Ryu, J., Wen, K., Twitchell, E., Bui, T., Ramesh, A., Weiss, M., Li, G., Samuel, H., Clark-Deener, S., Jiang, X., Lee, K., Yuan, L., 2016b. Increased and prolonged human Norovirus infection in RAG2/IL2RG deficient gnotobiotic pigs with severe combined immunodeficiency. *Sci Rep* 6, 25222.

Lei, S., Samuel, H., Twitchell, E., Bui, T., Ramesh, A., Wen, K., Weiss, M., Li, G., Yang, X., Jiang, X., Yuan, L., 2016c. Enterobacter cloacae inhibits human Norovirus infectivity in gnotobiotic pigs. *Sci Rep* 6, 25017.

Lei, S., Twitchell, E.L., Ramesh, A.K., Bui, T., Majette, E., Tin, C.M., Avery, R., Arango-Argoty, G., Zhang, L., Becker-Dreps, S., Azcarate-Peril, M.A., Jiang, X., Yuan, L., 2019. Enhanced GII.4 human norovirus infection in gnotobiotic pigs transplanted with a human gut microbiota. *J Gen Virol* 100(11), 1530–1540.

Lei, S., Yuan, L., 2019. Rice bran usage in diarrhea. In Watson, R., Preedy, V. (Eds.), *Dietary Interventions in Gastrointestinal Disease: Food, Nutrients and Dietary Supplements.* Academic Press, Elsevier Sciences, Cambridge, MA, 257–264.

Leung, B.P., Sattar, N., Crilly, A., Prach, M., McCarey, D.W., Payne, H., Madhok, R., Campbell, C., Gracie, J.A., Liew, F.Y., McInnes, I.B., 2003. A novel anti-inflammatory role for simvastatin in inflammatory arthritis. *J Immunol* 170(3), 1524–1530.

Lewis, H.M., Parry, J.V., Davies, H.A., Parry, R.P., Mott, A., Dourmashkin, R.R., Sanderson, P.J., Tyrrell, D.A., Valman, H.B., 1979. A year's experience of the rotavirus syndrome and its association with respiratory illness. *Arch Dis Child* 54(5), 339–346.

Li, C.Y., Lin, H.C., Lai, C.H., Lu, J.J., Wu, S.F., Fang, S.H., 2011. Immunomodulatory effects of Lactobacillus and Bifidobacterium on both murine and human mitogen-activated T cells. *Int Arch Allergy Immunol* 156(2), 128–136.

Li, X.Q., Zhu, Y.H., Zhang, H.F., Yue, Y., Cai, Z.X., Lu, Q.P., Zhang, L., Weng, X.G., Zhang, F.J., Zhou, D., Yang, J.C., Wang, J.F., 2012. Risks associated with high-dose *Lactobacillus rhamnosus* in an *Escherichia coli* model of piglet diarrhoea: Intestinal microbiota and immune imbalances. *PLOS ONE* 7(7), e40666.

Licciardi, P.V., Tang, M.L., 2011. Vaccine adjuvant properties of probiotic bacteria. *Discov Med* 12(67), 525–533.

Lievin-Le Moal, V., Sarrazin-Davila, L.E., Servin, A.L., 2007. An experimental study and a randomized, double-blind, placebo-controlled clinical trial to evaluate the antisecretory activity of *Lactobacillus acidophilus* strain LB against Nonrotavirus diarrhea. *Pediatrics*, 120(4), e795–803.

Ligtenberg, M.A., Rojas-Colonelli, N., Kiessling, R., Lladser, A., 2013. NF-kappaB activation during intradermal DNA vaccination is essential for eliciting tumor protective antigen-specific CTL responses. *Hum Vaccin Immunother* 9(10), 2189–2195.

Lin, S.C., Qu, L., Ettayebi, K., Crawford, S.E., Blutt, S.E., Robertson, M.J., Zeng, X.L., Tenge, V.R., Ayyar, B.V., Karandikar, U.C., Yu, X., Coarfa, C., Atmar, R.L., Ramani, S., Estes, M.K., 2020. Human Norovirus exhibits strain-specific sensitivity to host interferon pathways in human intestinal enteroids. *Proc Natl Acad Sci U S A* 117(38), 23782–23793.

Lindesmith, L., Moe, C., Lependu, J., Frelinger, J.A., Treanor, J., Baric, R.S., 2005. Cellular and humoral immunity following snow mountain virus challenge. *J Virol* 79(5), 2900–2909.

Link-Amster, H., Rochat, F., Saudan, K.Y., Mignot, O., Aeschlimann, J.M., 1994. Modulation of a specific humoral immune response and changes in intestinal flora mediated through fermented milk intake. *FEMS Immunol Med Microbiol* 10(1), 55–63.

Liu, F., Li, G., Wen, K., Bui, T., Cao, D., Zhang, Y., Yuan, L., 2010. Porcine small intestinal epithelial cell line (IPEC-J2) of rotavirus infection as a new model for the study of innate immune responses to rotaviruses and probiotics. *Viral Immunol* 23(2), 135–149.

Liu, F., Li, G., Wen, K., Wu, S., Zhang, Y., Bui, T., Yang, X., Kocher, J., Sun, J., Jortner, B., Yuan, L., 2013. *Lactobacillus rhamnosus* GG on rotavirus-induced injury of ileal epithelium in gnotobiotic pigs. *J Pediatr Gastroenterol Nutr* 57(6), 750–758.

Liu, F., Wen, K., Li, G., Yang, X., Kocher, J., Bui, T., Jones, D., Pelzer, K., Clark-Deener, S., Yuan, L., 2014. Dual functions of *Lactobacillus acidophilus* NCFM as protection against rotavirus diarrhea. *J Pediatr Gastroenterol Nutr* 58(2), 169–176.

Lozupone, C., Knight, R., 2005. UniFrac: A new phylogenetic method for comparing microbial communities. *Appl Environ Microbiol* 71(12), 8228–8235.

Macpherson, A.J., Harris, N.L., 2004. Interactions between commensal intestinal bacteria and the immune system. *Nat Rev Immunol* 4(6), 478–485.

Maffey, L., Vega, C.G., Mino, S., Garaicoechea, L., Parreno, V., 2016. Anti-VP6 VHH: An experimental treatment for rotavirus A-associated disease. *PLOS ONE* 11(9), e0162351.

Maier, E.A., Weage, K.J., Guedes, M.M., Denson, L.A., McNeal, M.M., Bernstein, D.I., Moore, S.R., 2013. Protein-energy malnutrition alters IgA responses to rotavirus vaccination and infection but does not impair vaccine efficacy in mice. *Vaccine* 32(1), 48–53.

Maitra, U., Gan, L., Chang, S., Li, L., 2011. Low-dose endotoxin induces inflammation by selectively removing nuclear receptors and activating CCAAT/enhancer-binding protein delta. *J Immunol* 186(7), 4467–4473.

Majamaa, H., Isolauri, E., Saxelin, M., Vesikari, T., 1995. Lactic acid bacteria in the treatment of acute rotavirus gastroenteritis. *J Pediatr Gastroenterol Nutr* 20(3), 333–338.

Mane, J., Pedrosa, E., Loren, V., Gassull, M.A., Espadaler, J., Cune, J., Audivert, S., Bonachera, M.A., Cabre, E., 2011. A mixture of *Lactobacillus plantarum* CECT 7315 and CECT 7316 enhances systemic immunity in elderly subjects. A dose-response, double-blind, placebo-controlled, randomized pilot trial. *Nutr Hosp* 26(1), 228–235.

Marionneau, S., Ruvoen, N., Le Moullac-Vaidye, B., Clement, M., Cailleau-Thomas, A., Ruiz-Palacois, G., Huang, P., Jiang, X., Le Pendu, J., 2002. Norwalk virus binds to histo-blood group antigens present on gastroduodenal epithelial cells of secretor individuals. *Gastroenterology* 122(7), 1967–1977.

Mathews, C.J., MacLeod, R.J., Zheng, S.X., Hanrahan, J.W., Bennett, H.P., Hamilton, J.R., 1999. Characterization of the inhibitory effect of boiled rice on intestinal chloride secretion in guinea pig crypt cells. *Gastroenterology* 116(6), 1342–1347.

Mathur, R.K., Awasthi, A., Wadhone, P., Ramanamurthy, B., Saha, B., 2004. Reciprocal CD40 signals through p38MAPK and ERK-1/2 induce counteracting immune responses. *Nat Med* 10(5), 540–544.

Mauroy, A., Gillet, L., Mathijs, E., Vanderplasschen, A., Thiry, E., 2011. Alternative attachment factors and internalization pathways for GIII.2 bovine noroviruses. *J Gen Virol* 92(6), 1398–1409.

Mausner-Fainberg, K., Luboshits, G., Mor, A., Maysel-Auslender, S., Rubinstein, A., Keren, G., George, J., 2008. The effect of HMG-CoA reductase inhibitors on naturally occurring CD4+CD25+ T cells. *Atherosclerosis* 197(2), 829–839.

Mayer, K.D., Mohrs, K., Crowe, S.R., Johnson, L.L., Rhyne, P., Woodland, D.L., Mohrs, M., 2005. The functional heterogeneity of type 1 effector T cells in response to infection is related to the potential for IFN-gamma production. *J Immunol* 174(12), 7732–7739.

McDonald, D., Price, M.N., Goodrich, J., Nawrocki, E.P., DeSantis, T.Z., Probst, A., Andersen, G.L., Knight, R., Hugenholtz, P., 2012. An improved greengenes taxonomy with explicit ranks for ecological and evolutionary analyses of bacteria and archaea. *ISME J* 6(3), 610–618.

McNeal, M.M., VanCott, J.L., Choi, A.H., Basu, M., Flint, J.A., Stone, S.C., Clements, J.D., Ward, R.L., 2002. CD4 T cells are the only lymphocytes needed to protect mice against rotavirus shedding after intranasal immunization with a chimeric VP6 protein and the adjuvant LT(R192G). *J Virol* 76(2), 560–568.

Medina, M., Vintini, E., Villena, J., Raya, R., Alvarez, S., 2010. *Lactococcus lactis* as an adjuvant and delivery vehicle of antigens against pneumococcal respiratory infections. *Bioeng Bugs* 1(5), 313–325.

Medzhitov, R., Janeway, C., Jr., 2000. The Toll receptor family and microbial recognition. *Trends Microbiol* 8(10), 452–456.

Meijerink, M., Wells, J.M., 2010. Probiotic modulation of dendritic cells and T cell responses in the intestine. *Benef Microbes* 1(4), 317–326.

Meng, X., Zhang, K., Li, J., Dong, M., Yang, J., An, G., Qin, W., Gao, F., Zhang, C., Zhang, Y., 2012. Statins induce the accumulation of regulatory T cells in atherosclerotic plaque. *Mol Med* 18, 598–605.

Messner, M.J., Berger, P., Nappier, S.P., 2014. Fractional poisson--A simple dose-response model for human Norovirus. *Risk Anal Off Publ Soc Risk Anal* 34(10), 1820–1829.

Meurens, F., Summerfield, A., Nauwynck, H., Saif, L., Gerdts, V., 2012. The pig: A model for human infectious diseases. *Trends Microbiol* 20(1), 50–57.

Meyer, F., Paarmann, D., D'Souza, M., Olson, R., Glass, E.M., Kubal, M., Paczian, T., Rodriguez, A., Stevens, R., Wilke, A., Wilkening, J., Edwards, R.A., 2008. The metagenomics RAST server - A public resource for the automatic phylogenetic and functional analysis of metagenomes. *BMC Bioinformatics* 9, 386.

Meyer, R.C., Bohl, E.H., Kohler, E.M., 1964. Procurement and maintenance of germ-free seine for microbiological investigations. *Appl Microbiol* 12, 295–300.

Meyers, S.N., Rogatcheva, M.B., Larkin, D.M., et al. 2005. Piggy-BACing the human genome II. A high-resolution, physically anchored, comparative map of the porcine autosomes. *Genomics* 86(6), 739–752.

Miettinen, M., Veckman, V., Latvala, S., Sareneva, T., Matikainen, S., Julkunen, I., 2008. Live *Lactobacillus rhamnosus* and *Streptococcus pyogenes* differentially regulate toll-like receptor (TLR) gene expression in human primary macrophages. *J Leukoc Biol* 84(4), 1092–1100.

Miyazaki, A., Kandasamy, S., Michael, H., Langel, S.N., Paim, F.C., Chepngeno, J., Alhamo, M.A., Fischer, D.D., Huang, H.C., Srivastava, V., Kathayat, D., Deblais, L., Rajashekara, G., Saif, L.J., Vlasova, A.N., 2018. Protein deficiency reduces efficacy of oral attenuated human rotavirus vaccine in a human infant fecal microbiota transplanted gnotobiotic pig model. *Vaccine* 36(42), 6270–6281.

Mohamadzadeh, M., Duong, T., Hoover, T., Klaenhammer, T.R., 2008. Targeting mucosal dendritic cells with microbial antigens from probiotic lactic acid bacteria. *Expert Rev Vaccines* 7(2), 163–174.

Mohamadzadeh, M., Olson, S., Kalina, W.V., Ruthel, G., Demmin, G.L., Warfield, K.L., Bavari, S., Klaenhammer, T.R., 2005. Lactobacilli activate human dendritic cells that skew T cells toward T helper 1 polarization. *Proc Natl Acad Sci U S A* 102(8), 2880–2885.

Monira, S., Nakamura, S., Gotoh, K., Izutsu, K., Watanabe, H., Alam, N.H., Endtz, H.P., Cravioto, A., Ali, S.I., Nakaya, T., Horii, T., Iida, T., Alam, M., 2011. Gut microbiota of healthy and malnourished children in Bangladesh. *Front Microbiol* 2, 228.

Murphy, W.J., Larkin, D.M., Everts-van der Wind, A., et al. 2005. Dynamics of mammalian chromosome evolution inferred from multispecies comparative maps. *Science* 309(5734), 613–617.

Nair, N., Feng, N., Blum, L.K., Sanyal, M., Ding, S., Jiang, B., Sen, A., Morton, J.M., He, X.S., Robinson, W.H., Greenberg, H.B., 2017. VP4- and VP7-specific antibodies mediate heterotypic immunity to rotavirus in humans. *Sci Transl Med* 9(395), eaam5434.

Narvaez, C.F., Angel, J., Franco, M.A., 2005. Interaction of rotavirus with human myeloid dendritic cells. *J Virol* 79(23), 14526–14535.

Nassar, C.A., Battistetti, G.D., Nahsan, F.P., Olegario, J., Marconato, J., Marin, C.F., Faccioni, D.M., da Costa, K.F., Kottwitz, L.B., Nassar, P.O., 2014. Evaluation of the effect of simvastatin on the progression of alveolar bone loss in experimental periodontitis--An animal study. *J Int Acad Periodontol* 16(1), 2–7.

Naylor, C., Lu, M., Haque, R., Mondal, D., Buonomo, E., Nayak, U., Mychaleckyj, J.C., Kirkpatrick, B., Colgate, R., Carmolli, M., Dickson, D., van der Klis, F., Weldon, W., Steven Oberste, M., PROVIDE study teams; Ma, J.Z., Petri Jr, W.A., 2015. Environmental enteropathy, oral vaccine failure and growth faltering in infants in Bangladesh. *EBiomedicine* 2(11), 1759–1766.

Nealon, N.J., Yuan, L., Yang, X., Ryan, E.P., 2017. Rice bran and probiotics alter the porcine large intestine and serum metabolomes for protection against human rotavirus diarrhea. *Front Microbiol* 8, 653.

Nelson, A.M., Walk, S.T., Taube, S., Taniuchi, M., Houpt, E.R., Wobus, C.E., Young, V.B., 2012. Disruption of the human gut microbiota following Norovirus infection. *PLOS ONE* 7(10), e48224.

Nguyen, T.V., Iosef, C., Jeong, K., Kim, Y., Chang, K.O., Lovgren-Bengtsson, K., Morein, B., Azevedo, M.S., Lewis, P., Nielsen, P., Yuan, L., Saif, L.J., 2003. Protection and antibody responses to oral priming by attenuated human rotavirus followed by oral boosting with 2/6-rotavirus-like particles with immunostimulating complexes in gnotobiotic pigs. *Vaccine* 21(25–26), 4059–4070.

Nguyen, T.V., Yuan, L., Azevedo, M.S., Jeong, K.I., Gonzalez, A.M., Iosef, C., Lovgren-Bengtsson, K., Morein, B., Lewis, P., Saif, L.J., 2006a. High titers of circulating maternal antibodies suppress effector and memory B-cell responses induced by an attenuated rotavirus priming and rotavirus-like particle-immunostimulating complex boosting vaccine regimen. *Clin Vaccin Immunol* 13(4), 475–485.

Nguyen, T.V., Yuan, L., Azevedo, M.S., Jeong, K.I., Gonzalez, A.M., Iosef, C., Lovgren-Bengtsson, K., Morein, B., Lewis, P., Saif, L.J., 2006b. Low titer maternal antibodies can both enhance and suppress B cell responses to a combined live attenuated human rotavirus and VLP-ISCOM vaccine. *Vaccine* 24(13), 2302–2316.

Nguyen, T.V., Yuan, L., Azevedo, M.S., Jeong, K.I., Gonzalez, A.M., Saif, L.J., 2007. Transfer of maternal cytokines to suckling piglets: In vivo and in vitro models with implications for immunomodulation of neonatal immunity. *Vet Immunol Immunopathol* 117(3–4), 236–248.

Nigro, G., Midulla, M., 1983. Acute laryngitis associated with rotavirus gastroenteritis. *J Infect* 7(1), 81–82.

Norazalina, S., Norhaizan, M.E., Hairuszah, I., Norashareena, M.S., 2010. Anticarcinogenic efficacy of phytic acid extracted from rice bran on azoxymethane-induced colon carcinogenesis in rats. *Exp Toxicol Pathol Off J Ges Toxikol Pathol* 62(3), 259–268.

Nubel, U., Engelen, B., Felske, A., Snaidr, J., Wieshuber, A., Amann, R.I., Ludwig, W., Backhaus, H., 1996. Sequence heterogeneities of genes encoding 16S rRNAs in *Paenibacillus polymyxa* detected by temperature gradient gel electrophoresis. *J Bacteriol* 178(19), 5636–5643.

O'Neal, C.M., Clements, J.D., Estes, M.K., Conner, M.E., 1998. Rotavirus 2/6 viruslike particles administered intranasally with cholera toxin, *Escherichia coli* heat-labile toxin (LT), and LT-R192G induce protection from rotavirus challenge. *J Virol* 72(4), 3390–3393.

O'Neal, C.M., Crawford, S.E., Estes, M.K., Conner, M.E., 1997. Rotavirus virus-like particles administered mucosally induce protective immunity. *J Virol* 71(11), 8707–8717.

O'Ryan, M., 2017. Rotavirus vaccines: A story of success with challenges ahead. *F1000Res* 6, 1517.

O'Ryan, M., Linhares, A.C., 2009. Update on Rotarix: An oral human rotavirus vaccine. *Expert Rev Vaccines* 8(12), 1627–1641.

O'Ryan, M.L., Hermosilla, G., Osorio, G., 2009. Rotavirus vaccines for the developing world. *Curr Opin Infect Dis* 22(5), 483–489.

Offit, P.A., Clark, H.F., 1985. Maternal antibody-mediated protection against gastroenteritis due to rotavirus in newborn mice is dependent on both serotype and titer of antibody. *J Infect Dis* 152(6), 1152–1158.

Offit, P.A., Dudzik, K.I., 1990. Rotavirus-specific cytotoxic T lymphocytes passively protect against gastroenteritis in suckling mice. *J Virol* 64(12), 6325–6328.

Oka, T., Katayama, K., Hansman, G.S., Kageyama, T., Ogawa, S., Wu, F.T., White, P.A., Takeda, N., 2006. Detection of human Sapovirus by real-time reverse transcription-polymerase chain reaction. *J Med Virol* 78(10), 1347–1353.

Okopien, B., Krysiak, R., Kowalski, J., Madej, A., Belowski, D., Zielinski, M., Labuzek, K., Herman, Z.S., 2004. The effect of statins and fibrates on interferon-gamma and interleukin-2 release in patients with primary type II dyslipidemia. *Atherosclerosis* 176(2), 327–335.

Olivares, M., Diaz-Ropero, M.P., Sierra, S., Lara-Villoslada, F., Fonolla, J., Navas, M., Rodriguez, J.M., Xaus, J., 2007. Oral intake of *Lactobacillus fermentum* CECT5716 enhances the effects of influenza vaccination. *Nutr (Burbank Los Angeles Cty Calif)* 23(3), 254–260.

Ouwehand, A.C., Salminen, S., Isolauri, E., 2002. Probiotics: An overview of beneficial effects. *Antonie Leeuwenhoek* 82(1–4), 279–289.

Paineau, D., Carcano, D., Leyer, G., Darquy, S., Alyanakian, M.A., Simoneau, G., Bergmann, J.F., Brassart, D., Bornet, F., Ouwehand, A.C., 2008. Effects of seven potential probiotic strains on specific immune responses in healthy adults: A double-blind, randomized, controlled trial. *FEMS Immunol Med Microbiol* 53(1), 107–113.

Palmer, C., Bik, E.M., DiGiulio, D.B., Relman, D.A., Brown, P.O., 2007. Development of the human infant intestinal microbiota. *PLOS Biol* 5(7), e177.

Pang, X., Hua, X., Yang, Q., Ding, D., Che, C., Cui, L., Jia, W., Bucheli, P., Zhao, L., 2007. Inter-species transplantation of gut microbiota from human to pigs. *ISME J* 1(2), 156–162.

Parreno, V., Hodgins, D.C., de Arriba, L., Kang, S.Y., Yuan, L., Ward, L.A., To, T.L., Saif, L.J., 1999. Serum and intestinal isotype antibody responses to Wa human rotavirus in gnotobiotic pigs are modulated by maternal antibodies. *J Gen Virol* 80(6), 1417–1428.

Patel, N.C., Hertel, P.M., Estes, M.K., de la Morena, M., Petru, A.M., Noroski, L.M., Revell, P.A., Hanson, I.C., Paul, M.E., Rosenblatt, H.M., Abramson, S.L., 2010. Vaccine-acquired rotavirus in infants with severe combined immunodeficiency. *N Engl J Med* 362(4), 314–319.

Pattekar, A., Mayer, L.S., Lau, C.W., Liu, C., Palko, O., Bewtra, M., Consortium, H., Lindesmith, L.C., Brewer-Jensen, P.D., Baric, R.S., Betts, M.R., Naji, A., Wherry, E.J., Tomov, V.T., 2021. Norovirus-specific CD8(+) T cell responses in human blood and tissues. *Cell Mol Gastroenterol Hepatol* 11(5), 1267–1289.

Patterson, L.J., Malkevitch, N., Venzon, D., Pinczewski, J., Gomez-Roman, V.R., Wang, L., Kalyanaraman, V.S., Markham, P.D., Robey, F.A., Robert-Guroff, M., 2004. Protection against mucosal simian immunodeficiency virus SIV(mac251) challenge by using replicating adenovirus-SIV multigene vaccine priming and subunit boosting. *J Virol* 78(5), 2212–2221.

Pearson, G.R., McNulty, M.S., 1977. Pathological changes in the small intestine of neonatal pigs infected with a pig reovirus-like agent (rotavirus). *J Comp Pathol* 87(3), 363–375.

Pedron, T., Mulet, C., Dauga, C., Frangeul, L., Chervaux, C., Grompone, G., Sansonetti, P.J., 2012. A crypt-specific core microbiota resides in the mouse colon. *mBio* 3(3), e00116-12.

Perdigon, G., Fuller, R., Raya, R., 2001. Lactic acid bacteria and their effect on the immune system. *Curr Issues Intest Microbiol* 2(1), 27–42.

Peroni, D.G., Morelli, L., 2021. Probiotics as adjuvants in vaccine strategy: Is there more room for improvement? *Vaccines (Basel)* 9(8), 811.

Perry, J.W., Wobus, C.E., 2010. Endocytosis of murine Norovirus 1 into murine macrophages is dependent on dynamin II and cholesterol. *J Virol* 84(12), 6163–6176.

Peterson, K.M., Buss, J., Easley, R., Yang, Z., Korpe, P.S., Niu, F., Ma, J.Z., Olortegui, M.P., Haque, R., Kosek, M.N., Petri, W.A., Jr., 2013. REG1B as a predictor of childhood stunting in Bangladesh and Peru. *Am J Clin Nutr* 97(5), 1129–1133.

Phelps, C.J., Koike, C., Vaught, T.D., Boone, J., Wells, K.D., Chen, S.H., Ball, S., Specht, S.M., Polejaeva, I.A., Monahan, J.A., Jobst, P.M., Sharma, S.B., Lamborn, A.E., Garst, A.S., Moore, M., Demetris, A.J., Rudert, W.A., Bottino, R., Bertera, S., Trucco, M., Starzl, T.E., Dai, Y., Ayares, D.L., 2003. Production of alpha 1,3-galactosyltransferase-deficient pigs. *Science* 299(5605), 411–414.

Plantinga, T.S., van Maren, W.W., van Bergenhenegouwen, J., Hameetman, M., Nierkens, S., Jacobs, C., de Jong, D.J., Joosten, L.A., van't Land, B., Garssen, J., Adema, G.J., Netea, M.G., 2011. Differential toll-like receptor recognition and induction of cytokine profile by *Bifidobacterium breve* and *Lactobacillus* strains of probiotics. *Clin Vaccin Immunol* 18(4), 621–628.

Plaza-Diaz, J., Gomez-Llorente, C., Fontana, L., Gil, A., 2014. Modulation of immunity and inflammatory gene expression in the gut, in inflammatory diseases of the gut and in the liver by probiotics. *World J Gastroenterol* 20(42), 15632–15649.

Pleasants, J.R., Johnson, M.H., Wostmann, B.S., 1986. Adequacy of chemically defined, water-soluble diet for germfree BALB/c mice through successive generations and litters. *J Nutr* 116(10), 1949–1964.

Premenko-Lanier, M., Rota, P.A., Rhodes, G., Verhoeven, D., Barouch, D.H., Lerche, N.W., Letvin, N.L., Bellini, W.J., McChesney, M.B., 2003. DNA vaccination of infants in the presence of maternal antibody: A measles model in the primate. *Virology* 307(1), 67–75.

Price, M.N., Dehal, P.S., Arkin, A.P., 2010. FastTree 2--Approximately maximum-likelihood trees for large alignments. *PLOS ONE* 5(3), e9490.

Rachmilewitz, D., Katakura, K., Karmeli, F., Hayashi, T., Reinus, C., Rudensky, B., Akira, S., Takeda, K., Lee, J., Takabayashi, K., Raz, E., 2004. Toll-like receptor 9 signaling mediates the anti-inflammatory effects of probiotics in murine experimental colitis. *Gastroenterology* 126(2), 520–528.

Ramesh, A., Mao, J., Lei, S., Twitchell, E., Shiraz, A., Jiang, X., Tan, M., Yuan, A.L., 2019. Parenterally administered P24-VP8* nanoparticle vaccine conferred strong protection against rotavirus diarrhea and virus shedding in gnotobiotic pigs. *Vaccines (Basel)* 7(4), 177.

Ramesh, A.K., Parreno, V., Schmidt, P.J., Lei, S., Zhong, W., Jiang, X., Emelko, M.B., Yuan, L., 2020. Evaluation of the 50% infectious dose of human Norovirus Cin-2 in gnotobiotic pigs: A comparison of classical and contemporary methods for endpoint estimation. *Viruses* 12(9), 955.

Ramig, R.F., 2004. Pathogenesis of intestinal and systemic rotavirus infection. *J Virol* 78(19), 10213–10220.

Ramirez, K., Wahid, R., Richardson, C., Bargatze, R.F., El-Kamary, S.S., Sztein, M.B., Pasetti, M.F., 2012. Intranasal vaccination with an adjuvanted Norwalk virus-like particle vaccine elicits antigen-specific B memory responses in human adult volunteers. *Clin Immunol* 144(2), 98–108.

Ray, B., Hutterer, C., Bandyopadhyay, S.S., Ghosh, K., Chatterjee, U.R., Ray, S., Zeittrager, I., Wagner, S., Marschall, M., 2013. Chemically engineered sulfated glucans from rice bran exert strong antiviral activity at the stage of viral entry. *J Nat Prod* 76(12), 2180–2188.

Reeck, A., Kavanagh, O., Estes, M.K., Opekun, A.R., Gilger, M.A., Graham, D.Y., Atmar, R.L., 2010. Serological correlate of protection against norovirus-induced gastroenteritis. *J Infect Dis* 202(8), 1212–1218.

Reed, L.J., Muench, H., 1938. A simple method for estimating fifty percent endpoints. *Am J Epidemiol* 27, 493–497.

Reid, G., Friendship, R., 2002. Alternatives to antibiotic use: Probiotics for the gut. *Anim Biotechnol* 13(1), 97–112.

Repa, A., Grangette, C., Daniel, C., Hochreiter, R., Hoffmann-Sommergruber, K., Thalhamer, J., Kraft, D., Breiteneder, H., Mercenier, A., Wiedermann, U., 2003. Mucosal co-application of lactic acid bacteria and allergen induces counter-regulatory immune responses in a murine model of birch pollen allergy. *Vaccine* 22(1), 87–95.

Robinson, C.M., Jesudhasan, P.R., Pfeiffer, J.K., 2014. Bacterial lipopolysaccharide binding enhances virion stability and promotes environmental fitness of an enteric virus. *Cell Host Microbe* 15(1), 36–46.

Rockx, B.H., Vennema, H., Hoebe, C.J., Duizer, E., Koopmans, M.P., 2005. Association of histo-blood group antigens and susceptibility to Norovirus infections. *J Infect Dis* 191(5), 749–754.

Rollo, E.E., Kumar, K.P., Reich, N.C., Cohen, J., Angel, J., Greenberg, H.B., Sheth, R., Anderson, J., Oh, B., Hempson, S.J., Mackow, E.R., Shaw, R.D., 1999. The epithelial cell response to rotavirus infection. *J Immunol* 163(8), 4442–4452.

Roth, A.N., Helm, E.W., Mirabelli, C., Kirsche, E., Smith, J.C., Eurell, L.B., Ghosh, S., Altan-Bonnet, N., Wobus, C.E., Karst, S.M., 2020. Norovirus infection causes acute self-resolving diarrhea in wild-type neonatal mice. *Nat Commun* 11(1), 2968.

Rothkotter, H.J., Kirchhoff, T., Pabst, R., 1994. Lymphoid and non-lymphoid cells in the epithelium and lamina propria of intestinal mucosa of pigs. *Gut* 35(11), 1582–1589.

Rowe, J.H., Ertelt, J.M., Way, S.S., 2012. Foxp3(+) regulatory T cells, immune stimulation and host defence against infection. *Immunology* 136(1), 1–10.

Ryan, E.P., 2011. Bioactive food components and health properties of rice bran. *J Am Vet Med Assoc* 238(5), 593–600.

Saif, L., Yuan, L., Ward, L., To, T., 1997. Comparative studies of the pathogenesis, antibody immune responses, and homologous protection to porcine and human rotaviruses in gnotobiotic piglets. *Adv Exp Med Biol* 412, 397–403.

Saif, L.J., Ward, L.A., Yuan, L., Rosen, B.I., To, T.L., 1996. The gnotobiotic piglet as a model for studies of disease pathogenesis and immunity to human rotaviruses. *Arch Virol Suppl* 12, 153–161.

Sato, M., Miyoshi, K., Nagao, Y., Nishi, Y., Ohtsuka, M., Nakamura, S., Sakurai, T., Watanabe, S., 2014. The combinational use of CRISPR/Cas9-based gene editing and targeted toxin technology enables efficient biallelic knockout of the alpha-1,3-galactosyltransferase gene in porcine embryonic fibroblasts. *Xenotransplantation* 21(3), 291–300.

Schloss, P.D., Handelsman, J., 2005. Introducing DOTUR, a computer program for defining operational taxonomic units and estimating species richness. *Appl Environ Microbiol* 71(3), 1501–1506.

Schmidt, B., Mulder, I.E., Musk, C.C., Aminov, R.I., Lewis, M., Stokes, C.R., Bailey, M., Prosser, J.I., Gill, B.P., Pluske, J.R., Kelly, D., 2011. Establishment of normal gut microbiota is compromised under excessive hygiene conditions. *PLOS ONE* 6(12), e28284.

Schmidt, P.J., 2015. Norovirus dose-response: Are currently available data informative enough to determine how susceptible humans are to infection from a single virus? *Risk Anal Off Publ Soc Risk Anal* 35(7), 1364–1383.

Schwiertz, A., Gruhl, B., Lobnitz, M., Michel, P., Radke, M., Blaut, M., 2003. Development of the intestinal bacterial composition in hospitalized preterm infants in comparison with breast-fed, full-term infants. *Pediatr Res* 54(3), 393–399.

Senok, A.C., Ismaeel, A.Y., Botta, G.A., 2005. Probiotics: Facts and myths. *Clin Microbiol Infect* 11(12), 958–966.

Shackelford, C., Long, G., Wolf, J., Okerberg, C., Herbert, R., 2002. Qualitative and quantitative analysis of nonneoplastic lesions in toxicology studies. *Toxicol Pathol* 30(1), 93–96.

Shawli, G.T., Adeyemi, O.O., Stonehouse, N.J., Herod, M.R., 2019. The oxysterol 25-hydroxycholesterol inhibits replication of murine Norovirus. *Viruses* 11(2), 97.

Sheflin, A.M., Borresen, E.C., Wdowik, M.J., Rao, S., Brown, R.J., Heuberger, A.L., Broeckling, C.D., Weir, T.L., Ryan, E.P., 2015. Pilot dietary intervention with heat-stabilized rice bran modulates stool microbiota and metabolites in healthy adults. *Nutrients* 7(2), 1282–1300.

Shen, J., Zhang, B., Wei, H., Che, C., Ding, D., Hua, X., Bucheli, P., Wang, L., Li, Y., Pang, X., Zhao, L., 2010. Assessment of the modulating effects of fructo-oligosaccharides on fecal microbiota using human flora-associated piglets. *Arch Microbiol* 192(11), 959–968.

Shornikova, A.V., Casas, I.A., Isolauri, E., Mykkanen, H., Vesikari, T., 1997. *Lactobacillus reuteri* as a therapeutic agent in acute diarrhea in young children. *J Pediatr Gastroenterol Nutr* 24(4), 399–404.

Shu, Q., Qu, F., Gill, H.S., 2001. Probiotic treatment using *Bifidobacterium lactis* HN019 reduces weanling diarrhea associated with rotavirus and *Escherichia coli* infection in a piglet model. *J Pediatr Gastroenterol Nutr* 33(2), 171–177.

Sierra, S., Lara-Villoslada, F., Olivares, M., Jimenez, J., Boza, J., Xaus, J., 2005. Increased immune response in mice consuming rice bran oil. *Eur J Nutr* 44(8), 509–516.

Silva, M., Jacobus, N.V., Deneke, C., Gorbach, S.L., 1987. Antimicrobial substance from a human Lactobacillus strain. *Antimicrob Agents Chemother* 31(8), 1231–1233.

Sindhu, K.N., Sowmyanarayanan, T.V., Paul, A., Babji, S., Ajjampur, S.S., Priyadarshini, S., Sarkar, R., Balasubramanian, K.A., Wanke, C.A., Ward, H.D., Kang, G., 2014. Immune response and intestinal permeability in children with acute gastroenteritis treated with *Lactobacillus rhamnosus* GG: A randomized, double-blind, placebo-controlled trial. *Clin Infect Dis* 58(8), 1107–1115.

Sinkora, M., Butler, J.E., 2009. The ontogeny of the porcine immune system. *Dev Comp Immunol* 33(3), 273–283.

Sironi, L., Banfi, C., Brioschi, M., Gelosa, P., Guerrini, U., Nobili, E., Gianella, A., Paoletti, R., Tremoli, E., Cimino, M., 2006. Activation of NF-kB and ERK1/2 after permanent focal ischemia is abolished by simvastatin treatment. *Neurobiol Dis* 22(2), 445–451.

Snydman, D.R., 2008. The safety of probiotics. *Clin Infect Dis Off Publ Infect Dis Soc Am* 46(Suppl 2), S104–111; discussion S144–151.

So, J.S., Kwon, H.K., Lee, C.G., Yi, H.J., Park, J.A., Lim, S.Y., Hwang, K.C., Jeon, Y.H., Im, S.H., 2008a. *Lactobacillus casei* suppresses experimental arthritis by down-regulating T helper 1 effector functions. *Mol Immunol* 45(9), 2690–2699.

So, J.S., Lee, C.G., Kwon, H.K., Yi, H.J., Chae, C.S., Park, J.A., Hwang, K.C., Im, S.H., 2008b. *Lactobacillus casei* potentiates induction of oral tolerance in experimental arthritis. *Mol Immunol* 46(1), 172–180.

Souza, M., Cheetham, S.M., Azevedo, M.S., Costantini, V., Saif, L.J., 2007a. Cytokine and antibody responses in gnotobiotic pigs after infection with human Norovirus genogroup II.4 (HS66 strain). *J Virol* 81(17), 9183–9192.

Souza, M., Costantini, V., Saif, L.J., 2007b. A human norovirus-like particle vaccine adjuvanted with ISCOM or mLT induces cytokine and antibody responses and protection to the homologous GII.4 human norovirus in a gnotobiotic pig disease model. *Vaccine* 25, 8448–8459.

Splichalova, A., Trebichavsky, I., Rada, V., Vlkova, E., Sonnenborn, U., Splichal, I., 2011. Interference of *Bifidobacterium choerinum* or *Escherichia coli* Nissle 1917 with Salmonella Typhimurium in gnotobiotic piglets correlates with cytokine patterns in blood and intestine. *Clin Exp Immunol* 163(2), 242–249.

Srivastava, V., Deblais, L., Huang, H.C., Miyazaki, A., Kandasamy, S., Langel, S.N., Paim, F.C., Chepngeno, J., Kathayat, D., Vlasova, A.N., Saif, L.J., Rajashekara, G., 2020. Reduced rotavirus vaccine efficacy in protein malnourished human-faecal-microbiota-transplanted gnotobiotic pig model is in part attributed to the gut microbiota. *Benef Microbes* 11(8), 733–751.

Stals, A., Baert, L., Botteldoorn, N., Werbrouck, H., Herman, L., Uyttendaele, M., Van Coillie, E., 2009. Multiplex real-time RT-PCR for simultaneous detection of GI/GII noroviruses and murine Norovirus 1. *J Virol Methods* 161(2), 247–253.

Steele, J., Feng, H., Parry, N., Tzipori, S., 2010. Piglet models of acute or chronic *Clostridium difficile* illness. *J Infect Dis* 201(3), 428–434.

Stepankova, R., Sinkora, J., Hudcovic, T., Kozakova, H., Tlaskalova-Hogenova, H., 1998. Differences in development of lymphocyte subpopulations from gut-associated lymphatic tissue (GALT) of germfree and conventional rats: Effect of aging. *Folia Microbiol* 43(5), 531–534.

Summerfield, A., McCullough, K.C., 2009. The porcine dendritic cell family. *Dev Comp Immunol* 33(3), 299–309.

Suzuki, T., Sundt, T.M., 3rd, Mixon, A., Sachs, D.H., 1990. In vivo treatment with antiporcine T cell antibodies. *Transplantation* 50(1), 76–81.

Swindle, M.M., Makin, A., Herron, A.J., Clubb, F.J., Jr., Frazier, K.S., 2012. Swine as models in biomedical research and toxicology testing. *Vet Pathol* 49(2), 344–356.

Szajewska, H., Wanke, M., Patro, B., 2011. Meta-analysis: The effects of *Lactobacillus rhamnosus* GG supplementation for the prevention of healthcare-associated diarrhoea in children. *Aliment Pharmacol Ther* 34(9), 1079–1087.

Tacket, C.O., Sztein, M.B., Losonsky, G.A., Wasserman, S.S., Estes, M.K., 2003. Humoral, mucosal, and cellular immune responses to oral Norwalk virus-like particles in volunteers. *Clin Immunol* 108(3), 241–247.

Takenaka, S., Itoyama, Y., 1993. Rice bran hemicellulose increases the peripheral blood lymphocytes in rats. *Life Sci* 52(1), 9–12.

Tamminen, K., Huhti, L., Koho, T., Lappalainen, S., Hytonen, V.P., Vesikari, T., Blazevic, V., 2012. A comparison of immunogenicity of Norovirus GII-4 virus-like particles and P-particles. *Immunology* 135(1), 89–99.

Tamura, M., Hori, S., Hoshi, C., Nakagawa, H., 2012. Effects of rice bran oil on the intestinal microbiota and metabolism of isoflavones in adult mice. *Int J Mol Sci* 13(8), 10336–10349.

Tan, M., Huang, P., Meller, J., Zhong, W., Farkas, T., Jiang, X., 2003. Mutations within the P2 domain of Norovirus capsid affect binding to human histo-blood group antigens: Evidence for a binding pocket. *J Virol* 77(23), 12562–12571.

Tan, M., Huang, P., Xia, M., Fang, P.-A., Zhong, W., McNeal, M., Wei, C., Jiang, W., Jiang, X., 2011. Norovirus P particle, a novel platform for vaccine development and antibody production. *J Virol* 85(2), 753.

Tan, M., Jiang, X., 2005a. Norovirus and its histo-blood group antigen receptors: An answer to a historical puzzle. *Trends Microbiol* 13(6), 285–293.

Tan, M., Jiang, X., 2005b. The p domain of Norovirus capsid protein forms a subviral particle that binds to histo-blood group antigen receptors. *J Virol* 79(22), 14017–14030.

Tan, M., Jiang, X., 2012. The formation of P particle increased immunogenicity of Norovirus P protein. *Immunology* 136(1), 28–29.

Tao, Y., Drabik, K.A., Waypa, T.S., Musch, M.W., Alverdy, J.C., Schneewind, O., Chang, E.B., Petrof, E.O., 2006. Soluble factors from *Lactobacillus* GG activate MAPKs and induce cytoprotective heat shock proteins in intestinal epithelial cells. *Am J Physiol Cell Physiol* 290(4), C1018–1030.

Tate, J.E., Burton, A.H., Boschi-Pinto, C., Parashar, U.D., World Health Organization-Coordinated Global Rotavirus Surveillance Network, 2016. Global, regional, and national estimates of rotavirus mortality in children <5 years of age, 2000–2013. *Clin Infect Dis* 62(Suppl 2), S96–105.

Tate, J.E., Burton, A.H., Boschi-Pinto, C., Steele, A.D., Duque, J., Parashar, U.D., World Health Organization-Coordinated Global Rotavirus Surveillance Network, 2012. 2008 estimate of worldwide rotavirus-associated mortality in children younger than 5 years before the introduction of universal rotavirus vaccination programmes: A systematic review and meta-analysis. *Lancet Infect Dis* 12(2), 136–141.

Teunis, P.F., Moe, C.L., Liu, P., Miller, S.E., Lindesmith, L., Baric, R.S., Le Pendu, J., Calderon, R.L., 2008. Norwalk virus: How infectious is it? *J Med Virol* 80(8), 1468–1476.

Thomas, J.W., Touchman, J.W., Blakesley, R.W., et al. 2003. Comparative analyses of multi-species sequences from targeted genomic regions. *Nature* 424(6950), 788–793.

Tu, N. Thi Kha, Thi Thu Hong, N., Ny, N. Thi Han, My Phuc, T., Tam, P. Thi Thanh, Doorn, H.R.V., Dang Trung Nghia, H., Thao Huong, D., An Han, D., Thi Thu Ha, L., Deng, X., Thwaites, G., Delwart, E., Virtala, A.K., Vapalahti, O., Baker, S., Van Tan, L., 2020. The virome of acute respiratory diseases in individuals at risk of zoonotic infections. *Viruses* 12(9), 960.

Tian, P., Jiang, X., Zhong, W., Jensen, H.M., Brandl, M., Bates, A.H., Engelbrektson, A.L., Mandrell, R., 2007. Binding of recombinant Norovirus like particle to histo-blood group antigen on cells in the lumen of pig duodenum. *Res Vet Sci* 83(3), 410–418.

Tlaskalova-Hogenova, H., Sterzl, J., Stepankova, R., Dlabac, V., Veticka, V., Rossmann, P., Mandel, L., Rejnek, J., 1983. Development of immunological capacity under germfree and conventional conditions. *Ann N Y Acad Sci* 409, 96–113.

To, T.L., Ward, L.A., Yuan, L., Saif, L.J., 1998. Serum and intestinal isotype antibody responses and correlates of protective immunity to human rotavirus in a gnotobiotic pig model of disease. *J Gen Virol* 79(11), 2661–2672.

Toki, S., Kagaya, S., Shinohara, M., Wakiguchi, H., Matsumoto, T., Takahata, Y., Morimatsu, F., Saito, H., Matsumoto, K., 2009. *Lactobacillus rhamnosus* GG and *Lactobacillus casei* suppress *Escherichia coli*-induced chemokine expression in intestinal epithelial cells. *Int Arch Allergy Immunol* 148(1), 45–58.

Tomov, V.T., Osborne, L.C., Dolfi, D.V., Sonnenberg, G.F., Monticelli, L.A., Mansfield, K., Virgin, H.W., Artis, D., Wherry, E.J., 2013. Persistent enteric murine Norovirus infection is associated with functionally suboptimal virus-specific CD8 T cell responses. *J Virol* 87(12), 7015–7031.

Torii, A., Torii, S., Fujiwara, S., Tanaka, H., Inagaki, N., Nagai, H., 2007. *Lactobacillus acidophilus* strain L-92 regulates the production of Th1 cytokine as well as Th2 cytokines. *Allergol Int* 56(3), 293–301.

Torres-Medina, A., Wyatt, R.G., Mebus, C.A., Underdahl, N.R., Kapikian, A.Z., 1976a. Diarrhea caused in gnotobiotic piglets by the reovirus-like agent of human infantile gastroenteritis. *J Infect Dis* 133(1), 22–27.

Torres-Medina, A., Wyatt, R.G., Mebus, C.A., Underdahl, N.R., Kapikian, A.Z., 1976b. Patterns of shedding of human reovirus-like agent in gnotobiotic newborn piglets with experimentally-induced diarrhea. *Intervirology* 7(4–5), 250–255.

Torres, A., Ji-Huang, L., 1986. Diarrheal response of gnotobiotic pigs after fetal infection and neonatal challenge with homologous and heterologous human rotavirus strains. *J Virol* 60(3), 1107–1112.

Treanor, J.J., Atmar, R.L., Frey, S.E., Gormley, R., Chen, W.H., Ferreira, J., Goodwin, R., Borkowski, A., Clemens, R., Mendelman, P.M., 2014. A Novel intramuscular bivalent Norovirus VLP vaccine candidate - Reactogenicity, safety and immunogenicity in a phase I trial in healthy adults. *J Infect Dis* 210(11), 1763–1771.

Turnbaugh, P.J., Ridaura, V.K., Faith, J.J., Rey, F.E., Knight, R., Gordon, J.I., 2009. The effect of diet on the human gut microbiome: A metagenomic analysis in humanized gnotobiotic mice. *Sci Transl Med* 1(6), 6ra14.

Twitchell, E.L., Tin, C., Wen, K., Zhang, H., Becker-Dreps, S., Azcarate-Peril, M.A., Vilchez, S., Li, G., Ramesh, A., Weiss, M., Lei, S., Bui, T., Yang, X., Schultz-Cherry, S., Yuan, L., 2016. Modeling human enteric dysbiosis and rotavirus immunity in gnotobiotic pigs. *Gut Pathog* 8, 51.

Uchiyama, R., Chassaing, B., Zhang, B., Gewirtz, A.T., 2014. Antibiotic treatment suppresses rotavirus infection and enhances specific humoral immunity. *J Infect Dis* 210, 171–182.

Ulivieri, C., Fanigliulo, D., Benati, D., Pasini, F.L., Baldari, C.T., 2008. Simvastatin impairs humoral and cell-mediated immunity in mice by inhibiting lymphocyte homing, T-cell activation and antigen cross-presentation. *Eur J Immunol* 38(10), 2832–2844.

Valdez, Y., Brown, E.M., Finlay, B.B., 2014. Influence of the microbiota on vaccine effectiveness. *Trends Immunol* 35(11), 526–537.

van Diest, P., van, Van Dam, P., Henzen-Logmans, S., Berns, E., Van der Burg, M., Green, J., Vergote, I., 1997. A scoring system for immunohistochemical staining: Consensus report of the task force for basic research of the EORTC-GCCG. European Organization for Research and Treatment of Cancer-gynaecological Cancer Cooperative Group. *J Clin Pathol* 50(10), 801.

Van Niel, C.W., Feudtner, C., Garrison, M.M., Christakis, D.A., 2002. Lactobacillus therapy for acute infectious diarrhea in children: A meta-analysis. *Pediatrics* 109(4), 678–684.

Van Overtvelt, L., Moussu, H., Horiot, S., Samson, S., Lombardi, V., Mascarell, L., van de Moer, A., Bourdet-Sicard, R., Moingeon, P., 2010. Lactic acid bacteria as adjuvants for sublingual allergy vaccines. *Vaccine* 28(17), 2986–2992.

VanCott, J.L., McNeal, M.M., Flint, J., Bailey, S.A., Choi, A.H., Ward, R.L., 2001. Role for T cell-independent B cell activity in the resolution of primary rotavirus infection in mice. *Eur J Immunol* 31(11), 3380–3387.

Varyukhina, S., Freitas, M., Bardin, S., Robillard, E., Tavan, E., Sapin, C., Grill, J.P., Trugnan, G., 2012. Glycan-modifying bacteria-derived soluble factors from *Bacteroides thetaiotaomicron* and *Lactobacillus casei* inhibit rotavirus infection in human intestinal cells. *Microbes Infect* 14(3), 273–278.

Vega, C.G., Bok, M., Vlasova, A.N., Chattha, K.S., Fernandez, F.M., Wigdorovitz, A., Parreno, V.G., Saif, L.J., 2012. IgY antibodies protect against human rotavirus induced diarrhea in the neonatal gnotobiotic piglet disease model. *PLOS ONE* 7(8), e42788.

Vega, C.G., Bok, M., Vlasova, A.N., Chattha, K.S., Gomez-Sebastian, S., Nunez, C., Alvarado, C., Lasa, R., Escribano, J.M., Garaicoechea, L.L., Fernandez, F., Bok, K., Wigdorovitz, A., Saif, L.J., Parreno, V., 2013. Recombinant monovalent llama-derived antibody fragments (VHH) to rotavirus VP6 protect neonatal gnotobiotic piglets against human rotavirus-induced diarrhea. *PLOS Pathog* 9(5), e1003334.

Verma, N., Rettenmeier, A.W., Schmitz-Spanke, S., 2011. Recent advances in the use of Sus scrofa (pig) as a model system for proteomic studies. *Proteomics* 11(4), 776–793.

Villabruna, N., Koopmans, M.P.G., de Graaf, M., 2019. Animals as reservoir for human Norovirus. *Viruses* 11(5), 478.

Villadangos, J.A., Cardoso, M., Steptoe, R.J., van Berkel, D., Pooley, J., Carbone, F.R., Shortman, K., 2001. MHC class II expression is regulated in dendritic cells independently of invariant chain degradation. *Immunity* 14(6), 739–749.

Villena, J., Aso, H., Kitazawa, H., 2014a. Regulation of toll-like receptors-mediated inflammation by immunobiotics in bovine intestinal epitheliocytes: Role of signaling pathways and negative regulators. *Front Immunol* 5, 421.

Villena, J., Chiba, E., Vizoso-Pinto, M.G., Tomosada, Y., Takahashi, T., Ishizuka, T., Aso, H., Salva, S., Alvarez, S., Kitazawa, H., 2014b. Immunobiotic *Lactobacillus rhamnosus* strains differentially modulate antiviral immune response in porcine intestinal epithelial and antigen presenting cells. *BMC Microbiol* 14, 126.

Villena, J., Kitazawa, H., 2014. Modulation of intestinal TLR4-inflammatory signaling pathways by probiotic microorganisms: Lessons learned from *Lactobacillus jensenii* TL2937. *Front Immunol* 4, 512.

Vitetta, L., Saltzman, E.T., Thomsen, M., Nikov, T., Hall, S., 2017. Adjuvant probiotics and the intestinal microbiome: Enhancing vaccines and immunotherapy outcomes. *Vaccines (Basel)* 5(4), 50.

Vitini, E., Alvarez, S., Medina, M., Medici, M., de Budeguer, M.V., Perdigon, G., 2000. Gut mucosal immunostimulation by lactic acid bacteria. *Biocell* 24(3), 223–232.

Vlasova, A.N., Paim, F.C., Kandasamy, S., Alhamo, M.A., Fischer, D.D., Langel, S.N., Deblais, L., Kumar, A., Chepngeno, J., Shao, L., Huang, H.C., Candelero-Rueda, R.A., Rajashekara, G., Saif, L.J., 2017. Protein malnutrition modifies innate immunity and gene expression by intestinal epithelial cells and human rotavirus infection in neonatal gnotobiotic pigs. *mSphere* 2(2), e00046-17.

Vodicka, P., Smetana, K., Jr., Dvorankova, B., Emerick, T., Xu, Y.Z., Ourednik, J., Ourednik, V., Motlik, J., 2005. The miniature pig as an animal model in biomedical research. *Ann N Y Acad Sci* 1049, 161–171.

Voltan, S., Castagliuolo, I., Elli, M., Longo, S., Brun, P., D'Inca, R., Porzionato, A., Macchi, V., Palu, G., Sturniolo, G.C., Morelli, L., Martines, D., 2007. Aggregating phenotype in *Lactobacillus crispatus* determines intestinal colonization and TLR2 and TLR4 modulation in murine colonic mucosa. *Clin Vaccin Immunol* 14(9), 1138–1148.

Walters, E.M., Prather, R.S., 2013. Advancing swine models for human health and diseases. *Mo Med* 110(3), 212–215.

Wang, H., Gao, K., Wen, K., Allen, I.C., Li, G., Zhang, W., Kocher, J., Yang, X., Giri-Rachman, E., Li, G.H., Clark-Deener, S., Yuan, L., 2016a. *Lactobacillus rhamnosus* GG modulates innate signaling pathway and cytokine responses to rotavirus vaccine in intestinal mononuclear cells of gnotobiotic pigs transplanted with human gut microbiota. *BMC Microbiol* 16(1), 109.

Wang, H.F., Zhu, W.Y., Yao, W., Liu, J.X., 2007a. DGGE and 16S rDNA sequencing analysis of bacterial communities in colon content and feces of pigs fed whole crop rice. *Anaerobe* 13(3–4), 127–133.

Wang, M., Donovan, S.M., 2015. Human microbiota-associated swine: Current progress and future opportunities. *ILAR J* 56(1), 63–73.

Wang, Q., Garrity, G.M., Tiedje, J.M., Cole, J.R., 2007b. Naive Bayesian classifier for rapid assignment of rRNA sequences into the new bacterial taxonomy. *Appl Environ Microbiol* 73(16), 5261–5267.

Wang, Y., Azevedo, M., Saif, L.J., Gentsch, J.R., Glass, R.I., Jiang, B., 2010. Inactivated rotavirus vaccine induces protective immunity in gnotobiotic piglets. *Vaccine* 28(33), 5432–5436.

Wang, Y., Vlasova, A., Velasquez, D.E., Saif, L.J., Kandasamy, S., Kochba, E., Levin, Y., Jiang, B., 2016b. Skin vaccination against rotavirus using microneedles: Proof of concept in gnotobiotic piglets. *PLOS ONE* 11(11), e0166038.

Ward, L.A., Rosen, B.I., Yuan, L., Saif, L.J., 1996a. Pathogenesis of an attenuated and a virulent strain of group A human rotavirus in neonatal gnotobiotic pigs. *J Gen Virol* 77(7), 1431–1441.

Ward, L.A., Yuan, L., Rosen, B.I., To, T.L., Saif, L.J., 1996b. Development of mucosal and systemic lymphoproliferative responses and protective immunity to human group A rotaviruses in a gnotobiotic pig model. *Clin Diagn Lab Immunol* 3(3), 342–350.

Ward, R.L., 2003. Possible mechanisms of protection elicited by candidate rotavirus vaccines as determined with the adult mouse model. *Viral Immunol* 16(1), 17–24.

Ward, R.L., McNeal, M.M., Farone, M.B., Farone, A.L., 2007. Reoviridae. In Fox, J.G., Barthold, S.W., Davisson, M.T., Newcomer, C.E., Quimby, F.W., Smith, A.L. (Eds.), *The Mouse in Biomedical Research.* Vol.2, Diseases. Academic Press, New York and London, 235–268.

Wei, C., Liu, J., Yu, Z., Zhang, B., Gao, G., Jiao, R., 2013. TALEN or Cas9 - Rapid, efficient and specific choices for genome modifications. *J Genet Genomics* 40(6), 281–289.

Wei, H., Shen, J., Pang, X., Ding, D., Zhang, Y., Zhang, B., Lu, H., Wang, T., Zhang, C., Hua, X., Cui, L., Zhao, L., 2008. Fatal infection in human flora-associated piglets caused by the opportunistic pathogen *Klebsiella pneumoniae* from an apparently healthy human donor. *J Vet Med Sci* 70(7), 715–717.

Weir, M.H., Mitchell, J., Flynn, W., Pope, J.M., 2017. Development of a microbial dose response visualization and modelling application for QMRA modelers and educators. *Environ Modell Softw* 88, 74–83.

Weiss, G., Rasmussen, S., Zeuthen, L.H., Nielsen, B.N., Jarmer, H., Jespersen, L., Frokiaer, H., 2010. *Lactobacillus acidophilus* induces virus immune defence genes in murine dendritic cells by a Toll-like receptor-2-dependent mechanism. *Immunology* 131(2), 268–281.

Wells, J.M., 2011. Immunomodulatory mechanisms of lactobacilli. *Microb Cell Factories* 10(Suppl 1), S17.

Wells, J.M., Mercenier, A., 2008. Mucosal delivery of therapeutic and prophylactic molecules using lactic acid bacteria. *Nat Rev Microbiol* 6(5), 349–362.

Wen, K., Azevedo, M.S., Gonzalez, A., Zhang, W., Saif, L.J., Li, G., Yousef, A., Yuan, L., 2009. Toll-like receptor and innate cytokine responses induced by lactobacilli colonization and human rotavirus infection in gnotobiotic pigs. *Vet Immunol Immunopathol* 127(3–4), 304–315.

Wen, K., Bui, T., Li, G., Liu, F., Li, Y., Kocher, J., Yuan, L., 2012a. Characterization of immune modulating functions of gammadelta T cell subsets in a gnotobiotic pig model of human rotavirus infection. *Comp Immunol Microbiol Infect Dis* 35(4), 289–301.

Wen, K., Bui, T., Weiss, M., Li, G., Kocher, J., Yang, X., Jobst, P.M., Vaught, T., Ramsoondar, J., Ball, S., Clark-Deener, S., Ayares, D., Yuan, L., 2016. B-cell-deficient and CD8 T-cell-depleted gnotobiotic pigs for the study of human rotavirus vaccine-induced protective immune responses. *Viral Immunol* 29(2), 112–127.

Wen, K., Li, G., Bui, T., Liu, F., Li, Y., Kocher, J., Lin, L., Yang, X., Yuan, L., 2012b. High dose and low dose *Lactobacillus acidophilus* exerted differential immune modulating effects on T cell immune responses induced by an oral human rotavirus vaccine in gnotobiotic pigs. *Vaccine* 30(6), 1198–1207.

Wen, K., Li, G., Yang, X., Bui, T., Bai, M., Liu, F., Kocher, J., Yuan, L., 2012c. CD4+ CD25- FoxP3+ regulatory cells are the predominant responding regulatory T cells after human rotavirus infection or vaccination in gnotobiotic pigs. *Immunology* 137(2), 160–171.

Wen, K., Li, G., Zhang, W., Azevedo, M.S., Saif, L.J., Liu, F., Bui, T., Yousef, A., Yuan, L., 2011. Development of gammadelta T cell subset responses in gnotobiotic pigs infected with human rotaviruses and colonized with probiotic lactobacilli. *Vet Immunol Immunopathol* 141(3–4), 267–275.

Wen, K., Liu, F., Li, G., Bai, M., Kocher, J., Yang, X., Wang, H., Clark-Deener, S., Yuan, L., 2015. *Lactobacillus rhamnosus* GG dosage affects the adjuvanticity and protection against rotavirus diarrhea in gnotobiotic pigs. *J Pediatr Gastroenterol Nutr* 60(6), 834–843.

Wen, K., Tin, C., Wang, H., Yang, X., Li, G., Giri-Rachman, E., Kocher, J., Bui, T., Clark-Deener, S., Yuan, L., 2014a. Probiotic *Lactobacillus rhamnosus* GG enhanced Th1 cellular immunity but did not affect antibody responses in a human gut microbiota transplanted neonatal gnotobiotic pig model. *PLOS ONE* 9(4), e94504.

Wen, X., Cao, D., Jones, R.W., Li, J., Szu, S., Hoshino, Y., 2012d. Construction and characterization of human rotavirus recombinant VP8* subunit parenteral vaccine candidates. *Vaccine* 30(43), 6121–6126.

Wen, X., Wen, K., Cao, D., Li, G., Jones, R.W., Li, J., Szu, S., Hoshino, Y., Yuan, L., 2014b. Inclusion of a universal tetanus toxoid CD4(+) T cell epitope P2 significantly enhanced the immunogenicity of recombinant rotavirus DeltaVP8* subunit parenteral vaccines. *Vaccine* 32(35), 4420–4427.

Wentzel, J.F., Yuan, L., Rao, S., van Dijk, A.A., O'Neill, H.G., 2013. Consensus sequence determination and elucidation of the evolutionary history of a rotavirus Wa variant reveal a close relationship to various Wa variants derived from the original Wa strain. *Infect Genet Evol* 20, 276–283.

West, N.P., Pyne, D.B., Peake, J.M., Cripps, A.W., 2009. Probiotics, immunity and exercise: A review. *Exer Immunol Rev* 15, 107–126.

Westerman, L.E., McClure, H.M., Jiang, B., Almond, J.W., Glass, R.I., 2005. Serum IgG mediates mucosal immunity against rotavirus infection. *Proc Natl Acad Sci U S A* 102(20), 7268–7273.

Wickert, L.E., Karta, M.R., Audhya, A., Gern, J.E., Bertics, P.J., 2014. Simvastatin attenuates rhinovirus-induced interferon and CXCL10 secretion from monocytic cells in vitro. *J Leukoc Biol* 95(6), 951–959.

Wiesel, M., Oxenius, A., 2012. From crucial to negligible: Functional CD8(+) T-cell responses and their dependence on CD4(+) T-cell help. *Eur J Immunol* 42(5), 1080–1088.

Winkler, P., de Vrese, M., Laue, C., Schrezenmeir, J., 2005. Effect of a dietary supplement containing probiotic bacteria plus vitamins and minerals on common cold infections and cellular immune parameters. *Int J Clin Pharmacol Ther* 43(7), 318–326.

Wu, S., Yuan, L., Zhang, Y., Liu, F., Li, G., Wen, K., Kocher, J., Yang, X., Sun, J., 2013. Probiotic *Lactobacillus rhamnosus* GG mono-association suppresses human rotavirus-induced autophagy in the gnotobiotic piglet intestine. *Gut Pathog* 5(1), 22.

Wyatt, R.G., James, W.D., Bohl, E.H., Theil, K.W., Saif, L.J., Kalica, A.R., Greenberg, H.B., Kapikian, A.Z., Chanock, R.M., 1980. Human rotavirus type 2: Cultivation in vitro. *Science* 207(4427), 189–191.

Xia, S., Du, J., Su, J., Liu, Y., Huang, L., Yu, Q., Xie, Z., Gao, J., Xu, B., Gao, X., Guo, T., Liu, Y., Zhou, X., Yang, H., 2020. Efficacy, immunogenicity and safety of a trivalent live human-lamb reassortant rotavirus vaccine (LLR3) in healthy Chinese infants: A randomized, double-blind, placebo-controlled trial. *Vaccine* 38(46), 7393–7400.

Yamane, H., Paul, W.E., 2013. Early signaling events that underlie fate decisions of naive CD4(+) T cells toward distinct T-helper cell subsets. *Immunol Rev* 252(1), 12–23.

Yang, K., Wang, S., Chang, K.O., Lu, S., Saif, L.J., Greenberg, H.B., Brinker, J.P., Herrmann, J.E., 2001. Immune responses and protection obtained with rotavirus VP6 DNA vaccines given by intramuscular injection. *Vaccine* 19(23–24), 3285–3291.

Yang, X., Twitchell, E., Li, G., Wen, K., Weiss, M., Kocher, J., Lei, S., Ramesh, A., Ryan, E.P., Yuan, L., 2015. High protective efficacy of rice bran against human rotavirus diarrhea via enhancing probiotic growth, gut barrier function, and innate immunity. *Sci Rep* 5, 15004.

Yang, X., Wen, K., Tin, C., Li, G., Wang, H., Kocher, J., Pelzer, K., Ryan, E., Yuan, L., 2014. Dietary rice bran protects against rotavirus diarrhea and promotes Th1-type immune responses to human rotavirus vaccine in gnotobiotic pigs. *Clin Vaccin Immunol* 21(10), 1396–1403.

Yang, X., Yuan, L., 2014. Neonatal gnotobiotic pig models for studying viral pathogenesis, immune responses, and for vaccine evaluation. *Br J Virol* 1, 87–91.

Yazdi, M.H., Soltan Dallal, M.M., Hassan, Z.M., Holakuyee, M., Agha Amiri, S., Abolhassani, M., Mahdavi, M., 2010. Oral administration of *Lactobacillus acidophilus* induces IL-12 production in spleen cell culture of BALB/c mice bearing transplanted breast tumour. *Br J Nutr* 104(2), 227–232.

Yuan, L., 2000. *Studies of Immunity to Human Rotavirus and Candidate Vaccines in a Gnotobiotic Pig Model.* PhD Dissertation. Ohio State University. The Ohio State University and OhioLINK, Columbus, OH.

Yuan, L., Azevedo, M., Zhang, W., González, A., Nguyen, T., Wen, K., Yousef, A., Saif, L., 2006. Impact of colonization by probiotic Lactobacilli on development of T cell responses in neonatal gnotobiotic (Gn) pigs infected with human rotavirus (HRV). Abstract W18-10. 25th Annual Meeting of American Society for Virology. July 15–19, 2006. University of Wisconsin, Madison, WI.

Yuan, L., Azevedo, M.S., Gonzalez, A.M., Jeong, K.I., Van Nguyen, T., Lewis, P., Iosef, C., Herrmann, J.E., Saif, L.J., 2005. Mucosal and systemic antibody responses and protection induced by a prime/boost rotavirus-DNA vaccine in a gnotobiotic pig model. *Vaccine* 23(30), 3925–3936.

Yuan, L., Geyer, A., Hodgins, D.C., Fan, Z., Qian, Y., Chang, K.O., Crawford, S.E., Parreno, V., Ward, L.A., Estes, M.K., Conner, M.E., Saif, L.J., 2000. Intranasal administration of 2/6-rotavirus-like particles with mutant *Escherichia coli* heat-labile toxin (LT-R192G) induces antibody-secreting cell responses but not protective immunity in gnotobiotic pigs. *J Virol* 74(19), 8843–8853.

Yuan, L., Geyer, A., Saif, L.J., 2001a. Short-term immunoglobulin A B-cell memory resides in intestinal lymphoid tissues but not in bone marrow of gnotobiotic pigs inoculated with Wa human rotavirus. *Immunology* 103(2), 188–198.

Yuan, L., Honma, S., Ishida, S., Yan, X.Y., Kapikian, A.Z., Hoshino, Y., 2004. Species-specific but not genotype-specific primary and secondary isotype-specific NSP4 antibody responses in gnotobiotic calves and piglets infected with homologous host bovine (NSP4[A]) and porcine (NSP4[B]) rotavirus. *Virology* 330(1), 92–104.

Yuan, L., Honma, S., Kim, I., Kapikian, A.Z., Hoshino, Y., 2009. Resistance to rotavirus infection in adult volunteers challenged with a virulent G1P1A[8] virus correlated with serum immunoglobulin G antibodies to homotypic viral proteins 7 and 4. *J Infect Dis* 200(9), 1443–1451.

Yuan, L., Iosef, C., Azevedo, M.S., Kim, Y., Qian, Y., Geyer, A., Nguyen, T.V., Chang, K.O., Saif, L.J., 2001b. Protective immunity and antibody-secreting cell responses elicited by combined oral attenuated Wa human rotavirus and intranasal Wa 2/6-VLPs with mutant *Escherichia coli* heat-labile toxin in gnotobiotic pigs. *J Virol* 75(19), 9229–9238.

Yuan, L., Kang, S.Y., Ward, L.A., To, T.L., Saif, L.J., 1998. Antibody-secreting cell responses and protective immunity assessed in gnotobiotic pigs inoculated orally or intramuscularly with inactivated human rotavirus. *J Virol* 72(1), 330–338.

Yuan, L., Saif, L.J., 2002. Induction of mucosal immune responses and protection against enteric viruses: Rotavirus infection of gnotobiotic pigs as a model. *Vet Immunol Immunopathol* 87(3–4), 147–160.

Yuan, L., Ward, L.A., Rosen, B.I., To, T.L., Saif, L.J., 1996. Systematic and intestinal antibody-secreting cell responses and correlates of protective immunity to human rotavirus in a gnotobiotic pig model of disease. *J Virol* 70(5), 3075–3083.

Yuan, L., Wen, K., Azevedo, M.S., Gonzalez, A.M., Zhang, W., Saif, L.J., 2008. Virus-specific intestinal IFN-gamma producing T cell responses induced by human rotavirus infection and vaccines are correlated with protection against rotavirus diarrhea in gnotobiotic pigs. *Vaccine* 26(26), 3322–3331.

Yuan, L., Wen, K., Liu, F., Li, G., 2013. Dose effects of LAB on modulation of rotavirus vaccine induced immune responses. In J.M. Kongo (Ed.), *Lactic Acid Bacteria - R & D for Food, Health and Livestock Purposes.* http://www.intechopen.com/books/lactic-acid-bacteria-r-d-for-food-health-and-livestock-purposes/dose-effects-of-lab-on-modulation-of-rotavirus-vaccine-induced-immune-responses.

Yuan, L.J., Jobst, P.M., Weiss, M., 2017. Gnotobiotic pigs: From establishing facility to modeling human infectious diseases. In Schoeb, T.R., Eaton, K.A. (Eds.), *Gnotobiotics.* Acamedic Press, Cambridge, MA, USA, 349–368.

Zaman, K., Dang, D.A., Victor, J.C., Shin, S., Yunus, M., Dallas, M.J., Podder, G., Vu, D.T., Le, T.P., Luby, S.P., Le, H.T., Coia, M.L., Lewis, K., Rivers, S.B., Sack, D.A., Schodel, F., Steele, A.D., Neuzil, K.M., Ciarlet, M., 2010. Efficacy of pentavalent rotavirus vaccine against severe rotavirus gastroenteritis in infants in developing countries in Asia: A randomised, double-blind, placebo-controlled trial. *Lancet* 376(9741), 615–623.

Zambrana, L.E., McKeen, S., Ibrahim, H., Zarei, I., Borresen, E.C., Doumbia, L., Bore, A., Cissoko, A., Douyon, S., Kone, K., Perez, J., Perez, C., Hess, A., Abdo, Z., Sangare, L., Maiga, A., Becker-Dreps, S., Yuan, L., Koita, O., Vilchez, S., Ryan, E.P., 2019. Rice bran supplementation modulates growth, microbiota and metabolome in weaning infants: A clinical trial in Nicaragua and Mali. *Sci Rep* 9(1), 13919.

Zambrana, L.E., Weber, A.M., Borresen, E.C., Zarei, I., Perez, J., Perez, C., Rodriguez, I., Becker-Dreps, S., Yuan, L., Vilchez, S., Ryan, E.P., 2021. Daily rice bran consumption for 6 months influences serum glucagon-like Peptide 2 and metabolite profiles without differences in trace elements and heavy metals in weaning Nicaraguan infants at 12 months of age. *Curr Dev Nutr* 5(9), nzab101.

Zegers, N.D., Boersma, W.J., Claassen, E. (Eds.), 1995. *Immunological Recognition of Peptides in Medicine and Biology.* CRC Press, Inc., Boca Raton, New York, London, and Tokyo, 105–123.

Zelaya, H., Tsukida, K., Chiba, E., Marranzino, G., Alvarez, S., Kitazawa, H., Aguero, G., Villena, J., 2014. Immunobiotic lactobacilli reduce viral-associated pulmonary damage through the modulation of inflammation-coagulation interactions. *Int Immunopharmacol* 19(1), 161–173.

Zeuthen, L.H., Fink, L.N., Metzdorff, S.B., Kristensen, M.B., Licht, T.R., Nellemann, C., Frokiaer, H., 2010. *Lactobacillus acidophilus* induces a slow but more sustained chemokine and cytokine response in naive foetal enterocytes compared to commensal *Escherichia coli*. *BMC Immunol* 11, 2.

Zhang, H., Wang, H., Shepherd, M., Wen, K., Li, G., Yang, X., Kocher, J., Giri-Rachman, E., Dickerman, A., Settlage, R., Yuan, L., 2014. Probiotics and virulent human rotavirus modulate the transplanted human gut microbiota in gnotobiotic pigs. *Gut Pathog* 6, 39.

Zhang, M., Zhang, M., Zhang, C., Du, H., Wei, G., Pang, X., Zhou, H., Liu, B., Zhao, L., 2009. Pattern extraction of structural responses of gut microbiota to rotavirus infection via multivariate statistical analysis of clone library data. *FEMS Microbiol Ecol* 70(2), 21–29.

Zhang, Q., Widmer, G., Tzipori, S., 2013a. A pig model of the human gastrointestinal tract. *Gut Microbes* 4(3), 193–200.

Zhang, W., 2007. *Effects of Probiotic Lactic Acid Bacteria on Innate and B Cell Responses to Rotavirus*. The Ohio State University, Wooster, OH.

Zhang, W., Azevedo, M.S., Gonzalez, A.M., Saif, L.J., Van Nguyen, T., Wen, K., Yousef, A.E., Yuan, L., 2008a. Influence of probiotic Lactobacilli colonization on neonatal B cell responses in a gnotobiotic pig model of human rotavirus infection and disease. *Vet Immunol Immunopathol* 122(1–2), 175–181.

Zhang, W., Azevedo, M.S., Wen, K., Gonzalez, A., Saif, L.J., Li, G., Yousef, A.E., Yuan, L., 2008b. Probiotic *Lactobacillus acidophilus* enhances the immunogenicity of an oral rotavirus vaccine in gnotobiotic pigs. *Vaccine* 26(29–30), 3655–3661.

Zhang, W., Azevedo, M.S.P., Wen, K., Gonzalez, A.M., Saif, L.J., Li, G., Yousef, A.E., Yuan, L., 2008c. Probiotic *Lactobacillus acidophilus* enhances the immunogenicity of an oral rotavirus vaccine in gnotobiotic pigs. *Vaccine* 26(29–30), 3655–3661.

Zhang, W., Wen, K., Azevedo, M.S., Gonzalez, A., Saif, L.J., Li, G., Yousef, A.E., Yuan, L., 2008d. Lactic acid bacterial colonization and human rotavirus infection influence distribution and frequencies of monocytes/macrophages and dendritic cells in neonatal gnotobiotic pigs. *Vet Immunol Immunopathol* 121(3–4), 222–231.

Zhang, X., Tao, Y., Wang, J., Garcia-Mata, R., Markovic-Plese, S., 2013b. Simvastatin inhibits secretion of Th17-polarizing cytokines and antigen presentation by DCs in patients with relapsing remitting multiple sclerosis. *Eur J Immunol* 43(1), 281–289.

Zhang, Z., Xiang, Y., Li, N., Wang, B., Ai, H., Wang, X., Huang, L., Zheng, Y., 2013c. Protective effects of *Lactobacillus rhamnosus* GG against human rotavirus-induced diarrhoea in a neonatal mouse model. *Pathog Dis* 67(3), 184–191.

Zhu, S., Regev, D., Watanabe, M., Hickman, D., Moussatche, N., Jesus, D.M., Kahan, S.M., Napthine, S., Brierley, I., Hunter, R.N., 3rd, Devabhaktuni, D., Jones, M.K., Karst, S.M., 2013. Identification of immune and viral correlates of Norovirus protective immunity through comparative study of intra-cluster Norovirus strains. *PLOS Pathog* 9(9), e1003592.

Zhu, Y.H., Li, X.Q., Zhang, W., Zhou, D., Liu, H.Y., Wang, J.F., 2014. Dose-dependent effects of *Lactobacillus rhamnosus* on serum interleukin-17 production and intestinal T-cell responses in pigs challenged with *Escherichia coli*. *Appl Environ Microbiol* 80(5), 1787–1798.

Index